T0100731

THE BOOK OF MINDS

ALSO BY PHILIP BALL

H2O: A Biography of Water

Bright Earth: The Invention of Colour

Critical Mass: How One Thing Leads To Another

Nature's Patterns: A Tapestry in Three Parts

The Music Instinct: How Music Works and Why We Can't Do Without It

Curiosity: How Science Became Interested in Everything

Serving the Reich: The Struggle for the Soul of Physics Under Hitler

Invisible: The History of the Unseen from Plato to Particle Physics

The Water Kingdom: A Secret History of China

Beyond Weird: Why Everything You Thought You Knew About Quantum Physics is Different

How To Grow a Human: Adventures in Who We Are and How We Are Made

The Modern Myths: Adventures in the Machinery of the Popular Imagination

THE BOOK OF MINDS

HOW TO UNDERSTAND OURSELVES AND OTHER BEINGS, FROM ANIMALS TO AI TO ALIENS

PHILIP BALL

The University of Chicago Press

The University of Chicago Press, Chicago 60637

Published 2022
Printed in the United States of America

31 30 29 28 27 26 25 24 23 22   1 2 3 4 5

ISBN-13: 978-0-226-79587-4 (cloth)
ISBN-13: 978-0-226-82204-4 (e-book)
DOI: https://doi.org/10.7208/chicago/9780226822044.001.0001

Published in the United Kingdom by Picador, an imprint of Pan Macmillan.

Library of Congress Cataloging-in-Publication Data

Names: Ball, Philip, 1962– author.
Title: The book of minds : how to understand ourselves and other beings, from animals to AI to aliens / Philip Ball.
Description: Chicago : The University of Chicago Press, 2022. | Includes bibliographical references and index.
Identifiers: LCCN 2021061676 | ISBN 9780226795874 (cloth) | ISBN 9780226822044 (ebook)
Subjects: LCSH: Cognition. | Consciousness. | Brain. | Cognition in animals. | Artificial intelligence.
Classification: LCC BF311 .B27 2022 | DDC 153—dc23/eng/20220114
LC record available at https://lccn.loc.gov/2021061676

∞ This paper meets the requirements of ANSI/NISO Z39.48-1992 (Permanence of Paper).

Contents

1. Minds and Where to Find Them 1
2. The Space of Possible Minds 39
3. All The Things You Are 61
4. Waking Up To the World 115
5. Solomon's Secret 165
6. Aliens On the Doorstep 231
7. Machine Minds 267
8. Out of This World 333
9. Free to Choose 397
10. How To Know It All 443

Acknowledgements 459

End notes 463

Bibliography 475

Index 495

CHAPTER 1

Minds and Where to Find Them

The neurologist and writer Oliver Sacks was an indefatigable chronicler of the human mind, and in the too-brief time that I knew him I came to appreciate what that meant. In his personal interactions as much as his elegant case-study essays, Sacks was always seeking the essence of the person: how does *this* mind work, how was it shaped, what does it believe and desire? He was no less forensic and curious about his own mind, which I suspect represented as much of a puzzle to him as did anyone else's.

Even – perhaps especially – for a neurologist, this sensitivity to minds was unusual. Yes, Sacks might consider how someone's temporal lobe had been damaged by illness or injury, and he might wonder about the soup of neurotransmitters sloshing around in the grey matter of the brain. But his primary focus was on the individual, the integrated result of all that neural processing: a person existing, as best they could, in the company of others, each trying to navigate a path amid other minds that they could never really hope to fathom and certainly never to experience. It was emphatically not the brain but the mind that fascinated him.

None more so, I imagine, than the mind he once encountered in Toronto, even though he was never able to make a case study of it. That would not have been easy, because this individual was not human. In the city zoo, Sacks had this briefest of exchanges with a female orangutan.

> She was nursing a baby – but when I pressed my bearded face against the window of her large, grassy enclosure, she put her infant down gently, came over to the window, and pressed her face, her nose, opposite mine, on the other side of the glass. I suspect my eyes were darting about as I gazed at her face, but I was much more conscious of her eyes. Her bright little eyes – were they orange too? – flicked about, observing my nose, my chin, all the human but also apish features of my face, identifying me (I could not help feeling) as one of her own kind, or at least closely akin. Then she stared into my eyes, and I into hers, like lovers gazing into each other's eyes, with just the pane of glass between us.
>
> I put my left hand against the window, and she immediately put her right hand over mine. Their affinity was obvious – we could both see how similar they were. I found this astounding, wonderful; it gave me an intense feeling of kinship and closeness as I had never had before with any animal. 'See,' her action said, 'my hand, too, is just like yours.' But it was also a greeting, like shaking hands or matching palms in a high five.
>
> Then we pulled our faces away from the glass, and she went back to her baby.
>
> I have had and loved dogs and other animals, but I have never known such an instant, mutual recognition and sense of kinship as I had with this fellow primate.

A sceptic might question Sacks' confident assertion of meaning: his supposition that the orangutan was expressing a greeting and was commenting on their physical similarities, their kinship. You don't know what is going on in an ape's mind! You don't even know that an ape *has* a mind!

But Sacks was making this inference on the same grounds that we infer the existence of *any* other mind: intuitively, by analogy with our own. All we know for sure, as René Descartes famously observed, is that we exist – by virtue of our own consciousness, our sense of *being* a mind. The rest is supposition.

No one has ever got beyond Descartes' philosophical conundrum: how can we be *sure* of anything but ourselves? Imagine, Descartes said, if there were some mischievous demon feeding our mind information that creates the perfect illusion of an external world, filled with other minds like ours. Maybe none of that is real: it is all just phantoms and mirages conjured by the demon as if by some trick of infernal telepathic cinematography.

This position is called solipsism, and widely considered a philosophical dead end. You can't refute it, yet to entertain it offers nothing of any value. If other people are merely a figment of my imagination, I cease to have any moral obligations towards them – but to assume without evidence that this is the case (rather than at least erring on the side of caution) would require that I embrace a belief my experience to date has primed me to regard as psychotic. It would seem to invert the very definition of reason. At any rate, since the rational solipsist can never be sure of her belief, it can't necessitate any change in her behaviour: it's an impotent idea. Sure, the demon (and where then did *he* come from?) might decide to end the game at any moment. But everything that transpires in my mind advises me to assume he will not (because I should assume he does not exist).

We'll hear more from Descartes' demon later, because scientific and technological advances since the seventeenth century have produced new manifestations of this troublesome imp. Let us return to the mind of Oliver Sacks's orangutan.

What, exactly, persuaded Sacks that the ape had a kindred mind? The familiar gestures and the soulful gaze of apes seem to insist on it. This intuition goes beyond anatomical similarities; apes, probably more than any other creatures (dog owners might demur), exhibit a deeply eloquent quality of eye contact. Those eyes are too expressively reminiscent of what we see in other humans for us to imagine that they are the windows to vastly different minds – let alone to no mind at all. In short, there is a great deal we import from encounters with other people into meetings with our more distant primate relatives.

It's the similarity of our own behaviour to that of other people which convinces us they too have 'somebody at home': that a mind like ours inheres within the other's body, guiding its outward actions. It is simply more parsimonious to suppose that another person is a being just like us than to imagine that somehow the world is peopled with zombie-like beings able, through some bizarre quirk of physics or biology, to mimic us so perfectly. How weird and improbable it would be if the inscrutable laws of zombiehood impelled these other beings to use language (say) in just the same way as we do, yet without any of the same intent. What's more, other people's brains produce patterns of electrical activity identical in broad outline to those in our own, and the same patterns correlate with the same behaviours. Such coincidences, if that is all they were, could surely only be the product of demonic design.

Far from being a leap of faith, then, assuming the existence of other minds is the rational thing to do – not just in people but also in orangutans. We'll see later that this argument for the reality of animal minds – the assumption that they are not merely complex automata, rather as Descartes supposed – can be made much more concrete.

But how far can this reasoning take us? I hope you're willing to grant me a mind – I can assure you (though of course you have only my word for it) that you'd be doing the right thing. I suspect most people are now happy, even eager, to accept that it is meaningful to say that apes have minds. The difficulty in going further, however, is not that we are habitually reluctant to attribute minds to non-human beings and entities, but that we do this all too readily. We have evolved to see minds every damned place we look. So some caution is in order.

Is this world not so glorious and terrible in its profuse and sublime extent that it must constitute evidence of a minded* Creator? That's what humankind has long believed, filling all corners of the world with entities that possess mind and motive. That wind? The spirits of the air are on the move. That crash of thunder? The storm god is restless. That creaking floorboard? The tread of a ghostly being.

It has become common in our increasingly secular age to treat all this animism either as a quirk of our evolutionary past that we need to outgrow or, worse, as evidence that we are still in thrall to pre-scientific delusions. I suggest our tendency to attribute mind to matter is a lot more complicated than that. What if, for instance, in stripping the world of mindedness we sacrifice our respect for it too, so that a river devoid of any animating spirit will eventually become no more than a resource to be exploited and abused? As we will see, some scientists today seriously argue that plants have minds, partly on the grounds that this should deepen our ecological sensitivity. It is surely no coincidence that the British scientist and inventor James Lovelock, having conceived of the entire Earth as a self-regulating entity with organism-like properties, accepted the suggestion of his neighbour, novelist and Nobel laureate William Golding, to personify that image with the name of the Greek earth goddess Gaia. For some, these ideas veer too far from science and too close to mysticism. But the point is that the impulse to award mindedness

* I will be using the word *minded* in a somewhat unconventional sense, to mean *imbued with mind.*

where it does not *obviously* reside – and thereby to valorize the minded entity – has not gone away, and we might want to consider if there are good reasons why that is so.

I doubt that even the hardest-headed sceptic of our instinct to personify nature, objects, and forces has not occasionally cursed the sheer perversity or bloody-mindedness of their computer or car. 'Don't do that!' we wail in futile command as the computer decides to shut down or mysteriously junks our file. (I feel that right now I am tempting fate, or the Computer God, even to say such a thing.) 'Why me?' we cry when misfortune befalls us, betraying the suspicion that deep down there is a reason, a plan, that the universe harbours. (That, in a nutshell, is the Book of Job, which in a generous reading warns against interpreting another's bad luck as an indication that God, displeased with them, has meted out their just deserts.)

If there's a flaw in our tendency too casually to attribute mind, it might better be located in the anthropocentric nature of that impulse. We can't resist awarding things *minds like ours*. As we'll see later, Christian theologians have striven in vain to save God from that fate, and it is no wonder: their own holy book undermines them repeatedly, unless it is read with great subtlety. (God seems to speak more regularly to Noah, Jacob and Moses than many company CEOs today do to their underlings.) Our habit of treating animals as though they are dim-witted humans explains a great deal about our disregard for their well-being; giving them fully fledged, Disneyfied human minds is only the flipside of the same coin. We'll see too that much of the discussion about the perils of artificial intelligence has been distorted by our insistence on giving machines (in our imaginations) minds like ours.

Still, it's understandable. As we are reminded daily, it's hard enough sometimes to fathom the minds of our fellow humans – to accept that they might think differently from us – let alone to imagine what a non-human mind could possibly be like. But that is what we're going to try to do here, and I believe the task is not hopeless.

Making minds up

First of all, we need to ask the central question: *What is a mind?*

There is no scientific definition to help us. Neither can dictionaries, since they tend to define the mind *only* in relation to the human: it is, for example, 'the part of a person that makes it possible for him or her to think, feel emotions, and understand things.' It's bad enough that such a definition leans on a slew of other ill-defined concepts – thinking, feeling, understanding. Worse, the definition positively excludes the possibility of mind existing within non-human entities.

Some behavioural researchers dislike the word altogether. 'Mind', say psychologists Alexandra Schnell and Giorgio Vallortigara, 'is an immeasurable concept that is not amenable to rigorous scientific testing.' They say that instead of talking about the 'dog mind' or the 'octopus mind', we should focus on investigating the mechanisms of their cognition in ways that can be measured and tested.

They have a point, but science needs vague concepts as well as precise ones. 'Life' too is immeasurable and untestable – there is no unique way to define it – and yet it is an indispensable notion for making sense of the world. Even words like 'time', 'energy', and 'molecule' in the so-called hard physical sciences turn out to be far from easy to define rigorously. Yet they are useful. So can the idea of mind be, if we are careful how we use it.

It's understandable yet unfortunate that much of the vast literature on the philosophy of mind considers it unnecessary to define its terms. Gilbert Ryle's influential book *The Concept of Mind* (1949) is so confident that we are all on the same page from the outset that it plunges straight into a discussion of the attributes that people display. Daniel Dennett, one of the most eloquent and perceptive contemporary philosophers of mind, presents a nuanced exploration of what non-human minds might be like in his 1996 book *Kinds of Mind*, and yet he too has to begin by suggesting that, 'Whatever else a mind is, it is supposed to be something like our minds; otherwise we wouldn't call it a mind. So our minds, the only minds *we* know from the outset, are the standard with which we must begin.'

He is right, of course. But this constraint is perhaps only because, in exploring the *Space of Possible Minds*, we are currently no better placed than the pre-Copernican astronomers who installed the Earth at the centre of the cosmos and arranged everything else in relation to it, spatially and materially. Our own mind has certain properties, and it makes sense to ask whether other minds have more or less of those properties: how close or distant they are from ours. But this doesn't get us far in pinning down what the notion I am referring to as *mindedness* – possessing a mind – means. One thing my mind has, for example, is memory. But my computer has much more of that, at least in the sense of holding vast amounts of information that can be recalled exactly and in an instant. Does that mean my computer exceeds me in at least this one feature of mind? Or is memory in fact not a necessary requirement of *mindedness* at all? (I shall answer this question later, after a fashion.)

In short, 'mind' is one of those concepts – like intelligence, thought, and life – that sounds technical (and thus definable) but is in fact colloquial and irreducibly fuzzy. Beyond our own mind (and what we infer thereby about those of our fellow humans), we can't say for sure what mind should or should not mean. We are not really much better off than what Ambrose Bierce implied in his satirical classic of 1906, *The Devil's Dictionary*, where he defined mind as

> A mysterious form of matter secreted by the brain. Its chief activity consists in the endeavor to ascertain its own nature, the futility of the attempt being due to the fact that it has nothing but itself to know itself with.

Yet I don't believe that a definition of mind need be impossible, so long as we're not trying to formulate it with scientific rigour. On the contrary, it can be given rather succinctly:

For an entity to have a mind, there must be something it is like to be that entity.

I apologize that this is a syntactically odd sentence, and therefore not easy to parse. But what it is basically saying is that a mind hosts an experience of some sort.

Some might say this is not a properly scientific definition because it invokes subjectivity, which is not a thing one can measure. I'm agnostic about such suggestions, both because there *are* scientific studies that aim to measure subjective experience and because a concept (like life) doesn't have to be measurable to be scientifically useful.

You might, on the other hand, be inclined to object that this definition of mind is tautological. What else could a mind be, after all?* But I think there is a very good reason for making it our starting point, which is this: the only mind we know about is our own, and *that* has experience. We don't know *why* it has experience, but

* All the same, other definitions exist. For example, neuroscientists Ogi Ogas and Sai Gaddam require of a mind only that it 'takes a set of inputs from its environment and transforms them into a set of environment-impacting outputs that influence the welfare of its body.' It's not hard to make machines that do more or less this – and indeed Ogas and Gaddam seem to consider machine minds to be an unproblematic notion. We will see in Chapter 7 how far my own definition can be extended to machines.

only that it does. We don't even know quite how to characterize experience, but only that we possess it. All we can do in trying to understand mind is to move cautiously outwards, to see what aspects of our experience we might feel able (taking great care) to generalize. In this sense, trying to understand mind is not like trying to understand anything else. For everything else, we use our mind and experience as tools for understanding. But here we are forced to turn those tools on themselves.

That's why there is something irreducibly phenomenological about the study of mind, in the sense invoked by the philosophical movement known as Phenomenology pioneered by Edmund Husserl in the early twentieth century. This tradition grapples with experience from a first-person perspective, abandoning science's characteristic impulse of seeking understanding from an impersonal, objective position. My criterion of mind is, I'd argue, not tautological but closer to phenomenological, and necessarily so when *mind* is the subject matter.

Since we can't be sure about the nature of other minds, we have to be humble in our pronouncements. I do not believe that a rock has a mind, because I don't think a rock has experience: it does not mean anything to say that 'being like a rock' is to be like anything at all. This, however, is an opinion. Some philosophers, and some scientists too, argue that there *is* something it is like to be a rock – even if that is only the faintest glimmer of 'being like'. This position is called panpsychism:* the idea that qualities of mind pervade all matter to some degree. It could be right, but panpsychists can't prove it.

* This word has sometimes been used with a derogatory implication, as if the idea it denotes is self-evidently absurd. It is not, and indeed the panpsychist position has enjoyed something of a recent resurgence, partly for reasons that we shall discover later.

Yet we need not be entirely mired in relativism. Scientists and philosophers who suspect there might be something it is like to be a rock don't say so because of some vague intuition, or because they cleave to an animistic faith. The claim is one arrived at by reasoning, and at least some of that reasoning can be examined systematically and perhaps even experimentally. We are not totally in the dark.

How about a bacterium? Is there something it is like to be a bacterium? Here opinions are more divided. Some invoke the notion of *biopsychism*: the proposal that mindedness is one of the defining, inevitable properties of all living things. We'll look at this position more closely later, but let's allow for now that it is not obviously crazy. Personally, I'm not sure I believe there is something it is like to be a bacterium.

Still, you can see the point. At some stage on the complexity scale of life, there appears some entity for which there is something it is to be like that organism. I imagine most people are ready to accept today that there is something it is like to be an orangutan. You might well consider there is something it is like to be a mouse, perhaps even a fly. But a fungus? Maybe that's pushing it.

This is why it makes sense to speak in terms of mindedness, which acknowledges that minds are not all-or-nothing entities but matters of degree. My definition notwithstanding, I don't think it's helpful to ask if something has a mind or not, but rather, to ask what qualities of mind it has, and how much of them (if any at all).

You might wonder: why not speak instead of consciousness? The two concepts are evidently related, but they are not synonymous. For one thing, consciousness seems closer to a property we can identify and perhaps even quantify. We know that consciousness can come and go from our brains – general anaesthesia extinguishes it temporarily. But when we lose consciousness, have we lost our mind too? It's significant that we don't typically speak of it in those terms. As we will see, there are now techniques for measuring whether a human brain is conscious; they are somewhat controversial and it's not entirely clear what proxy for consciousness they are probing, but they evidently measure *something* meaningful. What's more, even though we still lack a scientific theory of consciousness (and might never have such a thing), there is a fair amount we can say, and more we can usefully speculate, about how consciousness arises from the activity of our neurons and neural circuits.

Being minded, on the other hand, is a capacity that is both more general and more abstract: you might regard the condition as one that entails being conscious at least some of the time, but that more specifically supplies a repertoire of ways to feel, to act and simply to be.

We might say, then, that mindedness is a disposition of cognitive systems that can potentially give rise to states of consciousness. I say *states* because it is by no means clear, and I think unlikely, that what we call consciousness corresponds to a single state (of mind or brain). By the same token I would suggest that while there's a rough-and-ready truth to the suggestion that greater degrees of mindedness will support greater degrees of consciousness, neither attribute seems likely to be measurable in ways that can be expressed with a single number, and in fact both are more akin to qualities than quantities. Different types of mind can be expected to support different kinds of consciousness. Can mind exist without any kind of consciousness at all? It's hard to imagine what it could mean to 'be like' an entity that lacks any kind of awareness – but as we'll see, we might make more progress by breaking the question down into its components.

Colloquial language is revealing of how we think about these matters. 'Losing one's mind' implies something quite different to losing consciousness; here what is really lost is not the mind per se but the kind of mind that can make good (beneficial) use of its resources. Mind is a verb too, implying a sort of predisposition:

Mind out, mind yourself, would you mind awfully, I really don't mind. We seem to regard mind as disembodied: mind over matter, the power of the mind. As we'll see, there is probably on the contrary a close and indissoluble connection between mind and the physical structure in which it arises – but the popular conception of mind brings it adjacent to the notions of will and self-determination: a mind does things, it achieves goals, and does so in ways that we conceptualize non-physically.

By what means does a mind enact this functional objective? Philosopher Ned Block has proposed that the mind is the 'software of the brain' – it is, you might say, the algorithm that the brain runs to do what it does. He identifies at least two components to this capability: intelligence and intentionality. Intelligence comes from applying rules to data: the mind-system takes in information, and the rules turn it into output signals, for example to guide behaviour. That process might be extremely complex, but Block suggests that it can be broken down into progressively less 'intelligent' subsystems, until ultimately we get to 'primitive processors' that simply convert one signal into another with no intelligence at all. These could be electronic logic gates in a computer, made from silicon transistors, or they could be individual neurons sending electrical signals to one another. No one (well, hardly anyone) argues that individual neurons are intelligent. In other words, this view of intelligence is agnostic about the hardware: you could construct the primitive processors, say, from ping pong balls rolling along tubes.

But intelligence alone doesn't make a mind. For that, suggests Block, you also need intentionality – put crudely, what the processors involved in intelligence are *for*. Intentionality is *aboutness*: an intentional system has states that in some sense represent – are about – the world. If you stick together a strip of copper and one of tin and warm them up, the two metals expand at different rates, causing the double strip to bend. If you now make this a component in an electronic circuit so that the bending breaks the circuit, you have a thermostat, and the double strip becomes an intentional system: it is *about* controlling the temperature in the environment. Evidently, intentionality isn't a question of what the system looks like or what, of itself, it does – but about how it relates to the world in which it is embedded.*

This is a very mechanical and computational view of the mind. There's nothing in Block's formulation that ties the notion of mind to any biological embodiment: minds, you might say, don't have to be 'alive' in the usual sense.†

We're left, then, with a choice of defining minds in terms of either their nature (they have sentience, a 'what it is to be like') or their purpose (they have goals). These needn't be incompatible, for one of the tantalizing questions about types of mind is whether it is possible even to conceive of a sentient entity that does *not* recognize goals – or conversely, whether the origin of a 'what it is to be like' resides in the value of such experiential knowledge for attaining a mind's goals. Dennett suggests their key objective by quoting the French poet Paul Valéry: 'the task of a mind is to produce future.' That is to say, Dennett continues, a mind must be a generator of expectations and predictions:

> it mines the present for clues, which it refines with the help of the materials it has saved from the past, turning them into anticipations of the future. And then it acts, rationally, on the basis of those hard-won anticipations.

* Does this mean a thermostat has a mind? Philosophers have in fact debated this question, but have not reached a consensus.

† The question of what constitutes 'being alive' is not settled either, but that's another matter.

If this formulation is correct, we might expect minds to have certain features: memories, internal models of the world, a capacity to act, and perhaps 'feelings' to motivate that action. A mind so endowed would be able not only to construct possible futures, but also to make selections and try to realize them.

Dennett's prescription imposes a requirement on the *speed* with which a mind deliberates: namely, that must happen at a rate at least comparable to that at which significant change happens in the environment around it. If the mind's predictions arrive too late to make a difference, the mind can't do its job – and so it has no value, no reason to exist. Perhaps, Dennett speculates, this creates constraints on what we can *perceive* as mind, based on what we perceive as salient change. 'If', he says,

> our planet were visited by Martians who thought the same sort of thoughts as we do but thousands or millions of times faster than we do, we would seem to them to be about as stupid as trees, and they would be inclined to scoff at the hypothesis that we had minds . . . In order for us to see things as mindful, they have to happen at the right pace.

It's unlikely, as we'll see, that we are overlooking a tree mind simply because it works at so glacial a pace; but Tolkien's fictional Ents serve to suggest that relative slowness of mind need not imply its absence, or indeed an absence of wisdom. Or to put it another way: mindedness might have an associated timescale, outside of which it ceases to be relevant.*

* This illustrates one respect in which our technical devices effectively expand our range of mind. A hundred years ago it made no difference to electrons moving in atoms on attosecond timescales (10^{-21} seconds) whether we had minds or not. But today we can use ultrafast laser pulses to alter those motions intentionally and

Block's view would seem to make mind a very general biological property. If intelligence is a matter of possessing some information-processing capacity that turns a stimulus into a behaviour, while intentionality supplies the purpose and motive for that behaviour by relating it to the world, then all living things from bacteria to bats to bank managers might be argued to have minds.

Neuroscientist Antonio Damasio demands more from a mind. Organisms, and even brains, he says, 'can have many intervening steps in the circuits mediating between response and stimulus, and still have no mind, if they do not meet an essential condition: the ability to deploy images internally and to order those images in a process called thought.'

Here, Damasio does not necessarily mean visual images (although they could be); evidently it is not necessary to possess vision at all in order to have a mind. The imagery could be formed from sound sensations, or touch or smell, say. The point is that the minded being uses those primitive inputs to construct some sort of internal picture of the world, and act on it. Action, says Damasio, is crucial: 'No organism seems to have mind but no action.' By the same token, he adds, there are organisms that have 'intelligent actions but no mind' – because they lack these internal representations through which action is guided. (This depends on what qualifies as a representation, of course.)

But there's still some postponing of the question in this formulation. It teeters on the brink of circularity: a mind is only a mind if it thinks, and thinking is what minds do. It is possible already to build machines that seem to satisfy all of Damasio's criteria – they can, for example, construct models of their environment based on input data, run simulations of these models to predict the

---

with design: our minds can touch and impose their plans on processes that happen far faster than thought itself.

consequences of different behavioural choices, and select the best. This can all be automated. And yet no one considers that these artificial devices warrant being admitted to the club of minded entities, because we have absolutely no reason to think that there is any awareness involved in the process. There is still nothing that it is to be like these machines.

At least, that's what nearly all experts in AI will say, and I believe they are right. But it's not obvious how we could find out for sure. We can, in principle, know everything there is about, say, a bird brain, except for what it is like to be 'inside' it. My definition of mind therefore can't obviously be tested, verified or falsified, any more than can the scenario posed by Descartes' demon. And by the same token, it's not productive to fret too much about that. Rather than arguing over the question of whether other minds exist or not, we can more usefully ask: how does mindedness arise from the cognitive workings of our own brains? Which if any of these cognitive properties are indispensable for that to happen? How might these appear or differ in other entities that might conceivably be minded, and what might the resulting minds be like from the inside? If we have answers, can we design new kinds of mind? Will we? Should we?

Why minds?

Damasio's description of mind is incomplete in a useful way. For if an intelligent system is able to acquire all of these features and yet *still not be a mind*, why is anything more needed, or of any value? Given those capacities, why is it necessary for there to be a 'what it feels like' at all? It's not obviously impossible that our distant evolutionary ancestors evolved all the way to being Damasio's intelligent yet mindless beings, and then natural selection 'discovered' that there was some added advantage to be had by installing a mind amidst it all. This used to be a common view: that we humans are unique as beasts with mind and awareness, distinguished by the fact that we are not automata but willed beings. It can be found in Aristotle's categorization of living things as those with only a nutritive soul (like plants), those with also a sensitive soul (like animals), and those with also a rational soul or *nous* (us). This exceptionalism persisted in Descartes' claim that humankind alone possesses a soul in the Christian sense: an immortal essence of being. It's not clear how deeply Descartes was persuaded of that, however: his account of the human body presented it, in the spirit of his times, as a machine, a contraption of pumps, levers and hydraulics. He may have insisted on the soul as the animating principle partly to avoid charges of heresy in presenting in so mechanical a manner the divine creation that is humanity. (It didn't entirely save him from censure.) His contemporary, the Frenchman Julien Offray de La Mettrie, had no qualms in making us all mere mechanism, a position he maintained in his 1747 book *L'Homme machine*, which the church condemned as fit for burning.

You needn't be an anthropocentric bigot to take the view that mind was an abrupt evolutionary innovation. Maybe that leap happened for the common ancestors we share with great apes? Perhaps mind appeared with the origin of all mammals?

The proposition, however, seems unlikely. Evolutionary jumps and innovations do happen – but as we'll see, there's no sign in the evolutionary record of a transition to mindedness suddenly transforming the nature or behaviour of pre-human creatures. Nor is there any reason to think that the explosion, around forty to fifty thousand years ago, in the capabilities and complexities of the behaviour of *Homo sapiens* was due to the abrupt acquisition of mind itself. It looks much more probable that the quality that I propose to associate with mind arose by degrees over a vast span of evolutionary time. There's now a widespread view that it was present to some extent before our very distant ancestors had even left the sea. If so, there is perhaps nothing any more special about it than there is about having a backbone, or breathing air.

In either scenario, it's by no means obvious that mindedness need confer an adaptive benefit at all. Could it be that this attribute, which strikes us as so central to our being (and surely it is, not least in being the quality that allows us to recognize it and be struck by it), was just a side effect of other cognitive adaptations? In other words, could it be that, if we and other creatures are to have brains that are able to do what they do, we have no option to incur a bit of mindedness too?

If that seems an alarming prospect – that evolution was indifferent (initially) to this remarkable and mysterious property that matter acquired – so too might be the corollary: perhaps matter could develop all kinds of capabilities for processing, navigating and altering its environment while possessing no mindedness at all. After all, a great deal (not all!) of what we find in the characteristics of life on Earth is highly contingent: the result of some accident or chance event in deep time that affected the course of all that followed on that particular branch of the tree of life. Could it be that evolution might have played out just as readily on Earth to populate it with a rich panoply of beings, some as versatile and intelligent as those we see today – yet without *minds*?

These could seem like idle speculations, fantastical might-have-beens that we can never go back and test. But by exploring the Space of Possible Minds, we can make them more than that.

What's the brain got to do with it?

Neuroscience barely existed as a discipline when Gilbert Ryle wrote *The Concept of Mind*, but he doubted that the 'science of mind' then prevailing – psychology* – could tell us much beyond rather narrow constraints. It was no different, he said, from other sciences that attempt to categorize and quantify human behaviour: anthropology, sociology, criminology, and the like. There is no segregated field of mental behaviour that is the psychologist's preserve, he said, nor could it ever offer causal explanations for all our actions in the manner of a Newtonian science of mind. At root, Ryle's scepticism towards a 'hard-science' approach derives from the central problem for understanding the mind: we can only ever come at it from the inside, which makes it different from studying every other object in the universe. That's one way of expressing what is often called the 'hard problem' of consciousness: why a mind has anything there is to be like. We can formulate testable theories of why the brain might generate subjective experience, but we don't know even how to formulate the question of why a given experience is like *this* and not some other way: why red looks red, why apples smell (to us) like apples. (It might not, as we'll see, even be a question at all.)

Ryle is surely right to suggest that some problems of mind are irreducibly philosophical. But he threw out too much. To recognize that there are limits to what the brain and behavioural sciences can tell us about the mind is not the same as suggesting that they can tell us nothing of value. Indeed, to talk about minds without consideration of the physical systems in which they arise is absurd, akin to the sort of mysticism of mind that Ryle wanted to dispel.

So we need to bring the brain into the picture – but with care. The human brain is surely the orchestrating organ of the human mind, but the two concepts are not synonymous – for the obvious reason that the human mind didn't evolve solely for the sake of the brain, or vice versa. Minds as we currently know them belong to

* The discipline of course still exists, but now it overlaps considerably with what is commonly called cognitive science, and indeed with neuroscience.

living entities, to organisms as a whole, even if they are not distributed evenly throughout them like some sort of animating fluid.

I fear Ryle wouldn't like this perspective either. He derided the Cartesian division of mind from body as the 'ghost in the machine', and he argued instead that mind shouldn't be regarded as some immaterial homunculus that directs our actions, but is synonymous with what we do, and thus inseparable from the body. Yet he felt the problem was not so much that the two are intimately linked as that Descartes' dualism is a category error: minds are fundamentally different sorts of things from bodies. It is no more meaningful, he wrote, to say that we are made up of a body plus a mind than that there is a thing made up of 'apples plus November'. Descartes only bracketed the two together (Ryle says) because he felt duty-bound, in that age, to give an account of mind that was couched in the language of mechanical philosophy: the body was a kind of mechanism, and so the mind had to be something of that kind too, or related to it. Ryle would probably take the same dim view of modern neuroscience, which seeks to develop a mechanistic account of the human brain. Yet the simple fact is that no one can (or at least, no one should) write a book today about the question of minds while excluding any consideration of neuroscience, brain anatomy and cognitive science. Nor, for that matter, can they ignore our evolved nature. It is like trying to talk about the solar system without mentioning planets or gravity.

The brain, though, is a profound puzzle as a physical and biological entity. Compare it, say, with the eye. That organ is a gloriously wrought device,* including lenses for focusing light, a moveable

* Wrought, I hope it goes without saying, by the blind forces of evolution, which sift random changes in form for ones that improve function, in ways advantageous to survival and the propagation of offspring. The marvellousness of the eye has made it a favourite example for those who wish us to believe that it must be

aperture, photosensitive tissues to record images, delicate colour discrimination, and more. All of these components fit together in ways that make use of the physical laws of optics, and those laws help us to understand its workings. The same might be said of the ear, with its membraneous resonator and the tiny and exquisitely shaped bones that convey sound along to the coiled cochlea, capable of discriminating pitch over many orders of magnitude in frequency and amplitude. Physics tells us how it all functions.

But the brain? It makes no sense at all. To the eye it is a barely differentiated mass of cauliflower tissue with no moving parts and the consistency of blancmange, and yet out of it has come *Don Quixote* and *Parsifal*, the theory of general relativity and *The X Factor*, tax returns and genocide.

Of course, under the microscope we see more: the root network of entangled dendrites and their synaptic junctions, the mosaic of neurons and other cells, bundles of fibres and organized layers of nerves, bursts of electrical activity and spurts of neurotransmitters and hormones. But that in itself is of little help in understanding how the brain works: there's nothing here suggestive of a physics of thought in the same way as there is of vision and hearing. Conceivably, the microscopic detail just makes matters worse (at least at first blush), because it tells us that the brain, with its 86 billion neurons and 1,000 trillion connections, is the most complex object we know of, yet its logic is not one for which other phenomena prepare us.

What's more, the lovely contrivances of the ear and eye (and other facilitators of the senses) are in thrall to this fleshy cogitator. Though we can understand the physical principles that govern sight and sound, the brain can override them. It makes us see things that are patently absent, such as the light falling on the retina, and also

truly miraculous: that a divine intelligence lies behind this and other of nature's designs. But any need for such foresight has long since been proved otiose.

remain blind to things that imprint themselves there loud and clear. The output of the ear is like an oscilloscope trace of complex sonic waveforms: none of it comes labelled as 'oboe' or 'important command' or 'serious danger alert', and certainly none instructs us to feel sad or elated. That's the brain's job.

All this means that science can be forgiven for not understanding the brain, and deserves considerable praise for the fact that it is not still a total mystery. The best starting point is an honest one, such as can be found in Matthew Cobb's magisterial 2019 survey *The Idea of the Brain*, which states very plainly that 'we have no clear comprehension about how billions, or millions, or thousands, or even tens of neurons work together to produce the brain's activity.'

What we *do* know a lot about is the brain's anatomy. Like all tissues of the body, it is made up of cells. But many of the brain's cells are rather special: they are nerve cells – *neurons* – that can influence one another via electrical signals. It's easy to overstate that specialness, for many other types of cell also support electrical potentials – differences in the amount of electrical charge, carried by ions, on each side of their membranes – and can use them to signal to one another. What's more, neurons are like other cells in conveying signals to one another via molecules that are released from one cell and stick to the surface of another, triggering some internal change of chemical state. But only neurons seem specially adapted to make electrical signalling their *raison d'être*. They can achieve it over long distances and between many other cells by virtue of their shape: tree-like, with a central cell body sporting branches called dendrites that reach out to touch other cells, and an extended 'trunk' called an axon along which the electrical pulse (a so-called action potential) can travel (Figure 1.1). Each of these pulses lasts about a millisecond.

This 'touching' of neurons happens at junctions called synapses (Figure 1.2), and it doesn't require physical contact. Rather, there is

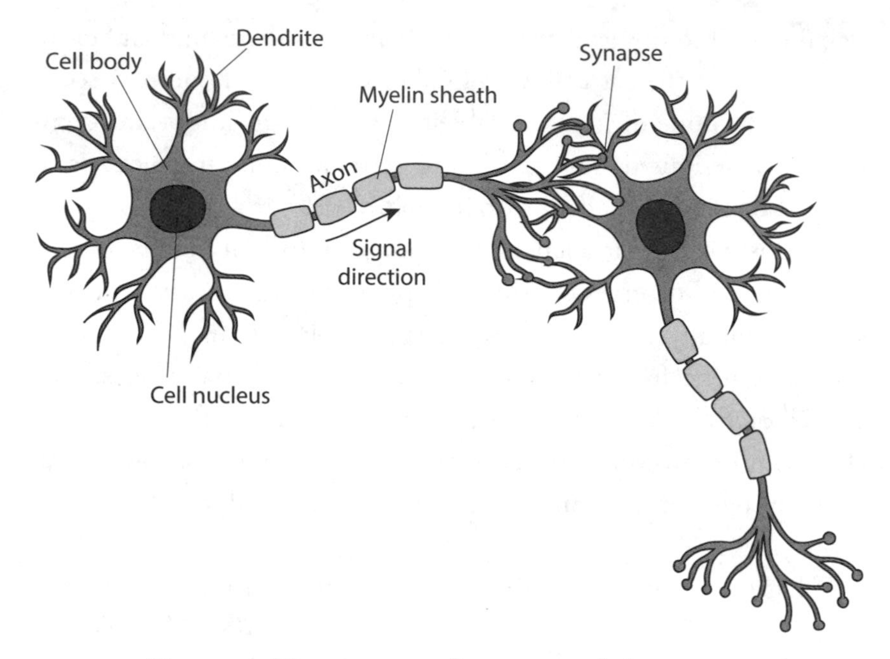

Figure 1.1. The structure of neurons and the synaptic junctions between them.

a narrow gap between the tip of an axon and the surface of another neuron's dendrite with which it communicates, called the synaptic cleft. When an electrical signal from the axon reaches the synapse, the neuron releases small biomolecules called neurotransmitters, which diffuse across the synaptic cleft and stick to protein molecules called receptors on the surface of another cell – these have clefts or cavities into which a particular neurotransmitter fits a little like a key into a lock. When that happens, the other cell's electrical state changes. Some neurotransmitters make the cell excited and liable to discharge a pulse of their own. Others quieten the cells they reach, suppressing the 'firing' of the neuron's distinctive electrical spike.*

* Different neurotransmitters are often described as though they have specific effects on the brain: serotonin is the 'well-being' signaller, suppressing violence

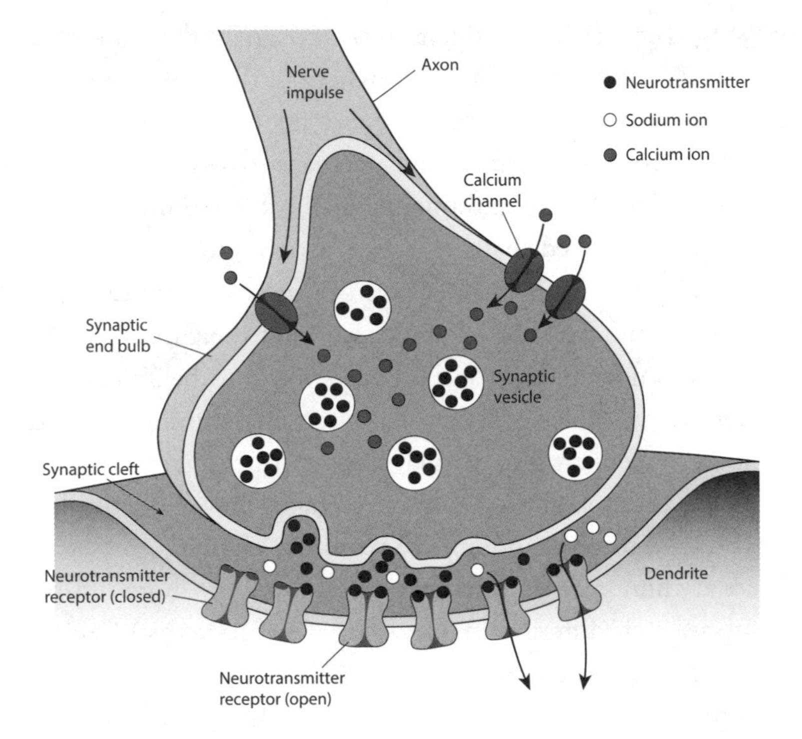

Figure 1.2. Close-up in the synaptic junction. Communication from one neuron to another happens when neurotransmitter molecules are released from the tip of the axon into the narrow gap between it and the dendrite of another neuron. These molecules diffuse across and bind to special 'receptor' molecules on the surface of the dendrite.

In this way, the network of neurons becomes alive with electrical activity. Typically the brain develops synchronized patterns of neural firing, and this synchrony seems central to coherent cognitive

---

and aggression, dopamine is the 'euphoria' signaller, and so on. But while it's surely true that different neurotransmitter molecules have different roles and effects, we should resist stereotyping them, in much the same way that we should resist labelling certain genes as being responsible for specific traits. Human-level experience and behaviour rarely if ever translate in a transparent way down to the level of molecules.

processing. The activity is influenced by the signals the brain receives from sensory organs and nerves elsewhere in the body – those, say, from the optic nerve connected to the retina of the eye. In this way the raw information registered by such sensory organs is somehow turned into mental images: thoughts, feelings, memories.

It's often overlooked that the brain's activity doesn't *rely* on such stimuli. It is intrinsically active: neurons are communicating with each other all the time. We don't know what they are 'saying', but the activity is not random, and it seems likely to be a vital part of cognition. In fact, the signals caused by sensory stimuli are typically very small by comparison, so that they can be detected by brain-monitoring technologies only after averaging away all the brain's noisy (but not merely random) 'background chatter'.

There are hundreds of different types (and many more sub-types) of neurons, each differing in size, shape and patterns of electrical activity. Some are excitatory: they stimulate others to fire – while others are inhibitory, suppressing activity in those to which they are connected. And brains are not just neurons. There are other cell types present too, in particular so-called glial cells, which outnumber neurons by a factor of about ten. Once considered just a kind of 'glue' tissue (that's what the name glia means) to bind the neurons together, glial cells are now known to play several vital roles in brain function, for example helping to maintain and repair neurons.

Too often overlooked also is the brain's energy economy. Our brains are big and expensive. Sheer physical size is not quite the issue – the sperm whale's brain is several times heavier (around 7.5kg) than ours, and men have a brain volume around 10 per cent larger on average than that of women without any difference in intelligence. What matters more is the ratio of brain mass to body mass, which is greater for humans than for any other animals. (Dolphins and great apes come next.) The brain accounts for about a quarter of the energy consumption in an adult (for newborn babies it is around 87 per cent), and the energy demand is almost twice as great when the brain is conscious and aware than when it is rendered unconscious by anaesthesia: thinking literally demands brain power. One of the marvels of the brain is that it doesn't simply fry itself with all this energy use: a computer with this density of components and interconnections would simply melt from the heat it would have to dissipate.

Although the general anatomy of the brain is as pre-determined as any other part of the body by the interaction and activation of genes and cells during development, the details of the wiring are shaped by experience. This happens in a process more akin to sculpture than to painting. Rather than the neural network growing gradually link by link, it starts as a profuse abundance of branches and junctions that gets pruned back. At birth, a baby's brain typically contains more than 100 billion neurons, which is somewhat more than the adult brain. Connections that are unused wither away, while recurring patterns of activity are reinforced. Neurons that are activated by a particular stimulus tend to forge connections with one another, creating circuits with specific functions.

This process continues throughout growth and life in a constant feedback between brain and environment. It's one reason why we can't easily or meaningfully separate what is innate in a brain from what is acquired by experience, for pre-existing traits may become amplified. A child with inherently good pitch perception or rhythmic coordination is encouraged to study music and thereby improves these traits still further, strengthening the relevant neural circuits. Any small gender-related differences in cognition – and what these are, or whether they exist at all, remains controversial* – are likely to

* Humans are in fact unusual among animals in having so few clear-cut gender differences in behavioural traits and abilities. Neuroscientist Kevin Mitchell

be amplified by cultural stereotypes, even in the most enlightened households. What's more, the traits we might recognize at a behavioural level, such as neuroticism or openness to new experiences, don't seem to have any clear correlates at the level of the neural circuitry – there aren't, say, 'conscientiousness' neurons. Personality is evidently a meaningful notion at the social level, but it's not obviously to be found in the wiring of the brain. Rather, these high-level, salient aspects of behaviour arise out of more 'primitive' characteristics, some of which *do* seem to have their origin in specific, identifiable aspects of the brain such as hormone levels or production of different neurotransmitters: aversion to harm, say, or ability to defer gratification.

Other traits and abilities are, however, more fundamental. The basic abilities that enable us to make sense of and navigate the world, such as vision, language, and motor skills, are each produced in specific regions of the brain. But not necessarily unique ones. Vision, for example, begins when light on the retina of the eye stimulates the optic nerve to send signals to the primary visual cortex (located, oddly, at the back of the brain). But as we'll see, for us to perceive an object visually it's not enough that its visual signal arrives here. The primary visual cortex communicates with other regions of the cortex that *interpret* what is seen, for example by separating it into different components (such as edges, textures, shadows, and movement) and interpreting them in terms of familiar objects. Vision defects can result from malfunctions at any of these stages.

---

speculates that perhaps some anatomical differences (on average) between male and female brains exist to *compensate* for the very clear physiological differences, for example in brain size or hormones, that might otherwise be expected to produce cognitive differences. Maybe there were good evolutionary reasons for human males and females to become more alike in their behaviour.

The brain has a mystique unequalled by any other part of the body. We commonly imagine that it holds the secrets of all that we are. There is marginally more justification for that notion than there is for the other modern biological myth of identity, which ascribes it all to the genome. For while our genes can dictate various dispositions in how the brain (as well as the rest of the body) develops, unquestionably influencing our traits, innate abilities and behaviours, the mantra that 'DNA is not destiny' can't be stressed enough.

It's a common view that personality and character traits arise through a combination of nature – genetic predispositions – and nurture, the slings and arrows of experience. Both play a role, but they are hard to disentangle and even harder to predict. Discomfiting though many find that idea, the clear fact is that genes play a significant, perhaps sometimes even dominant role, in guiding what we do. There is no human trait, from sexual orientation to a propensity for watching television or probability of divorcing, that doesn't appear to have a genetic component to it.* Yet environment or experience too may affect the way the brain gets wired up. Training for a skill, for example, can restructure the parts of the brain concerned, while deprivation or abusive treatment can leave long-lasting scars on the psyche – which necessarily means on the physical structure of the brain. Many environmental effects are hard to predict, however. Particular life events – a parental divorce, say – can have very different effects on different individuals, in part

* It sounds unlikely, even absurd, that there should be any genetic basis for television-watching, given that television is much too recent an invention for natural selection to have exerted any influence on watching habits. But as with so many other aspects of behaviour, the determining factors here are very general ones on which selection has long acted, such as an ability to maintain focused attention.

because of the innate characteristics of personality. Bluntly put, some people will weather trauma better than others.

There is a third influence, however, on the individual features of the brain (and the mind it supports) that is too often overlooked: neither nature (genes) nor nurture (experience), but chance events in both development and the environment. Genes give the relevant cells a programme of sorts for how to organize themselves into a brain structure, and they can bias certain aspects of that organization. But they cannot fully prescribe it, not least because the number of neural connections outweighs the number of genes involved in brain development by several billionfold. There is no blueprint for a brain in the genome. (In truth there's no blueprint for anything.)

The developmental process of growing a brain is thus astronomically complex, and tiny, random events during its course can have significant knock-on effects. If we have a particular talent, we love to say that it's in the genes, and seek for a relative to whom we can give the credit. Perhaps that's so; but it's often also possible that you just got lucky in the way your brain happened to wire up.

Is the brain a mind machine?

The brain's mystique surely stems from the notion that from it comes all that we are. The body, in this view, is just the housing: the mindless machinery that the brain controls. Thence comes the fascination with famous brains – as for example in the determination of some surgeons to seek the origin of Einstein's genius in slices of his grey matter, preserved against his wishes after his death, some of which circulated within medical networks like contraband.

There is a lot invested in the human brain. It is reasonably considered the 'engine of mind', provided that we remain alert to the shortcomings of such a mechanical metaphor. It both determines and constrains what a human mind can be like. But it does not embody the sum total of that issue, nor can it answer all the questions we might ask about mindedness. In exploring the Space of Possible Minds, we will need to perform a constant dance between the physical 'hardware' and the properties that emerge from it – a dialogue of mind and matter.

When we reason, think, feel, there are of course brain circuits underpinning those activities, and occasionally I will need to refer to them. But it's important to keep in mind that how the mind works and how the brain works are two distinct (though related) questions. Understanding one of them doesn't necessarily tell us about the other, and much of neuroscientific research today is aimed at trying to connect the two.

Neuroscience now has a wonderful battery of techniques for studying brain activity, most notably functional magnetic resonance imaging (fMRI), which reveals where blood flow has increased because of the demands for oxygen created by cell (neural) activity. These maps of brain activity can be revealing. If for example we discover that a specific cognitive task involves a part of the brain previously associated with a different function – if, say, listening to music activates regions linked to language processing – that can give us insight into the kind of mindfulness that is going on: the brain may be identifying a sort of syntax or grammar in the music. On the other hand, very often such regions have been labelled in the first place because of their activation in response to a particular task or behaviour, rather than because we know what's happening in said region. Unfortunately, brain-imaging methods don't come supplied with a convenient description of what the associated neural circuits are actually doing with the information they receive.

The danger is that the brain becomes simplistically divvied up into regions ascribed to specific functions, in much the same way as it was in the now discredited late nineteenth-century 'science' of phrenology. That tendency has given rise to simplistic and misleading tropes about brain anatomy: that the structure called the amygdala is said to be the 'fear centre', say. Such habits are encouraged by some neuroscientists' unfortunate tendency to confuse an explanation of a cognitive task with a mere list of the obscurely named parts of the brain that are active during it: the anterior cingulate cortex, the parahippocampal gyrus, and so on.

This returns us to the mind–brain problem: we can examine the brain as a biological organ all we like, but we still can't get inside the mind it helps create, and see what is going on. This is a tricky notion to grasp. Without the human brain, there is no human mind – we can be confident about that.* But does this mean that the brain is all there is to mind, or could it be analogous to saying that without the conductor, the orchestra does not play? The conductor is the organizer, the hub of all the activity – but heck, the conductor does not even play the music. Usually the conductor did not even compose the music. In fact, where *is* the music – on paper, in the vibrations of the air, or in the minds of the audience? What generates the music of the mind? And how many kinds of music can it play?

As we have seen, the brain–mind relationship is commonly couched today in terms of an analogy with computers: the brain is said to be the hardware (or perhaps, the 'wetware') and the mind is the software. In this picture – widely, if not universally, held in neuroscience – the brain is itself a kind of computer.

The analogy has existed since the earliest days of computers: the pioneer of information technology John von Neumann wrote a book titled *The Computer and the Brain* in 1957. Brains, as we'll see, have often been the explicit inspiration for work in computer

* Whether all minds require brains is another (and unresolved) matter, as we'll see.

science and artificial intelligence, to the point where, according to neuroscientist Alan Jasanoff, 'it can be difficult to tell which is the inspiration for which.' The computational view of mind is agnostic about where we might expect to find minds: it allows mindedness to arise in any system capable of the requisite computation. It represents one view of the philosophical perspective on mind called functionalism, which holds that a mental state or operation – a thought, desire, memory – doesn't depend on what the mind is made from, but on what it does – on its functional role in the mind-system, independent of the substrate in which it is supported. That view can be discerned in Thomas Hobbes' mechanistic view of the human mind, whereby it is all a kind of arithmetical processing: reason, he wrote in 1651, is 'nothing but *reckoning*, that is adding and subtracting.'

You can see the attraction of this idea. Just as computer circuits process information by passing digital (on/off) electrical signals through networks of vast numbers of interlinked 'logic gates' made from transistors, so the brain has neural networks made from interlinked neurons whose action potentials switch one another on and off. The brain, the argument goes, is much more powerful and versatile than the computer because it has many more components and connections; but the computer is often faster, because silicon devices can switch in nanoseconds (billionths of a second) while neurons fire only a few times per second.

One of the reasons the brain–computer analogy is so popular is that it offers what may be a false reassurance that the brain can be understood according to well-established engineering principles. It might be seen, Jasanoff suggests, as the modern equivalent of Descartes' separation of messy, material body from pristine, immaterial mind. The corollary is that mind doesn't need brain at all, but just information: a computer can simulate a brain in every detail, and thereby host a mind. 'Equating the organic mind to an inorganic mechanism might offer hope of a secular immortality', says Jasanoff. I'll return to this fantasy later.

But does the analogy hold up? History alone should make us wary: metaphors for the brain have always drawn on the most advanced technologies of the day. Once they were regarded as working like clockwork, or hydraulics, or electrical batteries. All of these now look archaic to us, and it's possible that the computer comparison will too, one day.

The most fundamental objection to the analogy is that brains and minds simply do not do what computers do. Gilbert Ryle considered it a common mistake to 'suppose that the primary exercise of minds consists in finding the answers to questions'. That's typically the task for a computer: it must compute an output based on the values input to a well-defined algorithm. It *looks* as though the brain does something like that too – that it enacts an algorithm for turning the input data of sensory experience into outputs such as actions. But the resemblance is superficial. Rather than information flowing through the brain in one direction from input to output, artificial-intelligence expert Murray Shanahan observes that 'waves of activation' can move back and forth between different regions until settling into a 'temporary state of mutual equilibrium.' The brain does not sit there waiting for questions to answer, but is in constant, active and structured conversation with itself and its environment. Frankly, we don't yet understand much about that discourse.

Moreover, the computer analogy says nothing about the role of sentience, which is surely one of the most salient aspects of our own minds. We can already make machines – assuredly non-sentient entities, at least at this stage of their development – that function as information-processing input–output devices that can decide on and execute actions. Conversely, our own minds seem capable of experiencing without acting: of being in states close to pure awareness, without any objectives or outputs to compute.

What's more, we can't in general, enumerate the reasons why we do what we do by mapping out the logical processes of mind in the way we can for a computer's circuits – identifying the input data, the key decision points and the outcomes of those processes, and so on. (We *do*, however, devote a lot of energy to constructing narratives of that sort – because coherent stories are something our minds appear to crave, for good adaptive reasons).

Yes, the mind has to make decisions conditioned on existing and new information – but those decisions are not what *make it a mind.* They are not what gives to a mind something that it is to be like. As we'll see, it looks possible that the subjectivity of awareness might be fundamental to what a mind does, not just an epiphenomenon of its problem-solving nature. Neither is the mind a kind of machine that goes to work on its environment; rather, it is something that emerges from the interaction of organism and environment. For this reason, it is neither some ethereal or abstract essence, nor a thing that can be written down as a kind of code or algorithm. It is a process, perpetually running, forever recreating and updating itself. The brain of a moving, active organism is part of a constant feedback loop: it controls and selects which stimuli it perceives, as well as adjusting to those stimuli. Typically, experiments on behaviour open up that loop artificially by selecting stimuli for the brain and watching the result. But ordinary behaviour is quite different, which is why cognition – what the brain actually does in the environment, an active and ongoing process – is not passive input–output computation. The 'input–output doctrine', writes neuroscientist Martin Heisenberg, 'is the wrong dogma, the red herring [in brain research].'

Even reflex actions – where stimulation of a given nerve or set of nerves automatically generates a particular response, like the patellar reflex where a tap of the knee produces an automatic leg jerk – are in fact controlled by the brain and not just passively enacted by them. What might look like a simple input–output response may actually be fine-tuned by the brain in response to the prevailing circumstances. Thus, some apparently instinctive and unmeditated responses in animals are actively altered and directed by the brain in subtle ways to suit the situation: the organisms are not acting with the predictability of a light switch.

What's more, brains do not need stimuli in order to be active. They are busy during sleep, for example, when our senses are mostly shut down. While external realities (coldness, or a full bladder) might sometimes leak into our dreams, most of their content is internally generated. Turn off all the stimuli to an awake and conscious person – place them in a sensory-deprivation chamber – and the mind will soon start to produce its own activity, which manifests as hallucinations. Nervous systems are dynamic and constantly active, whether or not the world feeds them sensory data. Minds exist *to mind.*

In contrast with a conventional computer algorithm, then, you cannot work out what a human brain will do when given a certain set of sensory inputs (which would have to include hormones and other bodily signals).* After all, the possible configurations and states that can be sustained in a typical neural network of a brain is larger than the number of particles in the observable universe. Yet just a tiny fraction of these states have been selected by evolution, and they are reliably produced in response to pretty much whatever stimuli the brain receives.†

* We'll see later that the approach generally used for today's artificial intelligence, called machine learning, also lacks transparent predictability in how inputs produce outputs. But we'll also see why this computational process differs from the way brains work.

† Are there any stimuli that the brain simply can't handle – for which it can't settle into one of its selected states? It would be intriguing, though perhaps hazardous,

That's the key reason why neuroscientists Gerald Edelman and Giulio Tononi argue that the brain can't be considered a computer. For even though many aspects of cognition and neural processing do closely resemble computation, it does not arrive at its internal states by logic operations, but by *selection*.

This might not be a coincidence. That's to say, it might be that the minds of evolved living beings *have* to be of this type, because purely logic-based machines won't work. Living organisms need, for one thing, to have some stability of behaviour – a reliable repertoire of possible actions adequate for most situations they will face, even though those are unpredictable and never identical. And they need to make decisions quickly, without laborious computation of all the input signals. But perhaps most of all: they are decision-making devices that *don't really have a well-defined input at all.* There's no moment (except perhaps in maths tests) where our mind is presented with a closed problem, characterized by uniquely specified, single-valued parameters, that we have to solve. The world is not a computer tape that feeds a binary digit string into our brain. 'Like evolution itself', say Edelman and Tononi, 'the workings of the brain [it might be better to say 'of the mind'] are a play between constancy and variation, selection and diversity.' Machines don't work this way – or perhaps it is better to say, we've never yet made machines that do. Computer scientist Josh Bongard and biologist Michael Levin turn the matter on its head, suggesting that perhaps living and minded entities show us 'machines as they could be'.

Yet my hunch is that no genuine mind will be like today's computers, which sit there inactive until fed some numbers to crunch.

---

to know. Might we indeed have a mental meltdown when faced with the cosmic infinitude of H. P. Lovecraft's Chthulu, or when fed the wisdom of super-advanced intelligences, as two hapless characters are in Fred Hoyle's sci-fi novel *The Black Cloud* (page 338)?

It's certainly notable that evolution has never produced a mind that works like such a computer. That's not to say it necessarily cannot – evolution is full of commitment bias to the solutions it has already found, so we have little notion of the full gamut it can generate – but my guess is that such a thing would not prove very robust. Edelman and Tononi postulate that logical computation (as in our artificial 'thinking devices') and selection (in evolved minds) might be the *only* two 'deeply fundamental ways of patterning thought'. If so, that would surely be a fundamental aspect of topography of the Space of Possible Minds. But *are* they the only two? We know of no others, but personally I'd be surprised if they don't exist. After all, we are just starting our journey into Mindspace, and we have very little conception of how big it is, and what lies out there.

CHAPTER 2

The Space of Possible Minds

In 1984 the computer scientist Aaron Sloman, of the University of Birmingham in England, published a paper arguing for more systematic thinking on the vague yet intuitive notion of mind. It was time, he said, to admit into the conversation what we had learned about animal cognition, as well as what research on artificial intelligence and computer systems was telling us. Sloman's paper was titled 'The structure of the space of possible minds'.

'Clearly there is not just one sort of mind', he wrote:

> Besides obvious individual differences between adults there are differences between adults, children of various ages and infants. There are cross-cultural differences. There are also differences between humans, chimpanzees, dogs, mice and other animals. And there are differences between all those and machines. Machines too are not all alike, even when made on the same production line, for identical computers can have very different characteristics if fed different programs.

Now an emeritus professor, Sloman is the kind of academic who can't be pigeon-holed. His ideas ricochet from philosophy to information theory to behavioural science, along a trajectory that is apt to leave fellow-travellers dizzy. Ask him a question and you're likely to find yourself carried far from the point of departure. He can sound dismissive of, even despairing about, other efforts to ponder the mysteries of mind. 'Many facts are ignored or not noticed,' he told me, 'either because the researchers don't grasp the concepts needed to describe them, or because the kinds of research required to investigate them are not taught in schools and universities.'

But Sloman shows deep humility about his own attempt four decades ago to broaden the discourse on mind. He thought that his 1984 paper barely scratched the surface of the problem and had made little impact. 'My impression is that my thinking about these matters has largely been ignored', he says – and understandably so, 'because making real progress is very difficult, time-consuming, and too risky to attempt in the current climate of constant assessment by citation counts, funding, and novel demonstrations.'

But he's wrong about that. Several researchers at the forefront of artificial intelligence now suggest that Sloman's paper had a catalytic effect. Its blend of computer science and behaviourism must have seemed eccentric in the 1980s but today it looks astonishingly prescient.

'We must abandon the idea that there is one major boundary between things with and without minds', he wrote. 'Instead, informed by the variety of types of computational mechanisms already explored, we must acknowledge that there are *many* discontinuities, or divisions within the space of possible systems: the space is not a continuum, nor is it a dichotomy.'

Part of this task of mapping out the space of possible minds, Sloman said, was to survey and classify the kinds of things different sorts of minds can do:

> This is a classification of different sorts of abilities, capacities or behavioural dispositions – remembering that some of the behaviour may be internal, for instance recognizing a face, solving a problem, appreciating a poem. Different sorts of minds can then be described in terms of what they can and can't do.

The task is to explain what it is that enables different minds to acquire their distinct abilities.

'These explorations can be expected to reveal a very richly structured space', Sloman wrote, 'not one-dimensional, like a spectrum, not any kind of continuum. There will be not two but many extremes.' These might range from mechanisms so simple – like thermostats or speed controllers on engines – that we would not conventionally liken them to minds at all, to the kinds of advanced, responsive, and adaptive behaviour exemplified by simple organisms such as bacteria and amoebae. 'Instead of fruitless attempts to divide the world into things with and things without the essence of mind, or consciousness', he wrote, 'we should examine the many detailed similarities and differences between systems.'

This was a project for (among others) anthropologists and cognitive scientists, ethologists and computer scientists, philosophers, and neuroscientists. Sloman felt that AI researchers should focus less on the question of how close artificial cognition might be brought to that of humans, and more on learning about how cognition evolved and how it manifests in other animals: squirrels, weaver birds, corvids, elephants, orangutans, cetaceans, spiders, and so on. 'Current AI', he said, 'throws increasing memory and speed and increasing amounts of training data at the problem, which allows progress to be reported with little understanding or replication of natural intelligence.' In his view, that isn't the right way to go about it.

What it is like

Although Sloman's concept of a Space of Possible Minds was stimulating to some researchers thinking about intelligence and how it might be created, the cartography has still scarcely begun. The relevant disciplines he listed were too distant from one another in the 1980s to make much common cause, and in any case we were then only just beginning to make progress in unravelling the cognitive complexities of our own minds. In the mid-1980s, a burst of corporate interest in so-called expert-system AI research was soon to dissipate, creating a lull that lasted through the early 1990s. The notion of 'machine minds' became widely regarded as hyperbole.

Now the wheel has turned, and there has never been a better time to consider what Sloman's 'Mindspace' might look like. Not only has AI at last started to prove its value, but there is a widespread perception that making further improvements – and perhaps even creating the kind of 'artificial general intelligence', with human-like capabilities, that the field's founders envisaged – will require a close consideration of how today's putative machine minds differ from our own. Understanding of animal cognition too has boomed in the past two decades, in part because of the new possibilities that neuroscience and information technologies have opened up (but frankly, mostly because of better behavioural experiments). Child-psychologists now routinely talk to roboticists and computer engineers, and neurologists to marine biologists. We have some of the conceptual and experimental tools to start mapping a landscape of minds.

To avoid raising false expectations, however, I must confess at once that neither I nor anyone else yet knows what the natural coordinates of Mindspace are – nor even whether they are well-defined at all. Minds seem unlikely to be the kinds of objects that one can represent by mathematical functions of precisely quantifiable variables: Mind = $xy^2 + qz^3$ or some such. Yet I do believe we can identify some of the likely *components of mindedness*: the qualities that minds seem to exhibit. What's more, we can anticipate that some minds are likely to be more richly imbued with these qualities than others. We can ask what kinds of minds, as a consequence, they are. Perhaps we can even try to imagine our way inside them.

To consider this Mindspace most fruitfully, I believe we will need to exercise our imagination. That facility is a vastly underrated tool in science, which (from the outside at least) can look like a pursuit built from strict logic and rigour, demanding precision of theory and observation. From the inside, on the other hand, I suspect there is no scientist who does not appreciate the value of imagination. Many like to cite Einstein's famous quote (which for once is genuine):

> I'm enough of an artist to draw freely on my imagination, which I think is more important than knowledge. Knowledge is limited. Imagination encircles the world.

What Einstein was referring to here is something like the leap of faith that enables the scientist to see just beyond the data, to construct new hypotheses for empirical testing, and to trust in the intuition that might be required to sustain their speculative ideas while no data exist to adjudicate on them.

The imagination that will help us to explore the Space of Possible Minds is of a somewhat different order, more akin to that required by a novelist: the ability to see through other eyes, to hear with other ears. With skill and research, it is possible to make a good fist of imagining oneself into the head of a Renaissance painter in Florence or a courtier of the Chinese Tang dynasty – or (if you are not already one yourself) into the viewpoint of an autistic boy, as in Mark Haddon's 2003 novel *The Curious Incident of the Dog in the Night-time*. On occasion, novelists have tried to go further: James Joyce projecting himself into early childhood at the start of *Portrait of an Artist as a Young Man*, Laline Paull's apian protagonist in *The Bees*, Orhan Pamuk making his narrator a coin or the colour red in *My Name is Red*. Yet there's little attempt in these latter works to make the characters anything more than humanized non-humans.

No dog, let alone a coin, can write a book, and no ghost-writer can help with that. One of the most gloriously quixotic attempts to install oneself in a non-human mind is Charles Foster's *Being a Beast* (2016), in which he tries to enter into the lifestyle and perspective of an otter, a fox, and a swift. In the end, Foster's book reveals how doomed an enterprise that is. But if there's one attribute that distinguishes the human mind from those of other creatures, it is our imagination: our capacity for fantastical 'what ifs', for mental metempsychosis. It's almost as if we have evolved to make these attempts to become what we are not. Of course we will fail if, say, we try to picture what it must be like to 'see' the world as a silicon-based neural network does, or a fruit fly. But what I hope to show you is that we can fail better – and to do so, we don't need to start from scratch, nor to set out totally blind into the landscape where our destination lies.

Anyone even vaguely familiar with the philosophy of mind will doubtless be wondering when Thomas Nagel's bat will flap into view. This seems to be its cue. In 1974 Nagel, a distinguished American philosopher, published what might be the most widely cited paper in the entire field of the philosophy of mind, titled 'What is it like to be a bat?'.

As you can see, the title reflects the very definition I have chosen for an entity to have mindedness: there is something it is like to be that entity. Nagel chose to pose the question for a bat partly because this creature is evolutionarily distant enough from us to seem rather alien, yet close enough for most people to believe that it does have some subjective experience. (We will look at whether that sort of belief is justified, and what it might mean, in Chapter 5.) But he also selected the bat because the modality of its sensory environment is so different from ours: although it has a visual system, it creates an image of its surroundings primarily using sonar echolocation, the reflection from surfaces of the acoustic waves it emits as shrill shrieks.

Bat sonar, Nagel wrote,

> though clearly a form of perception, is not similar in its operation to any sense that we possess, and there is no reason to suppose that it is subjectively like anything we can experience or imagine. This appears to create difficulties for the notion of what it is like to be a bat. We must consider whether any method will permit us to extrapolate to the inner life of the bat from our own case, and if not, what alternative methods there may be for understanding the notion.

He went on:

> It will not help to try to imagine that one has webbing on one's arms, which enables one to fly around at dusk and dawn catching insects in one's mouth; that one has very poor vision, and perceives the surrounding world by a system of reflected high-frequency sound signals; and that one spends the day hanging upside down by one's feet in an attic. *In so far as I can imagine this (which is not very far),* it tells me only what it would be like for me to behave as a bat behaves. But that is not the question. [My italics.]

It is, rather, the technique of the novelist, the modus operandi of, say, Richard Adams for imagining his rabbit protagonists in *Watership Down*. But, wrote Nagel,

> I want to know what it is like for a bat to be a bat. Yet if I try to imagine this, I am restricted to the resources of my own mind, and those resources are inadequate to the task. I cannot perform it either by imagining additions to my present experience, or by imagining segments gradually subtracted from it, or by imagining some combination of additions, subtractions, and modifications.

His conclusion that we cannot possibly hope – by our very nature, indeed essentially by our very definition – to know 'what it is like to be a bat' is hard to refute, and widely accepted. There's less consensus about the corollary Nagel presents, which is that consciousness itself – as he put it, the problem of how the mind arises from the body – is not accessible to scientific study:

> For if the facts of experience – facts about what it is like for the experiencing organism – are accessible only from one point of view, then it is a mystery how the true character of experiences could be revealed in the physical operation of that organism. The latter is a domain of objective facts par excellence – the kind that can be observed and understood from many points of view and by individuals with differing perceptual systems. There are no comparable imaginative obstacles to the acquisition of knowledge about bat neurophysiology by human scientists.

Even if we know all there is to know about the bat brain and physiology, that information can never tell us what it is to be a bat. By the same token, Nagel wrote that 'Martians might learn more about the human brain than we ever will' (he did not, I assume, actually believe there are Martians; they play the role of generic super-smart aliens) yet never know what human experience is like.

I don't want to give the impression that the only way to explore Mindspace is to project ourselves imaginatively into it and take a stroll. As Nagel says, we'd be fooling ourselves to think we could ever achieve that in other than a superficial and schematic manner. My point is that, merely by being open to the notion that there are other things it is to *be like* cuts some of the tethers that the philosophy of mind creates when it supposes not just that the human mind is the most interesting case to focus on, but that it is the alpha and omega of the subject.

The mind club

Discussions about varieties of mind have tended to be either freewheeling and amorphous or narrow and prescriptive. An example of the former is a 'taxonomy of minds' drawn up by Kevin Kelly, founding executive editor of the tech magazine *Wired*, which reads as an off-the-top-of-the-head list in which each item sounds like the premise of a sci-fi novel:

> Cyborg, half-human half-machine mind
> Super logic machine without emotion
> Mind with operational access to its source code
> Very slow 'invisible' mind over large physical distance
> Nano mind

and so on.

Alternatively, minds might be classified according to some single measure, most obviously 'intelligence'. Frankly, we don't know what we mean by intelligence – or rather, we don't all, or always, mean the same thing by it. Of course, we do measure *human* minds on an intelligence scale: the 'intelligence quotient' or IQ scale devised in 1912, allegedly the ratio of a person's mental to chronological age,* and which has been the subject of controversy ever since. Whatever their pros and cons for humans, IQ scales are certainly no good for other animals, or machines. If today's AI, or a dolphin, were to get a rotten score in an IQ test, we'd rightly suspect that it's not because they have 'low intelligence' (whatever that means) but because we're applying the wrong measure of intelligence – or more properly, of mind.

What's more, many tests of intelligence, like IQ scores, evaluate not cognition per se but performance. There's some value in doing that, but we know even from the way humans work that it can matter a great deal to the performance of a task how the question is asked, and in what circumstances. It is notoriously hard to find a fair test for all minds – trivially, I'd not fare at all well in an IQ test if it were written in Arabic. Performance alone might reveal rather little about the *kind* of mind being tested – about its modes of reasoning, of representing the environment, its intuitions and emotions (if any), its range of skills not probed by the test. A popular scheme for evaluating novel biologically inspired machine intelligence in the 2000s was called the Cognitive Decathlon, which tested skills such as recognition, discrimination, memory and motor control. It's too simplistic but nevertheless captures the spirit of the enterprise to say that the criterion of intelligence in this test was more or less whether

* IQ is generally considered a good proxy for the *g* factor, a measure of 'general intelligence' introduced around the same time by the English psychologist Charles Spearman. This was supposed to capture the observation that people (Spearman considered children) who perform well in one cognitive arena often do in others too. Like IQ, it has proved controversial – famously, the evolutionary biologist Stephen Jay Gould criticized it for reducing intelligence to a number, a 'single series of worthiness' that discriminated against disadvantaged groups.

the machine can play chess well. I don't think even Mensa would accept such a limited measure for humans, so why for machines?

While all those demonstrable skills matter a great deal for practical purposes, it seems equally important to have some sense of the kinds of cognitive realm these artificial 'minds' inhabit. In 1976, the psychologist Nicholas Humphrey argued that 'what is urgently needed is a laboratory test of "social skill"'. And what exactly is that? 'The essential feature of such a test', said Humphrey, 'would be that it places the object in a transactional situation where he [sic] can achieve a desired goal only by adopting his strategy to conditions which are continually changing as a consequence partly, but not wholly, of his own behaviour.' The principle is surely right, although such features of mind are likely also to be highly sensitive to context, and unlikely to be captured by a single measure.

What we really need to appreciate the varieties of mind is some way of plotting the dimensions that define the ways in which minds can work: more or less what Sloman's Space of Possible Minds posits. A simple yet ingenious way to begin that ambitious project was devised by American psychologists Daniel Wegner, Heather Gray and Kurt Gray (no relation) in 2007. They simply *asked* people about other minds: what attributes do we *think* they have? They canvassed the views of around 2,500 participants about the perceived mental capacities of humans, animals, and other entities such as robots, companies, and supernatural agents: ghosts and God. What, the volunteers were asked, do you imagine the minds of (say) chimpanzees and babies are like?

Surprisingly, the responses could be boiled down to a Space of Minds that had just two key attributes, which the researchers labelled *experience* and *agency*. Here 'experience' means not (as in colloquial usage) how *much* one has experienced, but rather the mind's innate capacity for an 'inner life': for feelings such as hunger, fear, pain, rage, pleasure, and joy. It really alludes to a sense that there is something it feels like to have such a mind; in other words, my working definition of mind demands non-zero measures of experience. 'Agency', meanwhile, connotes the ability to *do* things and accomplish goals, and to exercise control in doing so: to possess memory, say, or to plan and communicate.

So this Space of Possible *Perceived* Minds – what we humans imagine as the dimensions of mind – can be displayed as an ordinary two-dimensional graph on which each entity is represented as a data point with a certain amount of 'experience' and 'agency' (Figure 2.1). Those entities that have non-zero quantities of either or both attributes, suggested Wegner and colleagues, are the ones that we humans admit to the Mind Club.

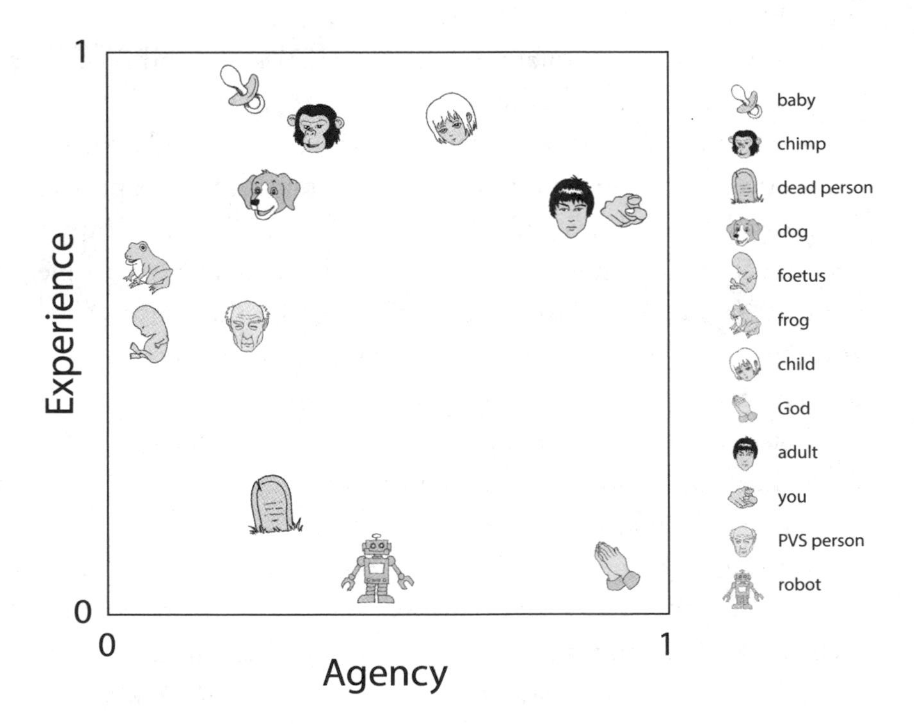

Figure 2.1. The Mind Club, according to Wegner, Gray and Gray: how we *think* about other minds.

It's something of a giveaway that adult humans are rated highly in both respects, and therefore appear at the top right of the graph. That's not to say we *don't* have a considerable amount of both experience (we feel, right?) and agency (we do stuff) – but it does suggest that all other minds are being assessed according to how they compare with us. That's hardly surprising. We have always thought in a human-centric way, and we shouldn't feel too bad about it. It is possibly in the nature, perhaps the very definition, of minds that they are at least somewhat egocentric and solipsistic. After all, when we started to map the cosmos we did that too from the human point of view: we put ourselves at the centre of all space. Literally, all else revolved around us – but what is more, it existed in relation to us and *for* us. It is us, not the beasts, that were deemed to be made in God's image and the primary focus of God's attention.

By degrees, our place in that cosmic expanse came to seem ever more contingent and insignificant. First we were on a planet like any other; then merely one of countless solar systems, embedded in countless galaxies – and perhaps even, some cosmologists think, inhabitants of one among countless universes. And just as the space of possible *worlds* looks very different today, so eventually will the Space of Possible Minds.

Humans are evidently regarded here as following a *path* in this space throughout the course of their lives. Babies were rated, on average, as slightly higher than adults on the ability to experience, but much lower on agency. Everything for a baby is intensely felt, eliciting happy gurgles or despondent howls; meanwhile, they can't get much done.* Children are intermediate between babies and

* Babies and even foetuses, like all complex organisms, have plenty of *biological* agency in the sense that I consider later (page 261) – but not in the more restrictive sense of the Mind Club.

adults: their agency steadily increases as they grow older, while their experience diminishes slightly.

Before the baby stage – when we are still foetuses – it's a different story. At that point we (are perceived to) have almost zero agency, but also less facility to experience. This stands to reason, for our brains and sensory systems are still forming: we don't yet have the full biological apparatus of mind. Presumably this trajectory begins, at conception, from close to the origin of the graph: zero agency, zero experience. At that point, we are deemed not much different to a sunflower, a bacterium, or maybe even a pebble. This is curious in the light of the moral controversy around embryos. If the general view is that they possess very little in the way of mind, what is it then that gives them a special moral status? As we will see, questions about the moral standing of entities are often tied to notions of their degree of consciousness, or as we might now say, of mindedness. Yet for humans it seems there is something else at play – some factor not represented in this particular Space of Possible Minds. It's not hard to see what that is: whether it is stated explicitly or not, most beliefs in the sanctity of the embryo stem from the notion of a soul – or as we might say in more secular terms, of *potential* for mindedness.

It's equally curious where the perceived human trajectory heads in cases of illness and death. The participants in the survey were asked to assess the minds of people in a permanent vegetative state (PVS): a comatose state in which the brain has more or less ceased to function, so that even basic body functions such as breathing need artificial assistance. Such people, who might be in that condition through illness or injury, have been the focus of some highly contentious legal battles about the 'right to live', a recent high-profile example being the case of Terry Schiavo in Florida. The Mind Club survey placed PVS patients at a similar level of experience to foetuses but with just a little more agency. They can't actually do significantly more than can foetuses, but one might intuit that their (generally) adult bodies are imbued with a *potentiality* for action. As Wegner and Kurt Gray write, 'the important [moral] question is whether the minds of [PVS] patients continue to exist, or at least whether they have minds that can be recovered.'

Even when we die, we are not deemed (in this survey) to return to the origin, so to speak. Dead people were attributed less experience than foetuses, *but not zero*. And strikingly, they were assessed as having rather more agency than people still alive but in permanent vegetative states. How the dead are regarded is obviously a cultural issue – the participants in the Mind Club survey were mostly white American Christians, and so a notion of life after death was presumably prevalent among them. But this conception demands that we entertain the idea of disembodied minds: in short, it seems Descartes's dualism is alive and well in the public perception.

If entities can feel – if they have experience – we tend to grant them moral rights: we consider that we have a duty to recognize and respect that experience. If, meanwhile, they have an ability to act – to cause change in the world – we are likely to give them some degree of moral responsibility: a duty to execute those actions in a way that respects moral rights. By this measure, adult humans incur both moral rights and moral responsibilities. (We afford children and animals rather more lenience, but devote some effort towards educating a sense of responsibility into them.)* Interestingly, we seem to perceive robots as having the latter but not the former: it matters, you might say, what they do to us but not what we do to them. And that perception stems from our view of what manner of mind they possess.

Look now at where 'God' is in this space of perceived minds: [S]he has high agency but hardly any experience. By this measure,

* I'm thinking of dogs, but not cats. We have resigned ourselves, I suspect, to the fact that cats refuse to recognize any moral responsibilities whatsoever.

God seems to be regarded almost as a super-powerful robot. That does not seem an entirely obvious result; I will return in Chapter 8 to what this implies about the Mind of God as it is perceived by us (which is, needless to say, all that can be said on that controversial matter).

Many animals, on this measure, are rated as having comparable moral rights to humans, because they have a similar degree of experience but less agency (in the case of 'simpler' organisms such as frogs, rather little). Here, perhaps, is where the relationship of agency to moral responsibility breaks down. Animals can certainly do things (including, in some cases, kill and eat us), but are they responsible for what they do? I don't suppose anyone feels the tiger ought to feel guilty about killing the monkey or the hare, or indeed the hapless person on incautious safari.

That's just one indication of how the Mind Club of Wegner and the Grays is not really a Space of Possible Minds, but rather a picture of how minds are colloquially viewed. One aspect of mind that has long exercised philosophers and theologians, and which surely features in our everyday considerations about behaviour, is not obviously included in this analysis: intention or volition, which traditionally has been labelled 'free will'. What makes all the difference between a tiger killing us and another person killing us is not how much agency or experience they possess, but what their intentions were. I will return to this in Chapter 9.

Wegner and the Grays included groups, institutions and nations in their survey.* 'Perhaps countries are conscious', they suggest, 'conversing with one another through us, just as we use our tongues to talk to other people.' People apparently deem them to be lodged, in these coordinates, somewhere between robots and God. Computer

* 'Is the United States conscious?' the researchers asked, and I suspect their answer was more than a little tongue-in-cheek: 'It's hard to know.'

scientist Daniel Hillis argues that nation-states and corporations are indeed 'hybrid intelligences' with their own emergent goals and modes of cognition. 'When talking about "What China wants" or "what General Motors is trying to do"', he says, 'we are not speaking in metaphors. These organizations act as intelligences that perceive, decide, and act.' We do, of course, speak colloquially in this fashion: 'China has ratified an agreement with India', 'Facebook is totally irresponsible', and so on. And countries, companies and committees make decisions that can't easily be attributed to any individuals – government ministers might all be trying to achieve one thing, say, while the outcome fails dismally to do so. It is more than metaphorical to say that companies act with intention and goals in mind; they certainly have agency, and they surely bear moral responsibility.

What is far less clear is whether they can, as a coherent entity, 'experience' anything. It does not matter that they have (as far as we can tell) no 'hive consciousness', for after all the agents that make them up certainly are conscious. Can the 'mindedness' of an aggregate of minds be more than the sum of its parts? It might help to think about the matter that way: to see that we cannot necessarily assume that the goals and motivations, the cognitive pathways and attributes, can be straightforwardly extrapolated from those of their constituents. For one thing, Hillis says, it seems quite likely that corporations might design the AI systems that they will increasingly use 'to have goals aligned with those of the corporation.' Frankly, I believe this is already apparent. It is possible that Mark Zuckerberg is as amoral and as indifferent to democracy as Facebook appears to be, but in fairness I doubt it. It is possible that Google's leadership took as vacuous and cosmetic a view of its unofficial motto 'Don't be evil' as the company as a whole has exhibited, but I'm prepared to believe the top executives were not really (quite) so cynical. Perhaps, then, we need rather urgently to understand how these collective pseudo-minds cognize.

Intellectology

Another attempt to plot out a Space of Minds has been made by neuroscientist Christof Koch of the Allen Institute for Brain Research in Seattle. Koch's Mindspace (Figure 2.2) is also two-dimensional, and has coordinates of 'intelligence' and 'consciousness'. Again, humans are considered to exhibit the greatest simultaneous complement of both attributes, while they decline in proportion to one another in other animals, with the most 'primitive' (such as jellyfish) displaying rather little of either. Today's computers and artificial-intelligence systems are deemed to be well endowed with intelligence but have negligible consciousness – and in Koch's view that is unlikely to change any time soon, or perhaps ever (see page 139).

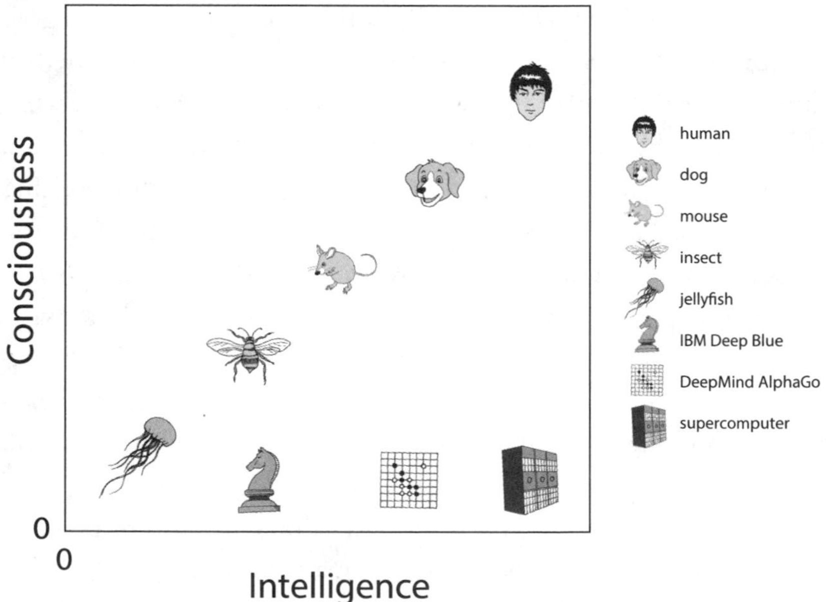

Figure 2.2. The Mindspace envisaged by neuroscientist Christof Koch.

Computer scientist Murray Shanahan has suggested another space of possible minds, in this case explicitly inspired by Sloman's idea, that is commendably honest about the anthropocentrism of the discussion. He suggests that one axis might measure the degree of similarity of other minds to ours – which again then naturally puts us at the top end of this scale (the H-axis, for human-likeness), with the maximal value arbitrarily assigned as 10.

The other axis in Shanahan's two-dimensional plot is much the same as Koch's: the 'capacity for consciousness' (the C-axis). This parameter (also assigned an arbitrary maximum value of 10) corresponds to 'the richness of experience' the entity is capable of. 'A brick scores zero on this axis', says Shanahan, 'while a human scores significantly more than a brick.'

In this plot, as on Koch's, animals of increasing cognitive complexity are asserted to appear more or less on the diagonal line extending from the origin (where the brick sits) to the datum point for humans (Figure 2.3*a*). Today's machines and computer algorithms vary a little in their human-likeness – Roomba, the domestic vacuum-cleaning robot, is considered more human-like than an AI algorithm such as the game-playing AlphaGo – but none is awarded any capacity for consciousness (Figure 2.3*b*).

Shanahan is willing to populate his H–C plane not just with entities we know about but with those we can imagine: more sophisticated forms of AI than we have today, such as 'artificial general intelligence' that has either a degree of human-like consciousness and none at all ('zombie AGI'), conscious exotic beings (whether biological, artificial or a bit of both) that are unlike us to the point of inscrutability, and 'mind children' we might one day make that have a superhuman capacity for consciousness while being similar enough to us for discourse to be possible (Figure 2.3*c*). Along the lower edge of the plot lies a 'void of inscrutability', where entities might reside that, even if they exhibit complex behaviour,

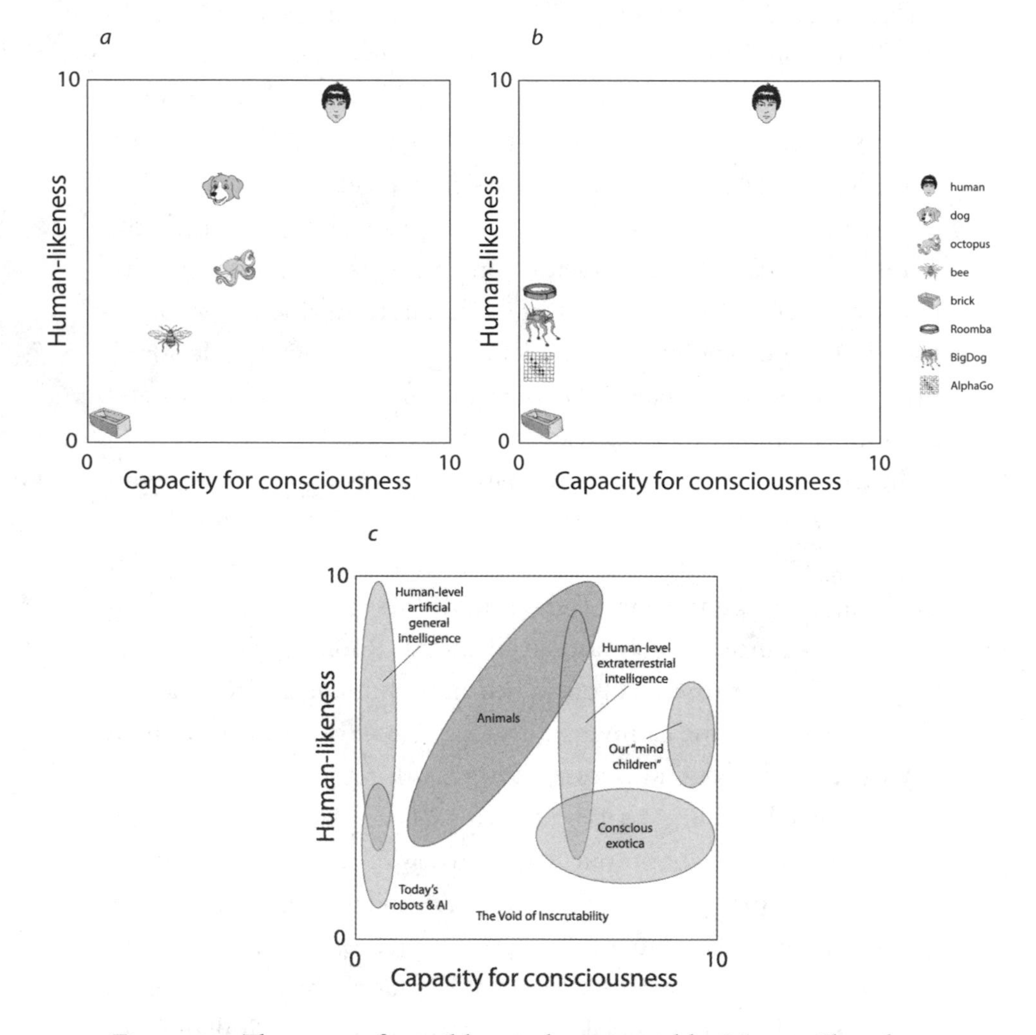

Figure 2.3. The space of possible minds envisaged by Murray Shanahan, characterized by the parameters of human-likeness (H) and capacity for consciousness (C). *a*, Humans compared to other animals. *b*, Humans and AI. *c*, Regions that might be populated by minds both known and unknown.

are so unlike us that we can't even meaningfully talk about a capacity for consciousness in terms that relate to our own. We would need other words or concepts to describe those dimensions of their minds.

Shanahan's plot illustrates the value of this exercise, however fanciful it might seem. It enables us to perceive and, to a rough extent, to quantify, similarities and differences between our minds and others. It lets us think inventively and yet coherently about the kinds of minds that might exist. It shows us which regions we know something about, which are a mystery, which might become populated by our near-term technologies, and which stand as a challenge to our imagination. 'To situate human consciousness within a larger space of possibilities strikes me as one of the most profound philosophical projects we can undertake', says Shanahan. 'It is also a neglected one. With no giants upon whose shoulders to stand, the best we can do is cast a few flares into the darkness.'

There is now a developing literature (of sorts) about the Space of Possible Minds, a discipline that computer scientist Roman Yampolskiy has proposed to call *intellectology*. It exists mostly in the grey zone where science meets unfettered futurist speculation about superintelligence and post-humanism. Sometimes it presents mind as little more than a capacity for abstract computation or manipulation of information represented by 'bit strings'. Since that position doesn't even seem obviously to connect to traditional ways of thinking about the human mind, it is not clear how productive it will be. Information theory surely has an important role to play in thinking about thinking and mindedness, but there's no reason to think our understanding of minds must (or should) be so arid, so reductively digitized.

Meanwhile, the philosophers' anthropocentric view of mind has preconditioned us to assume we are speaking of a conscious entity, most probably one like us. Sloman's wide-angle view of Mindspace questions that assumption, and certainly this book will admit into the debate some entities that we are sure lack this attribute of consciousness, such as today's artificial-intelligence systems. Questions can then arise. How much of Mindspace encompasses conscious minds? How sharp are its boundaries? Might consciousness be contingent on certain key characteristics of mind, and thus be restricted to a well-defined and unique region of the space? Or might there be many ways to become conscious, and indeed many different *forms* of consciousness? We don't yet have any of the answers, but the Space of Possible Minds creates an arena for formulating questions.

CHAPTER 3

All The Things You Are

The idea of a space, or at least a scale, of human minds would not have seemed unusual to Western thinkers from the Enlightenment to the Victorian age. They mostly took it for granted that there was a spectrum of human intelligence and intellect, sensitivity and taste, capacity for pain and pleasure, and indeed of sentience itself. These distinctions were reflected in society by the measures of race, class, and gender, with near-bestial savages at one end and white, well-bred males at the other.

With a legacy such as this, it is not surprising that both the sciences and the humanities have in recent times argued for the unity of human capabilities and experience: the idea that, while of course individuals differ in their specific mental attributes, there is no average distinction between the scope and skills of minds that can be correlated with race, class, gender, or any other societal labels. The survival of that idea of innate difference in a coded, pseudo-intellectual and racist discourse on 'human biodiversity' merely testifies to the value those obsolete hierarchies are deemed to have in the pursuit of certain social agendas.

And yet, just as the conceptual space of human minds was collapsed with the demise of one set of prejudices, it has been expanded by challenges to another. A greater understanding of neurodiversity – of significant differences in the way human minds can become wired, oriented, and attuned – has heightened our awareness that humans do after all experience, feel and reason on a 'spectrum', with different skill sets and different niches in which to contribute to society. Where some thinkers, such as Isaac Newton, Henry Cavendish, Paul Dirac, and Alan Turing, were once considered 'brilliant but odd', today we see ever less reason to stigmatize the aspects of their behaviour that set them apart from neurotypicals. Among the many demonstrations of how such behavioural characteristics have an essential part to play in the health of society, we need only contemplate how international awareness of the crisis of global heating has been galvanized by a young woman on the autistic spectrum.

To put it another way, our view of the space of human minds has become less that of a sphere of 'approved normality' with a scattering of unfortunate outliers, and more that of an irregular and multi-dimensional boundary that evolution and happenstance have shaped into a collective resource.

What are the coordinates of *this* part of Mindspace? We can recognize that most of us have some capabilities and attributes in common: for example, conscious experience, the ability to communicate with language (including sign), a capacity for mathematical abstraction, a need (of varying intensity) for companionship and affection, and so on. But can this plethora of behavioural and cognitive characteristics be condensed into a manageable number of dimensions? What shape are our minds?

The evolved mind

Although (or is it because?) we are so used to the sensation of looking out from our heads, it hasn't always been taken for granted that the mind is in the brain. For Aristotle, all sensation was produced in the heart – an idea that many learned people still accepted in the Middle Ages. The question that Shakespeare posed in *The Merchant of Venice* – 'Tell me where is fancy bred, Or in the heart or in the head?' – still resonates today in talk of being led by your head or your heart.

But even as the brain became accepted as the thinking organ during the Renaissance, the mind or psyche – what was once considered the soul – remained another matter. That Cartesian dualism was still alive and well when Gilbert Ryle attacked it as the ghost in the machine is not really surprising – for the mind has the astonishing capacity to project itself outwards from the body. The image in a photograph resides unambiguously within the object, in the particles and dyes of the photographic emulsion itself. But the image recorded on the retina does not seem to reside there, nor do we experience vision as happening 'in the brain'. It is weird to the point of being inconceivable that, as we look around us, what we see is happening inside the head; the mind somchow contrives to throw the experience back out into the world. It's the same with sound and smell: the brain reconstructs their source and presents that mental image as if it is no image at all but is *the thing itself*. Only occasionally – if we hallucinate, say, or when we look at optical illusions – are we reminded that the world does not necessarily correspond with our experience of it. This property of mind would seem almost magical if it were not so utterly, persistently familiar.

As with every aspect of living things, the first question to ask about why the mind is the way it is, is this: why might evolution have made it that way? (I say 'might' because it is a common error to suppose that all of life's grandeur is an evolutionary adaptation. Nonetheless, that's a good default position, and the most important thing to remember about the human mind is that it arises as a feature of an organism that evolved.)

So much of what evolution has produced seems so magical that it's understandable we could once not imagine it as being anything but the product of divine provenance and omnipotence. Whence the splendour of the peacock's plumage and the butterfly's wing, the unearthly glittering spectacle of schooling fish, the architectural precision of the honeybee or the web-spinning spider? None of this truly *is* magical at all, of course. Rather, it is a testament to what blind, Darwinian natural selection can achieve, given enough time.

The brain is one of the organs that enabled early humans to survive in a harsh environment, and its basic anatomy has not changed very much over tens of thousands of years. This does not mean (contrary to a popular trope) that we are now navigating our world with mental resources adapted to hunting and gathering on the savannah, for the whole point of the human brain is that it supports a phenomenally adaptable, versatile, fluid mind. A mind capable of sustaining itself in small societies in the inclement environment of the last Ice Age was inevitably going to be capable of so much more than that alone, in rather the same way that a technology capable of producing Excel spreadsheets, mp3 files, and Minecraft is never going to engender only those apps and file types. The brain of a person painting on the walls of the caves of Lascaux seventeen thousand years ago could most probably have supported a mind capable of using the internet. But such a mind was not yet called for, and those brains created the minds their circumstances demanded. This is a way of saying that we don't know all the capabilities and limits of the human mind, because our brains have only ever been exposed to certain mind-shaping environments.

We surely have a very different mental landscape to that of Tutankhamen or Shakespeare, let alone a person from the Upper Palaeolithic, and indeed my mental toolbox is different from that of a contemporary member of the Tsimané people of the Bolivian rain forest. It would be appropriate to say we have genuinely different kinds of minds, each appropriately attuned to its environment. My mind would not have been of much use during the Ice Age, and the mind of the Lascaux artist would not serve him well in central London.

Part of the misconception about our supposedly prehistoric intellectual endowment stems from this confusion between brain and mind. All humans, past and present, have slightly different brains, but it's not quite right to say that the differences between brains *cause* differences in mind. Rather, our brains have different 'mind-forming tendencies', and so do different environments. The organism survives by interacting with its environment; the mind mediates that interaction; and the brain is the part of the organism that coordinates it. When the environment changes, the mind adapts and evolves, and behaviour follows. 'Behaviour itself did not evolve', observes cognitive scientist Steven Pinker – 'what evolved was the mind.'

Here is the crucial feature of our evolved mind: while its architecture is partly (by no means entirely!) dictated by genes, it exists *to free us from our genes*: to allow us actions that are not preprogrammed. As we will see, minds probably began as an efficient way of coordinating body movements to achieve survival goals – but in humans (and many other animals) they are now a means by which we need not rely solely on genes to determine behaviour. The amazing thing about humans is not that our genes affect how we think and make choices, but that so much of our behaviour seems to escape their dominating influence. While we sometimes underestimate the flexibility of simple organisms to respond to their environment, more complex minds can create a much broader repertoire of ways in which we can fine-tune and adjust to our circumstances. If evolution wants (to put it much too teleologically) to give a creature many behavioural options to cope in a complex and unpredictable environment, it can either invest a lot into hard-wiring a response to every foreseeable circumstance, or – and this seems the most efficient route – it can build that creature a mind.

Shaped by society

Humans occupy a more diverse set of habitats than any other species on Earth. This follows directly from the fact that the intelligence we're endowed with is surprisingly broad. That's not a boast on behalf of my species; I'm the first to admit to the idiocies that humans routinely perpetrate. But it is frankly rather bizarre that a brain adapted on Darwinian tracks should be capable of producing the theory of general relativity. This is sometimes called Wallace's paradox, after the co-discoverer of natural selection Alfred Russell Wallace, who pointed out the apparent redundancy of a great part of human intelligence. Wallace figured that a brain with just one half more capacity than a gorilla's should have been fully adequate to give our hominid forebears the competitive edge they needed to thrive. Our brainpower seems like overkill.

It's unlikely that all this intelligence was acquired 'by accident' – that is, as a by-product of brain evolution for other abilities, such as better motor control for hunting and tool-making. Intelligence is a very costly trait: the human brain accounts for just 2 per cent of our total body weight, yet about a quarter of the amount of blood glucose we burn up and 20 per cent of the oxygen we inhale goes towards cogitation. No one knows why evolution made this investment, but one prominent idea is that it was motivated by the mental demands our ancestors experienced when living in community with others: that what distinguishes us most in the animal kingdom is *social intelligence*. To get along equably with our fellows, for example, we do well to anticipate how they will behave in response to our own actions: we need some mental model of what *their* minds are like. We need to remember who is who, and who did what: who is friendly and dependable, say, and who is not. We are (like many animals) inherently apt to cooperate, for evolutionary reasons: for example, to develop good hunting strategies or to benefit from the transmission of cultural knowledge in long-lasting social groups. One potential trigger for the social evolution of intelligence was the transition from a tree-based to a ground-based existence for our hominid ancestors, increasing their vulnerability to predation and therefore necessitating larger and more cohesive groups.

The real explosion of human culture and society, however, may have been entirely distinct from any major changes in our brains. Anatomically modern humans emerged in Africa around 200–300,000 years ago, and their first migration from that region occurred around 80,000 years ago. There is evidence of culture and artistry dating back around 75,000 years: that's the age, for example, of decorated beads and 'drawings' in ochre pigment found in the Blombos cave of South Africa, and there is a collection of bone weapons and tools (what its discoverers called a 'well developed bone industry') from the Western Rift Valley in Zaire dated to 90,000 BCE. But the archaeological record shows that it wasn't until around 40–50,000 years ago that early human culture became rich and diverse, for example with the creation of musical instruments. It looks as if dormant cognitive resources acquired much earlier were finally released as culture itself bloomed. We don't know why or how this happened; but it is not – as far as we can tell from cranial shape and capacity – accompanied by marked differences in the brain.

The evolution of intelligence due to social cooperation could have become a self-sustaining process. As our ancestors got smarter, they will have acquired an expanding repertoire of behaviours, rather than relying on a small and fixed set of responses to all eventualities. But then their minds needed to expand further to cope with this behavioural richness around them: as one group of researchers has put it, 'An individual's best behavioural option depends on what other group members are doing, which will vary from one hunt to the next, and even within the same hunt from moment to moment.'

This is not to say that cooperation between individuals in itself requires big brains, or even brains at all. We see it marvellously displayed, for example, in ant societies and even in bacterial colonies, for example in the way these microorganisms collaborate to produce resilient, slimy 'biofilms'. The adaptive benefits of cooperative and altruistic behaviour are typically explained by kin selection: gene mutations that (for example because of their effects on brain development) happen to promote beneficial collaborative behaviour among genetically related individuals will spread in a population. That seems to be enough to explain the sociality of ants and bees, in colonies of which the individuals are all related.

In more complicated societies where some individuals are closely related genetically and others are not, such altruistic behaviour might not be nuanced enough to accommodate these distinctions. It's simpler to cooperate with everyone in the group than to evolve some way of discriminating between more distant relatives and those who aren't meaningfully related at all. In principle, everyone benefits.

But cooperation that is not based on kinship can have drawbacks. It is vulnerable to free-riders, who exploit the cooperation of others for their own selfish gain. All the same, cooperative behaviour can still emerge from that balance. This is typically illustrated with a scenario – a 'game' – called the Prisoner's Dilemma, which places two individuals in a situation where they both have a better outcome in the long run by cooperating than by trying to outdo one another, even if cheating can yield short-term gains for the cheater.* In a

* It is the *Prisoner's* Dilemma because it was originally presented as a scenario in which two people accused of a crime are offered shorter sentences if they will strengthen the evidence for conviction by testifying against ('cheating on') the other.

single round of the game, one player does best if she cheats on the other rather than cooperating – but such selfishness is self-defeating if the game is played repeatedly, because the other player is likely to reciprocate, denying them both the rewards of cooperation. Many animals and other organisms show altruistic behaviour towards non-kin that can be explained on the basis of models like this.

It's possible to do well in Prisoner's Dilemma scenarios using extremely simple behavioural strategies that barely require any cognition at all. In one of these, called Tit For Tat, you do just what your opponent did in the last round. If they cooperated, so will you; if they cheated, that's what you do next. This is the tough way of dealing with cheats, but can end up locking the players into rounds of mutual recrimination, maybe due just to some initial error or misunderstanding. In some cases it can be better to follow a slightly more forgiving version of Tit For Tat that won't switch to cheating the moment your opponent does, but will only do so if they persist in this antisocial behaviour.

Rather simple rules like these can be followed by organisms with little intelligence. But to do well in such competitive Darwinian games in a society of cognitively more complex beings, you may need to get smarter to survive. The fact is that there is no single 'best' way to play the Prisoner's Dilemma – it all depends on what others are doing. And that, in a population of mobile creatures where you constantly encounter new 'players', is a demanding intelligence test. It requires flexibility – recognizing, say, that what worked against one player might not against another. It could, for example, pay to 'advertise' yourself as a trustworthy cooperator: to cultivate a reputation so as to encourage others to collaborate when they meet you. But you might also need to signal that you're no fool and won't stand for repeated exploitation. Or you might develop cunning, cheating only when no one else is looking. And so on.

In short, smartness brings rewards in such a society. It has been suggested that, because of considerations like this, cooperative behaviour among organisms tends to be bimodal: you find it either in organisms of low intelligence, like bacteria (where kin selection accounts for much of it), or among those with rather high intelligence, which show much more contingent behaviour rather than a small set of automatic, low-cognition responses.

Zoologist Luke McNally and his co-workers have shown that scenarios like the Prisoner's Dilemma can drive the evolution of intelligence in a simple 'model brain' of interconnected neurons that makes behavioural decisions. They started with simple networks of just a few connections (nodes), which supported quite simple strategies for a group of agents repeatedly playing two games among themselves. Both games gave the greatest mutual payoff for cooperative behaviour and the smallest payoff for mutual 'cheating' (defection). They allowed their agents to acquire extra nodes in their neural networks, allowing greater diversity and complexity in their behaviour – a crude measure of their intelligence. The researchers enacted this scenario on the computer in a Darwinian manner, such that the 'fitter' agents (those which get the biggest payoff in each round) will multiply and spread.

The researchers found that their population of artificial agents gradually evolved to become more 'intelligent' and complex in what they called a 'Machiavellian arms race' where the agents strive to outsmart one another. Although the amount of cooperation steadily increased too, this doesn't mean the agents were getting 'nicer'. They just got better at working the system for their own gain – in particular by developing a wider and more versatile repertoire of actions. This, say McNally and colleagues, supports the idea that high intelligence arose as a *social* ability.

That is not the only possibility. Others have suggested that the high intelligence of humans (and perhaps of other animals) arose from the demands of tool use: the adaptive advantages of being able to make and use tools favoured those individuals with the cognitive skills needed both for the requisite motor activities (such as knapping flints) and for the imagination to see how such objects might help them accomplish a task.

Others believe that intelligence had less to do with a generalized ability to socialize with others, and more with the challenge of finding a mate. This would make it the product of a variant of Darwinian selection called sexual selection. The idea is that superior cognitive skills would somehow boost an individual's ability to attract or select a mate. These characteristics don't help the individual to survive, for example by being good at predicting the patterns of behaviour of some animal that might make a nutritious meal; rather, they help the individual reproduce and pass on its genes. It could be that a complex repertoire of behaviours persuades a potential mate that this individual will be adept at the survival game too (which doesn't have to be true, but only persuasive!). In this view, intelligence might become like the peacock's tail or the stag's antlers: a trait that is liable to get blown out of all proportion relative to what is strictly needed, simply to convey an advantage in the mating game. If the idea that a sex drive made us more intelligent sounds surprising, just remember how much of our cultural and creative energy is still motivated by it.

Others think that the key to human intelligence is not so much biology but culture itself: that our intelligence took off in parallel with our ability to pass on learning, skills, and technologies to successive generations. There is surely something in this: arguably what distinguishes us most from other animals is not our basic cognitive skills but the complexity of our societies. Psychologist Cecilia Hayes argues that the human mind is a collection of 'cognitive gadgets' acquired not as pre-programmed instincts but through cultural learning in childhood. At birth, she argues, we are not so very different from chimpanzees.

Cultural evolution is surely a central determinant of our behaviour, and the habit in evolutionary psychology to reduce behaviour to 'primitive' biological urges too often skims off all that is important in it. As Daniel Dennett says, 'Every human mind you've ever looked at – including most especially your own, which you look at "from the inside" – is a product not just of natural selection but of cultural redesign of enormous proportions.'

Surely one of the biggest influences on this 'redesign' (if we wish to call it that; it's really just evolution by other means) is language. In particular, we can communicate complex ideas not just from person to person but across time, using language in verbal and symbolic form. Many suspect that it was with the advent of language that the human mind truly began to outstrip those of other animals. 'There is', writes Dennett,

> no step more uplifting, more explosive, more momentous in the history of mind design than the invention of language. When *Homo sapiens* became the beneficiary of this invention, the species stepped into a slingshot that has launched it far beyond all other earthly species in the power to look ahead and reflect.

Neuroscientist Euan Macphail believes that the very claim to have a mind at all – which in his view means to be self-aware and conscious – derives from language, and is therefore uniquely human. Without words to frame concepts and relations between them, he says, we cannot properly develop a notion of ourselves as beings in the world. It is an extreme position, according to which other primates and even human infants are reduced to little more than mindless automata – and in the end it imposes a rather arbitrary threshold of mindedness that overlooks the continuities of cognitive function between the adult human and other beings. Perhaps it is best to regard Macphail's view as showing how hard it is for us to imagine what other minds can be if they cannot articulate themselves to us.

While it is challenging to assess the merits of these ideas about the origins of human intelligence (and perhaps that of some other animals), collectively they imply that what we call intelligence is not some standard ingredient dished out by evolution, in quantities greater or lesser, to those beings that earn it. The word encompasses a cluster of abilities that have developed in response to specific selective pressures (that's to say, for specific 'purposes'). A mind shaped primarily by the demands of communal living is unlikely to be quite the same as one shaped by the demands of mate acquisition, or of capacity to make and deploy tools. (There is no reason to see these possibilities as mutually exclusive, but neither do they seem likely to be all equally important.)

If you worry that it is a terribly limiting view of the human mind to suppose that everything it does is ultimately geared towards Darwinian fitness – to enhance prospects of survival and reproduction – you're entirely correct. Just because those exigencies drove the evolution of our minds doesn't mean they wholly determine or constrain them, any more than the fact that balance and locomotion evolved for the same reasons means ballet is just about sex and survival. The capabilities of mind seem able to bootstrap themselves: they open up new possibilities that are, so to speak, unforeseen by evolution, and so the process continues. The central remarkable fact about the human mind is that it does so much else beyond what seems needful. The brand of over-reaching evolutionary psychology that reduces all artistic creativity to atavistic Darwinism (or an opportunistic offshoot thereof) dismally fails to grasp this.*

* I would rather draw an example from the more benign and thoughtful than from the dismal end of the spectrum, and E. O. Wilson's writings on the sources

Imagine that

Still, natural selection surely shaped the mental toolkit of all minded creatures. Minds arose to organize information about the creature's surroundings and its own body, acquired through the senses, to its own benefit. Minds are action-selection agents: given the current situation, what should the organism do? Seek food? Run or hide from danger? Do nothing and hope for the best?

Decisions per se don't need minds. The *if* function of computer logic will suffice: *if* there is input (stimulus) A alone, say, the output (action) is X, but *if* A occurs in conjunction with B, the output is Y. Many biologists think that plenty of organisms work with simple algorithmic logic like this: bacteria, plants, perhaps simple sessile (immobile) animals like sea urchins. They do not have a wide repertoire of actions, nor do they tend to encounter much variety and novelty in their environments, and so it is sufficient to link stimulus and response in an automaton-like fashion. Minds would be a mere encumbrance: too much expense for too little gain.

We will see later if things really are so simple for 'primitive' life. (Spoiler: that is by no means clear.) But when life gets more complex – when you're mobile in a rich ecosystem of predators and

---

of human creativity are surely the former. But when he writes that 'until a better picture can be drawn of prehistory, and by that means the evolutionary steps that led to present-day human nature can be clarified, the humanities will remain rootless', I think he is guilty of such over-reach. 'Why have the creative arts so dominated the human mind, everywhere and throughout history?' Wilson asks, before concluding that 'Because the creative arts entail a universal, genetic trait, the answer to the question lies in evolutionary biology.' But I don't feel that Wilson's book *The Origins of Creativity*, in which these quotes appear, makes that case by deepening our appreciation of or insight into the creative products of the human mind.

prey – such stimulus–response decision-making surely isn't going to be enough. You've got to make sense of much more complicated, changing and diverse information and predict ways of staying alive in the light of all that. Almost inevitably, the information you have will be incomplete: you can't know or see everything about what's around you, or where the dangers and opportunities lie. So your decision-making system needs to make a *best guess* about the situation and the required response. This, as we'll see, is the circumstance that probably prompted the emergence of the first minds, and it is what has guided their adaptation and evolution ever since.

In a complex environment, it's no longer a matter of computing outputs from inputs. Our own minds are rarely if ever confronted with so clear-cut a challenge – except, perhaps, when they are scientifically studied. It could be argued that what characterizes the human mind is not our capacity to perform specified actions according to precise instructions, but our potential instead to start reciting poetry on a whim, demand a cup of tea, suddenly remember leaving our keys in the car. Yet such responses are (for very good reasons) suppressed in the cognitive-science lab, where participants in experiments are typically asked to restrict themselves to certain actions in response to stimuli: 'when you see X, press this button'. The questions asked are typically of the yes/no variety, not 'can you say something funny?' Is it any wonder that a machine/computer metaphor for the mind typically emerges from such studies? Daniel Dennett points out that experiments in cognition and behaviour may thus distort our view of the human mind because, out of methodological necessity, they generally ask participants to behave as though they were 'transient virtual machines'.

What is truly striking about our minds, says Dennett, is not their ability to solve problems and perform tasks but the sheer open-endedness of our cognition, 'replete with [more] opportunities . . . than the lifeworld of any other living thing'. We can imagine ourselves not only into all manner of possible circumstances but into lots of *impossible* circumstances too: floating in the vacuum of space, living underwater, being an inch tall, travelling in time. What is the evolutionary point of such a freewheeling imagination?

Some neuroscientists and psychologists suspect that our overproductive imagination is not some quirk of evolution but a unique defining feature of the human mind: that we are as much *Homo imaginatus* as *Homo sapiens*. A mind that can conceive of possibilities beyond its own experience can prepare for the unexpected; better to over-anticipate than to be surprised. We seem able to do this by a kind of 'creative remembering'. A pattern of neural activity involving parts of the cortex along with the hippocampus (a brain region associated with memory), called the default mode network, is engaged both when we remember past events and when we imagine future ones. The default mode network has been shown to be activated in tasks that involve imagination and forethought, perspective-taking, and social scenarios. By drawing on and forging new associations among past experiences, we can project ourselves in our mind's eye 'anywhere in imaginary spacetime', as the neuroscientist Donna Rose Addis puts it – into medieval France, or Middle-earth, or the dystopian alternative universe of *The Matrix*. In that inner theatre, any performance can take place. As Theseus said in *A Midsummer Night's Dream*: 'Lovers and madmen have such seething brains, such shaping fantasies, that apprehend more than cool reason ever comprehends.'

The linguist Daniel Dor believes that it is for the 'instruction of the imagination', more than for simple communication, that we invented language. Language, he argues, 'must have been the single most important determinant of the emergence of human imagination as we know it.' He cites Ludwig Wittgenstein's remark in *Philosophical Investigations*: 'Uttering a word is like striking a note on the keyboard of the imagination.' We use words, Dor says, 'to communicate directly with our interlocutors' imaginations.' Through language we supply the basic ingredients – a kind of skeletal list or plan – that a listener can use to put together the *experience* of what is described. It's a way of passing actual experiences between us, and thereby to 'open a venue for human sociality that would otherwise remain closed.'

Thanks to language, imagination knows no boundaries. It doesn't just fantasize about tonight's meal, or for that matter fret about burning it. It conjures up grotesque monsters, galactic empires, fairy stories. Dor believes that this imaginative fecundity is unique to humans. While there is some evidence (as we will see) that other animals can 'mentally time-travel' – project themselves into past or future scenarios – there is no sign that they create such unconstrained narratives.

The invention of reality

This capacity for open-ended cognition equips us for what life has in store. For life is itself an open-ended process – even when it seems to involve starkly binary choices. Say you're selecting from the restaurant menu: will you have the beef or the duck? What choice could be more clear-cut? But while you peruse the menu, your mind might be much more focused on how your date is going: are you making a good impression? If a fire breaks out, you're not going to keep pondering your selection. And unlike most problems tackled by computers, there is rarely a best answer. I mean, there may be one that seems best for you, but on what criteria? You prefer beef, yes – but remember that amazing duck you had in Paris? What if they do it like that? Besides, you've resolved to cut down on red meat. And my goodness, that steak is expensive! Wait: your date has just ordered the beef. Will it look best to have what he's having, or to assert your independence of mind?

Put this way, the decision looks impossible: there are too many factors for any machine to compute a decision. *And yet we do*. It might take you just a few seconds to know what you'll order. Did you make the right decision? That (unless the duck is dreadful) is not a well-posed question. Yet the human mind, governed (but not as a dictatorship) by the brain, has evolved precisely to handle this kind of situation. Metaphorically speaking, menus were what the greater part of human existence has always been composed of: choices of *what to do*, and crucially, choices that *never end* and *might have no absolute right answer*. In fact, you might argue that we are often making decisions when we don't even know the question. Living is, then, not a computation but a process, a constant flow of inter-related decisions and actions and emotions, some big, some small, some conscious and others not, or less so. Our minds do, on occasion, generate specific outputs – we calculate our taxes, or decide what to cook for supper – based on more or less well-defined inputs. But living is not merely a string of such discrete calculations and decisions. The mind is more like a pilot than a computer.

The task of the pilot is not to 'solve reality' but to land the plane. What it needs to do that is a map of the world it must navigate. That representation need not be perfect, but merely good enough. Indeed, it *can't* be perfect, because the mind never has comprehensive information to base it on. What the eye receives, for example, is a two-dimensional pattern of bright, dark, and approximate wavelength information in the light that falls on the retina. (It doesn't even register the 'exact colour' of things, but detects light intensity in three bands of the spectrum – roughly, red, green and blue-violet – from which it must reconstruct all other visible colours.) Evolution has already elected to limit the information we receive, partly because of physical limitations on our receptor systems – it would be hard to make a stable and viable X-ray sensor from biological materials, say – and partly because it focuses on what is most salient for the organism's well-being. Some animals can see wavelengths of light that we cannot (both ultraviolet and infrared); some see the world at faster time resolution (at a higher 'frame rate', to speak cinematically); some make more use of odours, or hear in a different acoustic frequency band, or use sound as a kind of sight (here comes that bat again). Evolution has declared those capabilities not worth the investment for humans: the resources we'd need to devote to them would not be repaid in the value of the information they would provide.* We make our way viewing the world through a narrow little window, weaving what enters it into a good-enough picture to give us a fair chance of survival.

To put it another way: nothing in the world is intrinsically 'red' – our visual system just, rather arbitrarily, 'paints' some objects with that sensation. It's possible in principle to imagine a visual system that responds only to the 'red' band of the spectrum and yet divides it up into finer spectral gradations so that some of what we experience as different shades of red would look blue or green. 'We do not know, and it is improbable that we will ever know, what "absolute" reality looks like', says Antonio Damasio. It is more accurate, however, to say that there is nothing that 'absolute reality' looks like, because 'looks like' is a concept of mind, not of the world that minds inhabit. 'Perception is not simply, *I see the apple*', says author Jeanette Winterson. 'Perception is *I see the apple* in the way that I see it.'

* It's interesting to speculate whether humans would be able to detect, say, cosmic rays or dark matter, if there were some benefit to doing so. Throughout human history, we couldn't do much to hide from cosmic rays – but neither did they do us much harm, beyond inducing the occasional genetic mutation. So there was no advantage to be had from sensing them. The invisibility of dark matter, meanwhile, seems more a matter of fundamental physics. But still, might we have evolved the requisite detector apparatus, if it would have made a difference?

There's a great deal of potentially important information missing even in the visual data that we *can* access. One object might be partly obscured by another. One part of it might be in bright sunlight, the other in shade. We need to judge distance and depth, which are not evident from a single 2D image. Some of that information can be inferred from perspective, if say we assume that all those receding streetlights have the same height. But that's an added layer of inference beyond what the retinal image alone shows. The major clue to depth perception comes from our binocular nature: we can deduce it by comparing the two slightly different fields of view in each eye. But that demands a kind of mental calculation from the two flat images – and it means that what we experience does not exactly correspond to what the retina registers in either eye.

The character of visual perception is far more compromised even than all this. Whenever we fix our gaze on an object, our attention is constantly flickering from one spot to another in a series of involuntary eye movements called saccades that cause refocusing of our vision around three times every second. As the eye or head moves, the image blurs. And we blink roughly once every three or four seconds. If you add up all the duration of saccade-induced blurring and blink-induced darkness, it turns out that we are functionally blind for more than an hour every day. And yet we don't notice! You can stare at yourself for as long as you like in a mirror and neither see nor be aware of the saccades and blinking – because of course the eye can't register what it sees during either of them. But still our visual experience is smooth, sharp, and continuous, not the jerky strobe or blur that these phenomena would seem likely to produce.

Why not? Because what the retina registers is merely the raw data. *Seeing* involves the active intervention of the mind, guided by this data, to produce a reliable, stable best guess at the world. To that end, the brain extrapolates and makes assumptions to fill gaps. In the moment of blinking, the primary visual cortex that registers visual input to the eye falls momentarily silent (or nearly so) – but other brain areas also associated with processing vision compensate to sustain an image.

What's more, the vast majority of what we see comes from what is central to our field of view. At the peripheries, our eyes detect not only a fuzzy impression but one that lacks much information about colour. Yet we don't perceive any bleeding away of colour at the edges of vision, because the brain effectively assumes its presence, if only we could be bothered to turn and look at it directly. Using immersive virtual-reality technology, psychologist Michael Cohen and co-workers removed colour from the peripheral vision of their subjects, and found they didn't register this loss – in one case, even when less than 5 per cent of the visual scene remained in colour. 'Observers routinely failed to notice when color vanished from the majority of their visual world,' the researchers wrote, concluding that 'our intuitive sense of a rich, colorful world is largely incorrect.'

The brain also edits, selecting what (it assumes) matters in the sensory input and filtering out what doesn't. The auditory system will remove background noise that does not change – a persistent rumble or whistle, say – because if it has signalled nothing of consequence for some time, it is deemed not worth attending to. Selective attention may render invisible aspects of a scene that are perfectly evident on the retina, as in the famous 'invisible gorilla' illusion. Here people asked to count the number of passes in video footage of a basketball game may entirely fail to notice the person in a gorilla suit who wanders into frame, beats their chest, and wanders out again, because it is not what they have been told to attend to.

Perception is thus not a photographic reproduction of external stimuli, but what has been aptly described as a 'controlled hallucination'.* More properly, it frames a *hypothesis* about the world, based on the data available but subject to updates and improvements. The mind, says cognitive scientist Richard Shiffrin and colleagues, 'recovers the things in the world *as they would have to be* to explain our sensations' (my italics – but one might instead say 'as they are most likely to be'). This is a form of reasoning called *abduction*: finding the best explanation for a set of observations. Thus we may subtly change the world we perceive to conform with our top-down expectations – as in the optical illusion called the McGurk effect in which we hear an identical syllable in different ways ('fah', 'dah', 'bah') depending on the lip movements of a person apparently speaking it.† As David Hume put it, we 'gild and stain objects' in the external world 'with the colours borrowed from internal sentiment.'

This shaping of experience entails integrating all our sensory modalities and coming up with an interpretation that resolves them. If they conflict, the mind must choose. That's why it is disturbingly easy for one modality to override another and deceive us: even wine experts, faced with two glasses of white wine identical except that one has been turned ruby red with a tasteless food colouring, will swear they are different and might pronounce the 'red' to be a fine Côtes du Rhone. The mind's reality-building is as apt to be a competition as it is a collaboration of perceptions.

* Cognitive scientist Donald Hoffman argues that because perception necessarily hides reality behind our interface with it, we can conclude *nothing* about what it is really like, even whether objects persist at all when we do not perceive them. For most philosophers and neuroscientists this is a step too far towards solipsism: they figure we simply couldn't function or survive if the real world deviated too far from our perception of it.

† There are several examples online, such as https://www.youtube.com/watch?v=2k8fHR9jKVM

Mental models

The ultimate goal of all this is *action-selection*: what should we do? To make that decision, the first task is *accurate prediction*: anticipating what is to come and what the consequences of possible actions will be. By developing a model of the world, we can explore it 'offline', constructing imaginary scenarios and playing out the probable consequences without having to try it for real. In that way, it has been said, our imagined selves can 'die in our stead': we can figure out which decisions and actions are likely to be bad without suffering the consequences.

Much of this happens unconsciously. We don't think 'the trajectory of that falling brick will intersect with the position of my head' before we dodge it. Taking the time to consciously make such a prediction would not in that case be to our advantage. Sometimes, though, we *do* think things through in the discriminating theatre of consciousness – 'I'd better not say *that*, or she'll get mad at me.' Sometimes we might even make explicit reference to the psychological model we're using to predict someone's behaviour: 'She's always a bit grouchy in the morning in any case.' But even much of our social discourse is conducted instinctively, with the model itself out of sight, as it were. The brain has plenty of routes from model to prediction to action that do not involve a journey through consciousness, generally because we can't afford that time-consuming detour. We do, however, use our experience of outcomes to continually revise and refine our models, moment by moment – at least when our minds are working as we might hope they would.

It's astonishing how adept our minds are at this predictive model-building. For example, we're acutely sensitive to some *statistical* features of experience, without being aware of it: we seek out patterns and regularities and incorporate them into our expectations.

When you take pleasure in music, or indeed when you're unsettled or moved by it, the chances are that you're responding to the statistical regularities your mind has discerned and stored over the course of your musical experience so far. If you're hooked by the soaring melodic leaps at the start of 'Over the Rainbow' or 'Singing in the Rain', it's because you have subliminally learnt that their octave jumps ('Some-where . . . ') are rare in melodies and your attention is therefore gripped by this violation of expectation. If you delight in the wonky 'out-of-key' notes in the 'Peter' theme of Prokofiev's *Peter and the Wolf*, it's because they have challenged the harmonic model you probably never even knew you had. You didn't even register they were wonky? Ah, but you did – that's signalled by your emotional response of pleasure, which motivates and rewards your pattern-seeking behaviour.*

It might feel hard to accept that we're not experiencing some 'raw truth' about what is out there. But that's because the representation that we develop generally does a good job of assigning cause-and-effect and predicting future events and outcomes. It's when this faculty breaks down, sometimes in bizarre ways, because of brain dysfunction – illness, trauma or lesion, migraine, the action of drugs – that we're reminded how contingent our picture of the world is on what the mind does with the bare data.

We become particularly aware of this when we look at optical illusions. These persist in 'fooling us' even when we know the truth – that the lines are not parallel, the circles aren't spirals, the discs aren't moving – because we don't have conscious control over the brain's reality-modeller. Optical illusions are designed specifically to deceive this apparatus, and they work precisely because such

* Some violations of musical expectation can be detected by distinctive features in the electrical activity of the brain, such as a 'spike' of a characteristic shape arising in a specific region.

stimuli are rare or absent in the real world – for *that* is what the mind is designed to navigate. For this reason, when we see a Photoshopped image of cars with identical pixel-dimensions looking very different in size because of the perspective (Figure 3.1), our minds are not really playing tricks on us but are doing their job faithfully: if we ever saw a sight like this in real life, the bigger-looking car really would be bigger. It's the image, not the eye, that is 'faking reality.'

The mind even builds a model – a best-guess representation – of our own history. This is perhaps the most unsettling aspect of all. Surely we know at least what happened to us? At least . . . OK, sometimes we might remember things a little wrongly, but . . . But in fact your memory can be the most extraordinary fabulist. For several years, novelist Ian McEwan had a distinct memory of an

Figure 3.1. In this optical illusion, we're generally told that all the cars are actually 'identical' in size. But in what kind of real world would that be true?

'incredibly beautiful' novella that he'd written and then stowed somewhere in a drawer after he'd moved house. He looked all over the place for it. 'I saw it in my mind's eye, the folder, the pages, the drawer it was in,' McEwan says. But finally he had to accept the truth that he'd invented it. 'There was no gap in which this work could have been written,' he says. 'My time was fully accounted for.'

False memories like this are rather common, and can be rich in detail: not so much mistakes and misrememberings as elaborate, fantastical creations. I still 'remember' a book of piano pieces that I used to play in my youth – a compilation of tunes with the wistful romanticism of Chopin and Fauré. I can almost recall how some of them went, and I'd love to find that book again. But I know I won't, because I have had to gradually accept the truth that I made it up.

Experiments have shown that it is disturbingly easy to implant false memories in people's minds. All it takes is a slight nudge to initiate what the brain does anyway, which is to weave half-forgotten experiences into a coherent narrative: a *guess* at what happened. The baroque detail of some entirely false memories shows this model-building nature of our minds at its most inventive.

What can you do?

The constructive faculty of mind is not an end in itself. The brain does not (if we might put it rather perversely) say to the mind: 'Here is what I think is out there – now what do you want to do?' The interaction is much more fluid than that. You could say that the world the mind builds is *predicated* on the possibility of *doing something to it* (which includes doing something to or with one's own body). Thus the kind of world that the mind builds depends on the kind of actions it can take.

In other words, a central feature of the mind is that it is

embodied. It constructs a world based on assumptions and deductions about what interventions the organism might be capable of.

To the extent that the human mind arose as a survival engine, it looks for *affordances* to support that goal. This concept, first postulated by the psychologist James Gibson in 1966, refers to the potential of objects to serve useful roles. This pointy stick could extend my reach; this broad leaf could keep the rain off; this rock could be comfortably sat upon, but that one would be painful. As Murray Shanahan puts it, 'Cognition's trick is to open out the space of what the environment affords the creature beyond what is immediately apparent.'

Thinking about the world in terms of affordances creates flexibility and versatility. It does not impose fixed values or functions on aspects of the environment, but remains open to possibilities. A previously unknown plant offers the affordance of a nutritious meal, but perhaps at the risk of poisoning. A piece of flint offers the affordance of cutting or grinding; another hard stone suggests the affordance of knapping the flint to a keener edge.

The human mind is phenomenally good at spotting affordances. That's what has enabled us to spread throughout so many different environments and climate zones: we see affordances for finding food, keeping warm, collecting water resources, even in new and unfamiliar territories. Technology is one long history of affordance: recognition of the potential of materials and mechanisms to serve our needs.

Affordance is central to our ability to think laterally. Say you want to spread the butter on your toast but all the knives are being washed in the dishwasher. Ah well, a spoon can act as a makeshift knife: we recognize that it has the affordance to do so. We don't try to spread the butter with tinfoil (even though it's shiny like a knife) or with a cut-out picture of a knife, because we know they don't offer the right affordance.

In all probability, nowhere in our past did we learn that a spoon can be used as a knife, nor did anyone ever previously suggest it as a suitable replacement if the need arose. How do we know that a spoon will do the job? We might subconsciously picture ourselves scooping up a piece of butter and then using the flat handle of the spoon to spread it around. In other words, we play out that scenario in our minds.

This requires us to use some implicit knowledge: that the spoon is rigid, like a knife blade, and just about sharp enough to slide through the butter. We'd be unlikely to try the same thing with a chopstick, even though it too is long and rigid: it would be hard to pick up a knob of butter with it, and awkward to do the spreading.

The affordances we identify may be highly contingent on the nature of our bodies: we need to feel confident that we can function in the manner required for the affordance to be realized. An opposable thumb opens up new vistas of affordance not available to less dexterous creatures. A world of affordance vanished to our ancestors when they lost their prehensile tails and tree-climbing agility.

This, then, is the key aspect of the embodied mind: it selects courses of action predicated on a kind of mental rehearsal of the action we propose to take, and the expected outcomes, given our knowledge of what bodies can do and how the world responds. The human mind is forever imagining possible futures that leverage affordances, and choosing between them.

What enables these representations of possible scenarios to be so rich is not the specific cognitive abilities we have but the integration between them: what the archaeologist Stephen Mithen calls *cognitive fluidity*, which he considers to be the truly distinctive feature of the mind of anatomically modern humans. It's the difference between, say, learning and culturally transmitting how to make and use a tool by flint-knapping in the Palaeolithic (we will see that some other animals display similar capacity), and figuring out how to combine implicit and explicit knowledge for the complex process of smelting iron from its ore. It is the difference between taking care of business to keep oneself fed, warm and safe, and producing the imagined worlds of a cave painting. Stone Age art strikes us so deeply because it creates a connection to our modern minds in a way that stone tools do not. It speaks of a spiritual self, a sense of our place in a wider cosmos, an attempt not just to survive but to give our existence and survival meaning.

Learning to think

What are the mental building blocks that enable us to imagine things that have not happened yet, and perhaps never will? How do we abstract the knowledge required for that activity from the sights and events we experience? Much of what we now know about that process comes from watching the mind acquire these capabilities as it develops: by studying babies and young children.

Infants as young as two months behave in a way that shows they expect objects to follow smooth paths (rather than suddenly leaping to one side for no obvious reason), to persist rather than winking in or out of existence, to not interpenetrate, and not to affect one another at a distance without any obvious connection.* By six months they have some idea of material properties: that rigid objects behave differently from soft ones, and liquids do things that

* I'm not saying that all objects *do* do this. I'm saying that, for example, infants will be surprised to encounter magnets, precisely because most objects don't behave as magnets do, acting across empty space without contact. This is the kind of response that allows us to deduce what pre-verbal infants are 'thinking': differences in the nature of their attention – signified, for example, by head-turning or the duration of their gaze – reveal to us when things happen the way they expected, and when they do not.

solids don't.* At around a year old they have an intuitive sense of concepts like inertia – objects in motion tend to keep moving – containment, and collision. In due course children assimilate ideas about forces such as gravity and friction – not at the level of answering exam questions, for which we must be taught, but knowing that such influences exist and where they become manifest: when wooden-block towers will topple, when toy cars will stop rolling.

We use these intuitions to make predictions about what objects do. We know that a flat sheet of paper won't by itself hold up a book, that a glass tumbler will shatter on a stone floor while a plastic one will not. We develop what some researchers have called an 'intuitive physics engine'. This includes a capacity to deduce *causation*: to figure out generalizable reasons for why things happen as they do in chains of cause and effect. Children as young as six months typically figure out that for one object to move another by pushing, they have to be in physical contact: that's the basis of the causal mechanism. Somehow they deduce these things from very little explicit data: with just a few observed events to draw on (sometimes just a single one), they quickly identify the underlying causes.† People learn a lot from a little.

Some psychologists believe that children are innately inclined to seek out the 'rules' of causation – making them in effect wired with the impulse behind all scientific investigation. For example, researchers at the Massachusetts Institute of Technology presented two groups of four- to five-year-olds with a musical box activated by placing beads on it. One group was shown a box activated by any of four beads; the others were shown a box that required (any) *two*

* But oh, how easily they forget when it matters!

† All the same, we shouldn't underestimate the quantity of learning experiences infants routinely encounter: what to us is just getting on with life may to them be a new opportunity to figure out how life works.

beads to be placed on top before it played music. Encouraged by the researchers to play with beads to see what makes the machine work, the second group – who had been presented with a puzzle rather than a simple and unambiguous operating procedure – showed much more inclination to investigate and discover the 'rules' of the device. From a young age, we seem impelled to develop models of *how* the things we observe were produced.*

It may be that this intuitive physics leaves an imprint on our language. We think in terms of metaphors of place, time, force, trajectory: we're *on the right track*, but we're *taking our time*, *pondering* this *weighty* idea in the hope that eventually we will *understand*. In other words, we seem to find physical concepts useful tools not just for dealing with the physical world but for *thinking with*.

Harvard psychologist Elizabeth Spelke and her co-workers have proposed that the intuitions underlying cognition constitute a small number of 'core knowledge' systems, each of which works independently. One of these core systems gives us the ability to conceptualize objects: things that hold together as single entities and which persist over time, whether we can see them or not. We also have a spatial system that can understand distance and orientation: some things are nearer or farther than others, say, and sit in particular directions relative to us. (This is crucial knowledge for using landmarks to navigate with.) Other core systems deal with

* These aren't always reliable: magical thinking and conspiracy theories are examples of our tendency too readily to invent causal stories, and we are notoriously prone to mistaking correlation for causation: just because two things happen together doesn't mean one caused the other. While science surely draws on our innate drive to find causes and explanations, it has developed methods for helping us avoid false lures. Children might be 'natural scientists', but without training in critical thinking some might (and a few sadly do) end up as crackpots.

actions – our ability to affect things – and with 'number' and quantity.*

Spelke suggests that a fifth core system relates to social interactions: it supplies the basis for *intuitive psychology*, supporting the notion that there are other beings that act as *agents* with intentions and goals, as well as encoding the idea that there may exist agents whose behaviours and goals are aligned with ours (an in-group) and others for whom we cannot make that assumption (an out-group).†

Just as we use core concepts of 'physics' for thinking about our internal world, so we may project aspects of intuitive psychology onto the physical world: those wilting plants look 'sad', the storm clouds look 'angry', and of course humankind has always filled the natural world with agency: gods, demons, spirits.‡ Infants readily attribute goals and intentions to abstract and inanimate objects, for

* Our ability to conceive of numbers is more subtle than often appreciated. Despite common claims that very young children and some animals have an innate ability to 'count' using small numbers – perhaps up to four or five, and including zero – it seems more likely that such abilities rely on a capacity to discriminate between relative quantities. In that view, three objects are understood as being a greater quantity than two, but there is no concept of 'three' as a concept that can be applied to any set of objects. What such abilities demonstrate, then, is not *numeracy* but *numerosity*: a discriminating capacity on which the formal concept of number can, in humans, later be built.

† This discrimination – literally – can sound rather depressing: for example, very young infants show a same-race preference in their face-gazing activity. But this bias can be overcome by conditioning: by early exposure to diverse individuals. As is generally the case, what is 'innate' might be neither what is helpful or desirable for modern society nor set in stone.

‡ Reported encounters with supernatural events and agencies such as gods and spirits seem to depend on cultural models of the mind. They are more common in societies (typically non-Western) that consider the mind to be more 'porous' to the world and less hermetically sealed in a bounded self.

example distinguishing in their motions or interactions what they interpret as sociable or anti-social behaviour. If they see an object apparently obstructing or helping another (by being manipulated by an adult, say), they will be more inclined subsequently to play with the 'helpful' one, as if having decided it is friendly.

Central to intuitive psychology is what cognitive scientists call a Theory of Mind. That's not what perhaps it sounds like – a theory of how the mind works – but rather, the belief that other beings have minds like ours, with motivations and knowledge that is *personal to them*. Very young infants suppose that what they know is universal – it's a kind of ultimate (and generally endearing) egoism. The classic demonstration is to place some object – something desirable, like a grape – under one of three upturned cups in the sight of an adult experimenter (call her Alice). Alice then leaves the room, and another adult switches which cup the grape is under, in full view of the child. Up until around the age of four, the child will imagine that if Alice re-enters the room and looks for the grape, she will look first under the cup that now conceals it. Why wouldn't she? – for that is where it is! But of course Alice wouldn't do this if she didn't witness the switch – because now that means the child knows something Alice doesn't. When the child has developed a Theory of Mind, they will expect Alice to look under the original cup. Indeed, they will often delight in the deception: in their knowing what others don't.

Spelke's list of core systems may or may not be complete. I'd not be surprised, for example, if the mind does not also have a pattern-detecting core system, adept at spotting regularities in our sensory input so that we can make predictions, extrapolate from the specific to the general, and find simplifying order in the profusion.* Our

* If so, it is probably over-developed, leading for example to the phenomenon of pareidolia in which we see faces everywhere: in Martian rocks, burnt toast, the

ability to discern pattern may well speed up the process of language-learning, the rapidity and inferential nature of which has long been a source of dispute among linguists.

Spelke proposes that these 'mind modules' are innate, seeming to require virtually no learning: they are our brain's hard-wired toolkit for starting to find our way in the world. Modularity is surely an aspect of mind, and perhaps the biggest challenge is to understand how the modules all work so seamlessly (on the whole) together. Certainly, it makes sense to think of mind as arising *from* this integration of parts, and not as being some essence in which each of them is individually steeped. 'You can build a mind from many little parts, each mindless by itself', says computer scientist Marvin Minsky.

But we should distinguish the modularity of the mind from that of the brain: they are surely related, but not synonymous. Parts of the brain do particular things, but the wonder of this organ is that it can adapt and rewire, to some extent, to retain its functions: the features useful to the mind need not be wholly contingent on particular configurations of neurons. Blind people process Braille mostly in the part of the brain that handles vision in sighted people: these neurons are not intrinsically attuned to visual data alone. In a particularly striking example that could evidently not be ethically done with humans (and some may balk at hearing of it done to other animals), ferrets that were neurologically altered at birth to redirect visual stimuli to the auditory cortex retained their power of sight: the auditory cortex was just repurposed.

This basic toolkit of mind-functions is nowhere near enough to see us through life. We must also augment and modify our mental

---

radiators and headlights of cars. That's the price we pay. Science is, in part, a system that tries to make use of our pattern-finding nature without being in credulous thrall to it.

tools by learning. To the chagrin of those older folk among us, children are in many respects vastly superior learners. This is in some ways an inevitable feature of learning: once we've learnt something, it creates a bias about how we interpret what comes next, and so limits our ability to assimilate new information. Children are more open to learning from novel stimuli, but also for that reason potentially more credulous, easily misled, and prone to magical thinking.

Here again is why the argument about what is inherited/genetic in our behaviour and what is learned partly misses the point. Neither a pre-programmed cognitive device nor a blank-slate learning mechanism ready to be loaded with experience and facts would be a smart evolutionary solution for the task a brain (plus body) needs to do in constructing and supporting a mind. The brain comes preloaded with the basic abilities needed to improve itself.

Why do you care?

The mind's reality simulator constructs and evaluates possible futures based on its knowledge of its embodied resources: of what the body can do. This, however, goes beyond modelling the body as an abstract mechanism. Suppose you have just got back in from a late night at work. You are ravenous, and your kitchen is well stocked. You could cook yourself anything: a roast chicken with vegetables and all the trimmings. It would be delicious, and just what you need. But instead you open a tin of beans and put some bread in the toaster. What's stopping you from making something much more nutritious and satisfying? You know you're perfectly capable of it, and also that you have the resources to hand.

But the fact is, you can't be bothered. You're too tired and want the food too urgently. A 'better' solution is evidently available, but your gut wins out.

In other words, the decision you make is motivated not just by 'reason' but by 'emotion'. I use scare quotes here, because this is how we might often frame the issue – but it's highly debatable whether that's the right way to see it. It's not as though the situation is highlighting some design flaw in the evolution of the mind; the gut response is simply another aspect of mind.

For the gut has its reasons too. They might not be entirely defensible from a strictly rational point of view: you weren't about to faint with hunger if you didn't eat at once, and you won't be going to bed for a couple of hours anyway. But your gut is saying: this will do. It's good enough. If you're still hungry afterwards, you'll have some biscuits. The decision is neither right nor wrong – but it compels a particular action.

Of course, your actual gut itself didn't decide, any more than the heart sometimes trumps the head, or your legs make the decision about whether to go for that last lap. And yet, the gut *really is* involved.

For the mind includes in its representation of the world our own body and its current status. Your mind knows that your gut is empty, thanks to signals from the digestive system. That feeling of excited arousal you get when [insert name here] steps into the room: that's not *exactly* your heart sending signals to your brain, but rather, the effect of the hormones that raise your heartbeat. But it's the heartbeat that the mind registers: 'My pulse was racing!' If the body is not always the source of the emotional response, it can stand proxy for it.

Why should the mind need these (potentially confusing or conflicting) bodily sensations in making its decisions?

The fact is that brains are organs of the body, no more or less than are hearts, kidneys, and lungs. More: brains are *made for being in bodies*. The brain is not like Windows, which can run on a Hewlett Packard laptop or an IBM mainframe. It is 'engineered' specifically for use in a body – in our case, a human body.

'I don't think it's right to think of the brain as self-contained and self-actuating,' says neuroscientist Alan Jasanoff. 'Our decisions are always made with our bodies as well as with our brains. Every single thing we do is done against a backdrop of physiological signals in our body that are constantly coupling things going on in our brain with things going on within us.' It's a two-way process: the brain doesn't just respond to bodily signals but modifies them – for example, chronic mental stress can incapacitate some of the cells of the immune system and make us more vulnerable to infection, or anxiety can alter the regulation of sex hormones. And this is no more or less than what all organs of the body do.

In short, emotions are a part of the embodied mind. That being so, it is absurd that we have come so often to think of them as obstructions that *cloud* the mind. Evolution evidently had a use for them. One of their roles is to equip us with drives and motivations. The power of emotion to motivate actions to attain a meal or a mate (not necessarily in that order) is very clear. 'The emotions are mechanisms that set the brain's highest-level goals', says Steven Pinker. For David Hume, 'passions' rather than reason lie at the root of all action: we never do anything unless moved by some passion. Even restraint from imprudent or immoral acts, he said, involve the dominance of one passion (albeit a calming one) over another. Rationality itself cannot tell us what to do, but only how to do what passion recommends.

This is not simply a matter of giving an added emotional push to hard-headed pragmatism; it can serve to cut through slow or intractable deliberations of logic. How might you have decided, on that late-night return, whether to rustle up some quick beans on toast or prepare a full-blown roast, if it were not for the message of your 'gut'? There's no obvious 'logical' solution to the dilemma, and one can imagine a computer-like mind frozen into indecision like Buridan's famous ass, dying of indecisive starvation and dehydration at the precise midpoint between a pail of water and a bale of hay.* Emotions say, 'Cut the crap and do something.'

They *can* also act as a reinforcement mechanism: when reason advises a particular action, emotion adds a motivating, 'Hell yes!' As cognitive scientist Daniel Bor puts it,[†] 'Emotions put meat on the bones of what is beneficial or harmful.' Minds like ours – blessed with some appreciable capacity for reasoning but faced also with environments and situations where that will not suffice to guarantee effective action – may *require* emotions as a kind of tie-breaking or amplifying influence. Emotion is, then, not a distraction but a crucial part of the mental toolkit, and emotionality might be loosely considered to be one of the dimensions of Mindspace.

We might wonder whether there is some minimal and essential emotional toolkit. The 'basic' emotions are typically said to include – perhaps exhaustively – fear, anger, sadness, happiness, disgust and surprise. People of all cultures on Earth exhibit these six: they appear to be innate in humans, even if the triggers for them are culturally specific. There's reason to think that most if not all of them 'map onto'[‡] other primates (at least) such as chimpanzees: they might be a part of the repertoire of brain states for animals closely related to us.

* This scenario was posed as a satirical embodiment of the notion voiced by the medieval French theologian Jean Buridan that 'Should two courses be judged equal, then the will cannot break the deadlock, all it can do is to suspend judgement until the circumstances change, and the right course of action is clear.' The paradox ultimately stems from Aristotle.

† It's inevitably risky quoting Daniel Dor and Daniel Bor in the same chapter, but they are not the same person distinguished by a typo.

‡ This is a cautious way of avoiding the claim that other primates 'feel happy' and so forth – the word is only well defined for humans. It seems fair to say, however, that there is something like happiness or joy that chimps feel – but we can't know quite what that feeling is like.

Then there are so-called 'social emotions', such as envy and jealousy, pride, and embarrassment, which arise in relation to others. (If we feel embarrassment in private, it is generally in relation to an imagined or remembered social situation.) Given their complex social structures, it seems that other primates feel something akin to these too.

Might there be kinds of mind that experience some of these emotions but not others – that's to say, do they come as a necessarily complete set or as independent features? Certain brain disorders do seem to eliminate fear. For example, in 2010 researchers at the University of Iowa described a woman who lost the function of her amygdala in late childhood because of a rare genetic disease. Although she could feel other emotions such as happiness or sadness, she felt no fear in situations that would unsettle most people: watching a scary horror film, say, being threatened with a knife (which happened to her in real life), or encountering large spiders and snakes. Although professing to 'hate' those latter creatures, when she was taken to see some at an exotic pet store, she found to her own surprise that she had an impulse to touch them.

If emotions are an intrinsic aspect of how the human mind works, an emotionless but otherwise human-like person, like *Star Trek*'s Mr Spock, is a phantom – or at least, their mind will not work well or efficiently. (The writers of *Star Trek* often hinted, of course, that this might be so.) One of the chilling features of some psychopaths is their emotional disconnection from their actions. And differences from expected emotional responses are often one of the attributes noted in people on the autistic spectrum: not an absence of emotion by any means, but perhaps a quite different set of triggers for and reactions to them that leave them sometimes struggling to cope in our neurotypically biased human society.

Emotions are not terribly well understood from a cognitive or neuroscientific point of view, however. They are often measured in terms of arousal: some heightened mental states can be correlated with physiological markers such as increased skin conductance or pupil dilation. These are crude measures at best, as is notoriously evident in the unreliable and sometimes blatantly pseudoscientific technologies of lie detectors.

Some researchers think we have got the role of emotions completely backwards. The neuroscientist Joseph LeDoux argues that they don't *drive* behaviour at all, but rather, are human-specific *responses* to subconscious mental processes. Take the amygdala: although this brain region is often described as the 'fear circuit', the sensation of fear is in fact a higher cognitive process that arises in the cortex. Signals coming from the amygdala are somehow associated with it, but their role is to form a swift link between sensory information and the motor systems, without the intervention of higher cognition and out of sight of consciousness: those signals get us moving, perhaps away from some threat. LeDoux thinks that what we experience as emotions are side-effects of these physiological reactions: subjective narratives that our brain creates, caused by instinctive responses that are in themselves devoid of emotion. They rely on language and culture, he says: we consciously label these bodily reactions 'fear'.

In support of this view, he cites examples of blindsight, where people with a vision-processing impediment don't consciously register a stimulus even though the 'brain has seen it'. In these cases, they might feel no fear when, say, shown a snake – but still the brain triggers a bodily response, so that the heart beats faster. Some intimation of LeDoux's position is more familiar: many people have been in perilous situations in which they acted almost without thought or feeling, and only afterwards, when they have been able to reflectively process events, did they experience the fear and shock appropriate to the predicament.

If emotions are indeed, as LeDoux argues, not in themselves

'primitive' responses, but rather, sophisticated cognitive ones that depend on our labelling them as such – 'I feel happy/scared' – then it's not clear that animals would feel them at all. Sure, they might show all the physiological responses we associate with emotion – a dog in a dangerous predicament might howl and look wide-eyed – but that is not accompanied by a *feeling* of fear.

Anyone familiar with Descartes' reputation as a vivisectionist (see page 215) might feel some discomfort with this view. But LeDoux is not advocating anything of that sort. He merely cautions that we should not confuse the behaviour associated with emotions with the experience of them.

Whether or not LeDoux's view of emotions is correct (it is disputed), it touches on the inevitable challenge for understanding other minds: if all we have to go on is external behaviour (or measurements of brain activity or physiological signals), how can we conclude anything about subjective experience? There are hazards in both directions: we might too readily assume that equivalence of behaviour implies equivalence of experience, but we might equally too readily deny it. If we can feel no genuine emotion without the capacity to label it as such, then small infants will not feel fear: their bawling at a loud noise is just (say) an instinctive response to alert an adult to the danger. But wouldn't it be strange if babies and non-human animals can sometimes so convincingly mimic behavioural and physiological aspects of a familiar emotion while lacking what seems to be for us the central component of it: an ability to feel it? It seems to me that the problem with LeDoux's hypothesis is not that it is obviously right or wrong but that it makes a mental attribute too absolute, definitive and binary a thing – something that we either have or don't – rather than acknowledging that minds are complex blends of ingredients mixed in very different proportions and flavours, of which ours come from just one shelf.

The human mind-cloud

Efforts to understand how the human mind assembles and organizes itself are typically posed in terms of theme and variations – or rather, an answer and deviations from it. That's to say, the notion is propagated that humans generally have a *kind of mind*, and that we discover what kind of mind this is from aberrations where this and that aspect have been removed by the vagaries of biology or happenstance. And it is surely true that, say, the cognitive consequences for people who have experienced damage to specific areas of the brain may tell us much about what that part of the brain does.*

But this normative model of the human mind has been challenged in recent years. It's now clear that there is not a single point in Mindspace occupied by the human mind, from which individuals may be displaced by dysfunctions or quirks of their brain. The point central of the neurodiversity approach to human cognition, in particular with regard to people with autistic-spectrum conditions, is that there are appreciably different ways that the human mind might naturally structure itself, and that we should think more in terms of differences than deficits.

It's important not to be simplistic or romantic here. For many people on the autistic spectrum, everyday life can be an immense challenge. Some have cognitive and behavioural challenges that they themselves wish to be regarded as impairments that create special needs. But some of the difficulties faced by people with autism are surely socially rather than neurologically instantiated. If it is

* This approach needs caution: if a cognitive function disappears because of some brain lesion, this doesn't mean the function was generated solely or even primarily in the affected region, any more than that locomotion of a car is caused by its spark plugs.

correct to say that much of what makes the human mind distinct from that of other animals, especially our close relatives among the great apes, is that it has evolved for sophisticated social cognition, then it is not surprising that human minds not attuned to the complexities of social interactions will struggle to 'fit in'. Some people on the autism spectrum attest to feeling like visitors to another realm – as Temple Grandin, an animal behaviourist with autism, famously told Oliver Sacks, 'Much of the time I feel like an anthropologist on Mars.' And in a sense that is what people like Grandin are, for they come from other places – adjacent neighbourhoods, that is all – in Mindspace.

We're not talking about a tiny minority here. As Steve Silberman says in his 2015 survey of autism, *NeuroTribes*, in the USA there are roughly as many people on the autism spectrum as there are Jewish people.* You probably know some, or are one yourself. Some of these people struggle in this world designed for 'neurotypicals'; others find niches that can cater for minds a little different from the majority. Academic science seems particularly accommodating: several famous scientists and technologists, such as Paul Dirac, Steve Jobs, Thomas Edison, Nikola Tesla, and perhaps Isaac Newton, Albert Einstein, and Charles Darwin, may have been on the spectrum.† And why, after all, would it not be socially and adaptively valuable to see what cognition can produce in a different region of Mindspace? 'We have to learn to think more intelligently about people who think differently', says Silberman.

What parameters of mind are different in autistic people? The symptoms are very diverse, ranging from what might seem like a few random eccentricities (such as slightly obsessive interests) to a

* One estimate puts the incidence at 1 out of every 68 people.

† So too might computer scientist John McCarthy, who coined the term 'artificial intelligence'.

near-total inability to engage in verbal or indeed any sort of interpersonal interaction. But if a generalization might be made, it is that people with autism have difficulty in finding (or even thinking of looking for) a useful intuitive psychology. They may misread social cues, often leading to what seems to neurotypicals like a blunt, direct, and clumsy way of interacting. Some people with autism are not unaware of this; indeed, it can make social contact very stressful and distressing for them. As Hans Asperger, the psychiatrist who coined the term 'autistic' in the 1940s, said of one of his patients, 'He can recognize the facts, but cannot invent what may lie between them.' The formulation is extremely apt, for it points to what is truly weak or absent for many people with autism: they don't miss sensory data, but lack that dimension of mind that *creates* an effective model of what caused the data insofar as it relates to social behaviour.

Autistic-spectrum conditions are not (as is sometimes suggested) new and somehow caused by civilization or modernity. Autism has a genetic component, and many of the mutations are in genes associated with brain development that will have existed throughout human evolution.* That these mutations have persisted and spread widely suggests they might in fact confer some adaptive benefit. Cognitive scientist Simon Baron-Cohen has proposed that genes that are correlated with autism overlap considerably with those that are associated with our brain's pattern-recognizing capacity. His studies confirm the intuition that some people excel more in empathic skills and others in perceiving and developing schemes for

* We should be wary of supposing that there is any clear relationship between the implicated genes and the phenotype – the behavioural characteristics – associated with autism. The latter might have more to do with the way the developing brain converges on particular schemes of wiring and organization in response to lower-level genetic changes in the cellular processes involved.

classifying and systematizing.* The latter, he argues, tend to be better at spotting patterns of the kind on which science depends – for example, to advance from cause-and-effect relationships to hypotheses about interventions in the world: *if* X is true, *and* I do Y, *then* the likely outcome is Z. People with autism are, he argues, extreme cases of the systematizing, pattern-seeking brain – which is why these types tend to be over-represented in the sciences, maths, and engineering, and among the inventors and entrepreneurs of Silicon Valley. Baron-Cohen suggests that people on the autistic spectrum might therefore have played a key role in driving human discovery and invention. Of itself this is a scarcely testable hypothesis, but it surely makes an important point: autistic traits long labelled as odd, antisocial, or nerdy have always been a significant and productive aspect of the diversity of human minds, and we risk under-valuing and under-utilizing this resource if we structure society to marginalize people with such minds, in favour of valuing social skills. Autism appears to be an extreme version of a normal feature of the human Mindspace.

The issues are complex. While it is indisputable that people with autism have often suffered greatly from being written off as mentally deficient, 'weird', and socially burdensome, Silicon Valley and the IT industry has itself become notorious for its lack of empathy, its solipsism and disregard of social responsibility, not to mention its blatant misogyny. And while social skills are increasingly valued

* Baron-Cohen suggests that his scale between empathizers and systematizers indicates that there are five types of brain (one might perhaps better say, of mind), ranging from extreme to mild empathizers, balanced, to mild and extreme systematizers. These divisions, however, are arbitrary boundaries on a continuum – and moreover, classifying minds this way is too simplistic, because the questions on which the categorization are based only probe empathic and systematizing tendencies and so only admit those dimensions of mind.

by employers, jobs that require empathy (such as social care) have been traditionally undervalued and poorly paid in comparison with the fortunes made by number-crunching IT speculators and financial analysts. If Baron-Cohen is right to associate (on average) systematizing minds with maleness and empathizing minds with femaleness,* it's all too clear where the balance of power has lain. Even today there are some who argue (in the face of evidence) that the 'male brain' is better suited to the sciences and engineering, so that the continuing gender imbalance in those disciplines – which inevitably sets up still more biases and inequalities – is just 'nature's way'.

Some researchers, such as neuroscientist Henry Markram, believe that autism results not so much from a cognitive deficit as a surfeit: an excessive richness of information created by a mind that has in some sense 'too much awareness'. Faced with that potential overload, people with autism devote great energy to imposing structure and pattern on it all. In this 'intense world' theory of autism, social interactions might be actively shunned because they are just so information-rich and complex, and hard to reduce to obvious patterns. So far, however, there seems to be rather little hard evidence to support this intriguing view.

Autism shows the human mind operating differently from the norm, but we should be wary of turning it into a different kind of mind. 'I have myself long been attracted to the idea of meeting a genuinely different mind, but I am not so sure now that autism is

* It's a big and controversial 'if'. In its support, Baron-Cohen offers evidence that systematizing brains are generated by greater prenatal exposure to testosterone. There may be something in this, but the details are unclear. For one thing, others have asserted that the characteristics of autism in females are often rather different from the 'classic' symptoms displayed more commonly in males, and so the condition is under-diagnosed in females.

the place where I might be able to meet this mind', says cognitive scientist Uta Frith, an expert on autistic-spectrum disorders. The prodigious talents shown by some people with autism do not necessarily locate them in entirely distinct regions of Mindspace from the rest of us. Although it has been estimated that between 10 and 30 per cent of people with autism have some kind of exceptional, often highly specific, talent, Frith says that 'deeper scrutiny of these talents often disappoints. They tend to be narrow, and even shallow.'

All the same, the space of human minds clearly encompasses a range of skills and aptitudes, which should certainly not all be placed on a single scale between 'empathic' and 'system-seeking'. None is any better than another in any general sense, and it is increasingly recognized that diverse teams function best for problem-solving and innovative tasks. Even in science, where systematizing skills are traditionally seen to be at a premium, team members with good social skills play a vital (and at present, insufficiently acknowledged or rewarded) role. There are plenty of examples where collaborations between brilliant egotists have quickly fallen apart, not to mention Nobel laureates with notable deficits in empathy.

The real obstacle to acceptance of neurodiversity (which encompasses also conditions such as dyslexia and attention-deficit/hyperactivity disorder, ADHD) is that many societies are by their nature normative: set up not just to expect but often to enforce behaviour close to the mean. 'Many of the challenges [people with autism] face daily', writes Silberman, 'are not "symptoms" of their autism, but hardships imposed by a society that refuses to make basic accommodations for people with cognitive disabilities.' Since cognitive diversity is very likely to be adaptive in a population, this might seem odd – but as Sigmund Freud pointed out, it is intrinsic to any civilized culture (and I don't equate that with technological sophistication) that it imposes constraints on our minds and demands adjustments to behaviour that run contrary to our natures to some degree or another. Yet there is no reason why this should require us to judge other minds in a narrow and conformist manner. We have never really tried to imagine how societies might be constructed to better fit the landscape of human minds. But there's every reason to think that if we could improve that match, the benefits could be abundant.

To judge from experience, human minds tend to be conservative: to feel most comfortable among others of like mind. This is natural: if a Theory of Mind has played a key part in enabling us to coexist, it depends on an ability to project our own mental ontology into the heads of others. If we can't fathom another's mind, we find it challenging to know how to live alongside them. (No doubt this partly explains why we so relentlessly anthropomorphize and 'Disnefy' other animals.) As Olaf Stapledon pointed out in his 1930 novel *Last and First Men*, the 'rich diversity of personal character' has 'entailed endless wasteful and cruel personal conflicts' – but it also 'enabled every individual of developed sympathy to enrich his [sic] spirit by intercourse with individuals whose temperament, thought and ideals differed from his own.'

The extended mind

What is *in* your mind and what's *outside* it? We typically imagine that the skull marks the boundary – or, if we accept the notion of an embodied mind, then the skin. But some philosophers of mind dispute that.

Let's say that Betty and Jai have planned to meet at Tony's Diner on Eighth Avenue. Betty knows New York well, and although she lives a mile from the venue she can get there just by consulting her memory of the route. Jai is visiting from Denver, and so to walk from his hotel he has to consult Google Maps from time to time. Betty is navigating with her mind alone, but Jai is using an artificial aid.

But what if Jai too is a native New Yorker, but has Alzheimer's and can no longer navigate from memory? Then he needs the device for guidance because that part of his mind is impaired.

We're constantly and increasingly doing this even if we have no such impairment. Once we might have memorized phone numbers; now they're stored on our smartphones. One recent study found that people navigating the dense, small streets of London's Soho showed much more activity in their hippocampus – a brain region involved in memory and spatial navigation – when having to do it unaided than when they could use GPS guidance. Some researchers are concerned that there's a use-it-or-lose-it aspect to brain activity, and suggest that using it might postpone neurodegenerative decline. Outsourcing our mind actually alters it.

But there's another view. What, really, is the difference between Betty and Jai looking up information in order to make navigational decisions, except that one looks it up internally and the other externally? Might we ask if the information in the GPS device has *become* a part of Jai's mind, as readily accessible now as any memorized information?

It seems an unlikely proposition at first encounter. Why consider things outside the head to be a part of the mind itself? But consider how some elderly people who are taken from a care home into hospital become more confused and less mentally able. The loss of familiar landmarks doesn't just affect how well they can get around, but might also impact, say, their ability to get dressed or their awareness of when to get fed. They use their surroundings as cognitive aids. 'Taking them out of their homes', says Daniel Dennett, 'is literally separating them from large parts of their minds – potentially just as devastating a development as undergoing brain surgery.'

By the same token, evolutionary biologists Hilton Jaypassú and

Kevin Laland have argued that spiders' webs are a part of their 'embodied' cognitive system. The web works as a sort of sensorium: the spider sits and waits to pick up vibrations, through sensors in their legs, that signal the location of captured prey. The spider tunes the web to improve the value of the information it delivers, for example adjusting the tension in the strands so that information is channelled and filtered, rather as the sensory systems of the body do. And the web can act like a memory: if one section of it catches more prey, the spider might enlarge that part.

'We shouldn't accept a neural or even a biological model of persons or minds,' says philosopher Andy Clark, one of the prime advocates of this notion of an 'extended mind' that includes prosthetic aids to cognition and other aspects of the environment. 'The machinery of mind is not all in your head . . . it can be spread across brain, body and the world.' His position is in some ways a logical extension of Gilbert Ryle's: to regard the web formed by brain, body, and world as being under the control of some disembodied agency – rather than as the *mind itself* – is again to invoke a ghost in what now might be very literally 'the machine'. Clark argues that to destroy the artifacts with which a person with impaired cognition constructs a sense of self would be as grievous a crime as to harm a part of their brain.

As technologies that can interface and fuse with the mind become more complex, it certainly becomes ever less clear where mind stops. Suppose you had a chip implant that allowed your eye to collect visual information from a camera placed on a busy street: in a sense part of your mind is transported there too. What else is remote surgery – in which a machine operates on the patient controlled by the movements of a surgeon in another geographical location – doing but importing the *consciousness* of the surgeon to the location of the operation? In this view, it becomes hard to say what is mind at all – does it embrace everything we encounter, including other people? Clark and his student David Chalmers argue that to qualify for inclusion in one's mind, information has to be reliable and easily accessible. But those are rather hazy criteria. It's also unclear whether this information amounts to mere extension of memory, or whether the extended mind also performs some sort of *cognition*.

Feminist philosopher of technology Donna Haraway famously proposed in her 1985 essay 'A Cyborg Manifesto' that the distinctions and boundaries of human and machine were blurred even back then. That was surely true for physicist Stephen Hawking, all of whose interactions with the world through his body paralysed by amyotrophic lateral sclerosis were mediated by technology. In her 2012 book *Hawking Incorporated*, philosopher Hélène Mialet called him a 'collective' being, more machine than man and with agency that was wholly dependent on assistants. Some considered this a shocking, dehumanizing suggestion – but Mialet regarded Hawking's situation as only an extreme case of our own. Anyone whose will is exerted through the agency of other beings or objects is a collective, she said: we are all 'incorporated' in a complex nexus of machines and human beings. If mind doesn't just enable but actually arises from agency, might it be considered to permeate all that agency draws upon?

Whatever your view of these suggestions, they remind us that a Space of Possible Minds is not a map of types of brain, nor even of types of organism. Minds have a function that is defined, shaped, and arguably embodied by their environments.

Noble in reason?

What a piece of work, then, is the human mind? Unsurprisingly, many of its characteristics are likely to be shared by any mind developed through biological evolution to enable life on this planet. It seeks to sustain the host organism in its given environment, having instinctive urges to address the basic evolutionary needs – food and warmth (that is, homeostasis: keeping our bodies within the bounds of optimal operation), sex, escaping predators – and the sensory processing capacity to navigate its surroundings accordingly. These demands create criteria of *value*: the mind assesses experience in terms of actual and predicted losses and gains of value for us as individuals, and guides responses accordingly. Such value judgements don't occur on any single scale, nor do they always concur with one another, within or between minds.

The values and meanings that our mind ascribes to experience create shortcuts to decisions about actions: our conscious lives are conducted largely by rules of thumb (which we are experts at justifying post hoc). To that end, our sensory systems limit the amount of information about the environment to a fraction of what is potentially out there, focusing on what has proved most significant (to the organism) in the evolutionary past. Even the gross imperfection of our memories might be in part a necessary filter against overwhelm; Jorge Luis Borges' short story 'Funes the Memorious' warns us of how incapacitating total recall would be.

As we've seen, such a mind cannot really be a computer, because it doesn't know what it is supposed to compute, except that it must in the end do *something*. It does not understand its own urges, far less control or optimize them. It may end up creating without any purpose it can articulate, or destroying on the same basis.

Our urges to communicate and collaborate, to classify and demarcate and find order, were surely the engines of civilization and innovation. In this way, culture bootstrapped its way to the bewildering edifice it has become, with its subtle codes and hierarchies, norms, and institutions. Our minds demonstrate how a capacity for mental model-building, coupled to rather basic evolutionary drives, opens up a Mindspace that becomes its own justification, where the expansion of mind is not really *for* anything any more but takes on the aspect of a natural process, like population growth or star formation. In the process, notions of value may become dissociated from any real costs and benefits for survival and reproduction. In short, it is futile to suppose that all we do, all we decide, can be interpreted in terms of evolutionary pressures, although some surely can. To put it another way: when other animals do things that seem dumb, it is generally because of the limitations of their cognitive processing. But we have developed uniquely sophisticated ways of doing things that, from an evolutionary perspective, are pointless or dumb. That's the price we pay for these often remarkable minds of ours.

We have no idea where our endowment of intelligence – however you might choose to measure that – ranks in the scheme of the possible. Perhaps it is exceedingly modest. Yet it is adequate to pose such questions, and to start us wondering about our coordinates in the Space of Possible Minds.

CHAPTER 4

Waking Up To the World

René Descartes gets a raw deal these days. The Nobel laureate and physicist Steven Weinberg had scathing words for the seventeenth-century philosopher: 'For someone who claimed to have found the true method for seeking reliable knowledge, it is remarkable how wrong Descartes was about so many aspects of nature.' Antonio Damasio titled a book on the human mind *Descartes' Error*, and went on to ask '*which* error of Descartes' do I mean to single out?' Murray Shanahan says that 'It has become customary, in contemporary cognitive science, to blame Descartes for almost everything.'

But Descartes must have been doing something right – for many of the books about the mind and the brain, whether they are philosophical or scientific, feel compelled to begin with what he had to say on the issue, even if it is only to rough him up a little for saying it. If he was *that* wrong, you'd imagine his words would have been forgotten by now.

The truth is that Descartes highlighted (it could hardly be called a discovery) the crucial feature of the human mind: it knows about itself. That's to say, it is aware of its own existence. Descartes argued that this is the single thing we can know for sure: I am, because I am capable of knowing it. 'The one thing, the only thing, that is given, is my experience,' says neuroscientist Christof Koch. 'That's Descartes' central insight.'

It might sound trivial to the point of tautology. But on the contrary, Descartes' *cogito ego sum* expresses what might be the most profound fact about the universe that we know of: it has given rise to the possibility of knowing itself. The mind awakens our bodily matter to its own existence. No matter how much we understand about the mechanisms and algorithms and heuristics of the human mind, we won't fully appreciate its nature until we parse consciousness: the fact that there is something it is like to be us.

But are we capable of that? Some regard consciousness as the holy grail of neuroscience, while others see it as a distraction, a chimera that threatens to drag us into mysticism. For many years, consciousness was almost a dirty and disreputable word in the study of the brain; you admitted an interest in it at your peril. The pioneering cognitive psychologist George Miller wrote in 1962 that

> Consciousness is a word worn smooth by a million tongues . . . Maybe we should ban the word for a decade or two until we can develop more precise terms for the several uses which 'consciousness' now obscures.

Miller was more or less heeded, and for decades thereafter a scientist risked ridicule or disdain if they pronounced on the matter. In 1990, philosopher John Searle attested that

> As recently as a few years ago, if one raised the subject of consciousness in cognitive science discussions, it was generally regarded as a form of bad taste, and graduate students, who are always attuned to the social mores of their disciplines, would roll their eyes at the ceiling and assume expressions of mild disgust.

Today, in contrast, you can hardly move for theories of how consciousness arises. Countless books promise to give us 'consciousness explained' or 'consciousness demystified'. Plenty of them have interesting things to say on the matter, but none settles it, and it's not even clear whether there can be a truly scientific theory that ever will. Some people insist that consciousness isn't even real, but is a mere illusion that will evaporate once we understand the brain sufficiently. Others believe that consciousness pervades the universe, or at least obtains in every living thing within it. Some maintain that consciousness is a uniquely biological property, perhaps even unique to humans; others are convinced that we will have conscious machines before very much longer. Yet although there is not even remotely a consensus about any of this, we can now at least make a start towards understanding what is required for a mind to be conscious. At this point, I suggest that you approach theories of consciousness a little like book reviews: to be evaluated and appreciated not according to whether they are right (we barely know how to judge that), but whether they are intellectually and imaginatively stimulating. I rather doubt that consciousness, any more than a serious work of literature, can be 'explained' by a single, unquestionably correct interpretation or theory – so perhaps it is no bad thing that the theories proliferate. Right now we seem likely to gauge the shape of consciousness only by looking at it from many angles. So let's make the most of those we have.

Beyond thought

Some philosophers think that the puzzle of consciousness will be for ever beyond our ken, precisely because we can only ever see it from the inside. To comprehend it is rather like hoping that we will be able to see ourselves through our own eyes – not a reflection, not a photograph or image, but ourselves as we are. Yet that's the one thing we *can't* see. Descartes' *cogito* expresses this as a problem of where to begin in our knowledge of the world: the conscious self has to be a given, an axiom, and as such is beyond the reach of inspection. Descartes had to suppose that what we'd call consciousness or awareness – the mind or soul, if you prefer – is therefore a quality that exists outside of the physical world, yet which infuses the brain and acts on matter. 'I knew I was a substance whose whole essence or nature is simply to think', he wrote in *Discourse on the Method* (1637), 'and which does not require any place, or depend on any material thing, in order to exist.' (6:32–3)

It is this idea that Damasio considers Descartes' gravest error, for it promulgated the split between mind and body known as dualism. For Damasio, the *cogito* has it back to front: our physical being, the possession of a body in an environment, comes first, and thinking follows from that. The conscious mind is built out of the demands of the body.

If, on the other hand, we persist in separating consciousness from the body, we can perform all kinds of dubious intellectual legerdemain with it. Not least, we can deny that such an insubstantial thing exists at all. For some philosophers of mind, notably the husband-and-wife team Paul and Patricia Churchland, the experience we call consciousness is a delusion, a product of 'folk psychology', derived from naive and flawed thinking. All we experience and all we do can be (or one day will be) explained in neurobiological terms, and there is no function left for the conscious self that we fondly imagine to be in control of it all. This perspective is commonly called eliminative materialism.

Philosopher Galen Strawson dismisses it with impatience, almost contempt: 'If there is any sense in which these philosophers are rejecting the ordinary view of the nature of [conscious experiences] like pain . . . their view seems to be one of the most amazing manifestations of human irrationality on record.' It is more rational,

Strawson says, to believe in divine intervention than to imagine *that*.

This is more a battle of words than of ideas. The Churchlands are concerned that we don't *reify* consciousness, making it a distinct 'thing' akin to Descartes' immaterial soul that has somehow to be explained by theory. Strawson, meanwhile, worries that in our eagerness to demystify we risk denying the evidence of our own minds. Consciousness might one day be explained but it will never be *explained away*. That simply has no meaning. It results in a futile regression:

> 'Why do I feel this way?'
> —'Ah, you only *think* you feel that way.'
> 'Well then, why do I *think* that I feel this way?'
> —'Ah, you only *think* you think you feel that way . . .'

It would be like telling a pregnant woman in labour that she does not really feel pain; there are just nerve signals travelling to parts of her brain that process touch, emotion and so forth. (Seriously, don't ever try this.)

Let's be clear about what counts as illusory. It's often said that the solidity of objects is illusory, for example, because the atoms that constitute them are mostly empty space. If all the actual matter in my desk were to be collapsed so that the space between the atomic nuclei were filled, it would occupy about the volume of a pollen grain. But that claim is specious, because the property of solidity *is* the fact that my finger cannot penetrate my desk. That is what it *means*. It's the same for consciousness: it can only be illusory if we imagine it is some kind of ethereal fluid, not a property generated by a macroscopic brain. Simply, 'consciousness is experience', says Koch. We don't *think* we have experience; the experience of thinking we have experience is a part of what experience *is*.

So I shall assume that consciousness is a real phenomenon that warrants an explanation. Can we hope to say anything about what that is?

Some philosophers despair of that too. Colin McGinn is a leading advocate of a position often called 'mysterian', which sounds more mystical than it is: it simply posits that the human brain has not evolved the cognitive resources to understand so profound a problem. This remains possible, but contrasts with a popular view in neuroscience that consciousness must somehow 'fall out' of a computational theory of mind. Thomas Nagel's 'What Is It Like to Be a Bat' essay was intended largely to expose what he regarded as the hermetic subjectivity of consciousness. 'Without consciousness the mind–body problem would be much less interesting', he wrote. 'With consciousness it seems hopeless.'

While not regarding the quest quite as despairingly as that, Searle has argued that we can't hope to understand consciousness using a computational theory of mind. In 1980 he presented a scenario that he believed illustrates the gulf between mind as computation and our inner experience. He imagined a man confined to a closed room whose job it is to answer questions about a story that are submitted to him by people outside the room by posting them on paper through a slot. The man's answers are posted out through another slot. The problem is that both the story and the questions are written in Chinese characters, and the man must answer in the same form – yet he does not read, speak, or understand a word of Chinese. But he does not need to. He can consult a big book of rules that tell him what symbols to write down in response to any of the symbol combinations he receives.

To the people outside the room, it looks as if the man is fluent in Chinese and understands perfectly what is being asked of him. But all he is doing is carrying out a kind of computational algorithm to convert input into output, without any understanding of what he is saying.

If, however, the mind itself is purely computational, in what sense does the device of the 'Chinese room' lack any capacity the brain has?

You might say that the book of rules is cheating, since the scenario is then no longer the stark 'man in a room' – you might as well give him Google Translate. Yet what if the man possesses a prodigious memory, so that he can remember all those rules in his head? What, indeed, if he could do it with such speed that you could just write down your question and he'll instantly write the answer? How could this seem any different from a man who truly understands (written) Chinese?

Searle's point was that running an algorithm to manipulate symbols in a computer-like manner can be done without any comprehension at all. What is missing is any grasp of meaning. For Searle, this is closely linked to consciousness – if by that we mean to act with a purpose of which we are aware. The man in the Chinese room is, we might say, not *conscious* of his responses in Chinese – in that regard, he is like a zombie, even though he might have a rich inner world in other respects, for instance imagining what he will have for supper as he scribbles his answers in total ignorance of their meaning.

Many theorists of mind reject Searle's critique, although if it is wrong there is no consensus about why. Some say that there *is* consciousness and meaning at play in the process of translation, but that it is not located in the man himself – he is just one element of a conscious *system* that includes the way the book of answers was compiled. Others say that Searle has just rendered consciousness unrecognizable. Steven Pinker argues that, in the computational theory of mind, consciousness *does* consist in converting inputs to outputs, but that Searle has slowed it down to the point of invisibility. It's a bit like saying that neurons or the hippocampus have no 'understanding' or consciousness, Pinker says. The man in the Chinese room is acting like the brain's language module, which is not inherently conscious but may contribute to the neural process in which consciousness is constructed.

Others argue that consciousness is not to be located in any computational input–output problem but is exhibited in the fact that our mental processing is accompanied by a feeling of what it is like – what philosophers call *qualia*, the subjective experience triggered by a stimulus.* It would be a straightforward engineering problem to create a robot that reacts with apparent aversion to a plate of food that smells like sewage, even to the extent of grimacing: we'd need the requisite chemical sensors, motor controllers, and so forth, but you can imagine how it might be done. Why, then, has evolution not simply made us fleshy devices of this kind, which would perform equally well in avoiding things we should not eat and even signalling to others that they are bad? Why was it necessary also to evoke a subjective (and in this case deeply unpleasant) experience too?

Philosopher David Chalmers has said that understanding the engineering aspects of this scenario – how the brain processes smell, and how the outcome of that operation triggers behaviour – is the easy part of understanding the mental phenomenon we call consciousness. Understanding qualia, meanwhile, is the 'hard problem' – because we can't even see where to start. Experience seems irreducible: it is what it is. It's certainly not enough to say that the smell we experience from sewage is bad so that we avoid it, while the smell of baking is good because we will benefit from being attracted to it. That is merely a tautology: we like what smells good, and things smell good to us so that we will like them. And in any

* Or by, for example, a dream; not all qualia are stimulus-driven.

case, that doesn't explain why fresh bread smells like it does and not like caramel or coffee, which would have the same consequence.

Yet again, there's no consensus that this is the right way to look at the matter. Daniel Dennett says that instead of trying to answer the 'hard problem', we would do better to consider 'hard questions': in particular, what is gained by an item entering conscious awareness? It's not, he says, as though we need to *explain* why coffee smells like *this* and not *that*, for what after all is this 'smell'? It's not a property intrinsic to the volatile chemical constituents of coffee beans, but an arbitrary output of our olfactory system, just as redness is an arbitrary response of our visual system to light of a certain wavelength. In this sense, he argues that qualia don't really exist: not as properties of the physical world. Once we understand all the brain is doing as it processes the smell of coffee, Dennett argues, the 'hard problem' will evaporate. 'The elusive subjective conscious experience – the redness of red, the painfulness of pain – that philosophers call qualia?' he writes. 'Sheer illusion.' Cognitive neuroscientist Stanislas Dehaene believes that qualia, like the concept of vitalism, will one day be seen as 'a peculiar idea of the prescientific era'.

On the other hand, there's a good argument for why qualia are not merely (mysterious) products of the brain. Damasio argues that they must be linked to our bodily sensory mechanisms in much the same way and for the same reasons that the body is central to emotions and feelings. Qualia are then not entirely arbitrary, but depend on the kinds of bodies we have. They will not be explained by any abstract 'theory of the mind' that fails to (literally) incorporate the body.

At any rate, it's not clear that qualia can be pinned to specific neural states of the brain – because they are themselves always embedded in other, simultaneous experiences. We won't track down the 'what it is like to experience red' brain state, Daniel Bor points out, because we're never in the pure state of 'experiencing red'. All we have is, say, 'experiencing a shade of red on a computer screen in a rather bland testing lab and feeling a little bored as well as queasy after having bacon and eggs for breakfast'. And this is not a complication of consciousness but part of its very definition: it can't be atomized but is always necessarily experienced as a unified whole, a single, evolving, patchworked 'instant' of qualia-flavoured sensations.

This need not, however, insulate consciousness from reductive investigation. Cognitive scientist Anil Seth argues that we shouldn't worry about 'hard' and 'easy' questions here, but should find *real* ones: that is, ones for which we can pose specific questions about why particular experiences are the way they are, and not some other way, and which we can investigate empirically. We should, he says, focus on 'distinguishing different aspects of consciousness, and mapping their phenomenological properties onto underlying biological mechanisms.' Perhaps, he suggests, by trying to address the hard problem head on, or by sweeping it under the carpet, or by pretending that qualia don't even exist, we have made consciousness a more complicated problem than it need be. There is at this point no reason to suppose that consciousness – specifically, the subjective, experiential properties of it – cannot be a target of scientific enquiry.

The readiness of neuroscientists to engage with consciousness today is partly due to advances in experimental methods. Techniques for probing neural activity in the brain, such as functional magnetic resonance imaging (fMRI) and electro- and magnetoencephalography, which record electrical and magnetic signals from the brain using sensors on the cranium, have made it possible to compare what is happening in our grey matter with what people *say* they are experiencing. This has made 'introspection' – subjective reports – respectable, where once they had been more or less outlawed from the behavioural sciences. It's not that people are invariably 'right' about what they report as perceiving or remembering – we fool ourselves all the time – but we can now test whether, when people say they have experienced something ('yes, I perceived the flash/word/sound'), there's some signal in the brain that correlates with and thus corroborates the *experience*. Another person's conscious experience is no longer just a matter of trust.

What is consciousness made of?

Some of the disputation that consciousness studies provoke comes from a lack of clarity about definitions. We recognize consciousness as something we possess – we are (usually) aware of our existence – but we also speak of being conscious of specific things, experiences, events. The two are not the same. There is plenty that our senses receive, and which affects our behaviour, but of which we are not conscious in the sense that we could report back on it if questioned. What we describe as the content of our conscious mind seems sometimes to depend on what we are asked and how the answer is solicited.

Some studies on animals and small children equate consciousness with self-knowledge: a recognition that the self exists. This is often assessed in tests using a mirror, in which the infant or animal sees themselves with a mark – a smear of paint, say – on their forehead. Those capable of self-recognition will generally attempt to remove the mark from their own bodies.

But there's nothing in this response that guarantees it comes from *awareness* of self in the sense that we experience it. Sure, the mirror test seems to be passed by animals that we already suspect for other reasons to have some degree of consciousness – great apes, elephants, dolphins, for example. But others, such as pigeons, can be *trained* to pass the test. This surely can't mean they are trained into consciousness; cognitive scientist Stanislas Dehaene calls them 'mirror-using automata'.*

Perhaps consciousness is just an ability to access information in the brain – what some call access-consciousness? Some researchers believe that all we then need to do is identify which parts of the brain are involved in this retrieval (that is, the neural correlates), and how they work. That is Dennett's view; as Steven Pinker expresses it, 'Once we have isolated the computational and neurological correlates of access-consciousness, there is nothing left to explain.'

Here consciousness is regarded as the *construction* of experience: all it requires is the neural apparatus that produces perceptions. So, for example, the 'primary consciousness' of mental images arises in creatures that can create inner mental representations from stimuli: that, for example, possess camera-like eyes connected to neural circuits for vision, mapping the light from the external world into a photograph-like 'topographic map' in the brain. Meanwhile, 'affective consciousness' – emotions and feelings – comes from neural circuits that attach valence to experience: is this thing good or bad? And 'interoceptive consciousness', or awareness of body states, likewise requires only the respective neural links between viscera and brain that produce map-like representations of the body. Once we have identified the circuits responsible, what else can be missing?

Yet this picture doesn't seem to explain why consciousness feels the way it does: how these various elements that represent and evaluate the world get bound together into a unified subjective experience, a 'what it is like'. Some researchers talk about this subjective feeling as 'sentience', a kind of minimal consciousness that manifests as subjective feeling, or what Damasio calls 'core

* This might be unfair to pigeons. They are in some respects not the brightest of birds, but they have some remarkable cognitive faculties (for navigation, say), and 'bird consciousness' seems highly likely in at least some avian species: see page 207.

consciousness'. This, he thinks, arises from the moment-by-moment engagement of an internal representation of the body – the proto-self – with its environment, involving input from all the senses. We have a sense of our presence in the scene that we build from the warmth of the room, the sound of traffic passing on the wet street outside, the sight of the November gloom through the window. We assemble all these pieces of experience into a 'wordless narrative', which constitutes the foundation of consciousness.

Animals with core consciousness perceive objects and their environment, as well as bodily states, and respond to them with feelings that act to guide decisions and actions. But, Damasio says, there is no reason to believe that they have an image of themselves doing these things: an internal dialogue along the lines of 'Here I am, trying to write a book'. Damasio calls this 'autobiographical consciousness': a uniquely human ability to build and sustain a picture of ourselves as autonomous entities with a history and identity. That internal dialogue is literal, even if not articulated in precise words; Damasio claims that autobiographical consciousness 'relies extensively on language' and so must be exclusively human by definition.

This distinction feels right, for we have no reason to suppose that other species have a self-image of themselves as beings with dreams, frustrations, a concrete past and a richly imagined future, and so on, living in community with other like-minded souls. All the same, such a dissection of consciousness must be fluid and rather elusive, since we are by no means constantly inhabiting this autobiographical self. Much if not most of what we do makes little reference to it; it's there just as a kind of background buzz that we can bring to attention if the circumstances require. Yet the mind maintains it. When we wake up, we might sometimes have a moment of wondering where and when we are, but we don't generally have to work at recalling *who* we are. We're right there, ready to go.

Autobiographical consciousness seems for us to be more of a resource than a condition: a type of awareness we can consult if we wish.

Where does this sense of self come from? The great philosopher of selfhood Derek Parfit believed that it stems from memory: we build our self-image out of the things we have experienced. What matters for a sense of self, then, is psychological continuity with the past. But this is more of a metaphysical definition than a scientific description, and does not say much about the neurological basis of how we weave our remembered world into a self.* Damasio, in contrast, posits that the conscious sense of self is built from four cognitive elements. First, we recognize a local perspective from which we map the objects around us: we interpret sensory data from a knowledge of being *here*, *now*. Second, we have a feeling of ownership of our mind: this representation is *mine*, and mine alone. And we feel we have agency: I could, if I wanted, manipulate those objects (but only in certain ways, consistent with my intuitive physics; I can't make them vanish, say). Fourth, we have primordial feelings of embodiment that aren't connected to the objects and surroundings: I am an entity in a living body.

When these images of the 'self aggregate' are 'folded together with images of non-self objects', says Damasio, 'the result is a conscious mind' – by which he means an *autobiographically* conscious mind, one that has a concept of the self persisting through time and is capable of 'self-oriented deliberation'. When Damasio defines

* It's possible that Parfit's view was influenced by the struggles he experienced in maintaining an inner representation of his own world. He was apparently unable to form mental images (a condition called aphantasia), and could not call to mind his own wife or house unless they were there in front of him, although he had no difficulty recognizing them. Parfit was an accomplished photographer of haunting images, often devoid of people; this too seems no coincidence.

consciousness as 'a state of mind in which there is knowledge of one's own existence and of the existence of surroundings', he is setting a high bar (although a great deal is elided in that unexplored word *knowledge*). This is why he believes that consciousness is not synonymous with mind, and that minds need not be conscious. It is not, he says, the same as mere wakefulness, although we often use it colloquially in that way ('Is she conscious yet?'). Nor is it the same as being conscious *of* something, in the sense that a thing that has appeared in our field of view has come to our attention.

The sense of self might also require an ability to articulate it: to categorize and label objects and feelings, and use such labels for organizing experience, intention, and prediction. If that's so, Damasio's autobiographical consciousness must be not only an intrinsically human attribute but one that emerges during our development and is not available in any significant degree to very young infants. That might seem unkind to babies, but not if we recognize that this form of consciousness is by definition an emergent feature of the social human, not some fundamental and absolute property of our mind.* It is not then an attribute like Chinese nationality that you either have or you don't, but more like a sense of humour: a question of degree, and not identical in all minds.

Building a workspace

Such a sophisticated form of consciousness, though, far exceeds the criterion of there being something it is to be like. How first does the

* The production of a sense of selfhood becomes a more problematic issue if it is tied to claims of moral status – as in the suggestion of philosopher Michael Tooley that only when we develop a concept of a continuing self, perhaps around the age of three months, do we acquire full human rights to life. 'The brain deserves a better philosophy [than that]', says Dehaene, with justification.

brain build the subjective awareness that is the hallmark of Damasio's core consciousness?

This involves a purposeful *bringing into mind*: a kind of focusing of the attention on what is salient in our experience, and in the process imbuing it with significance for us – with subjective meaning. It demands selection among our sensory data, filtering some of it out and boosting other components. In that way we can hold onto a piece of information, an idea, or an intention, so that it can motivate a range of behaviours. If we're hungry, we salivate, but we might also decide what to eat, get up, go downstairs and get it from the cupboard. If we can't hold 'hunger' in our awareness during that journey, we may find ourselves wondering why we're standing in the kitchen. (Yes, I know . . .)

This awareness filter is rather aggressive: it permits only three or four items to sit within our working memory at any moment and command our attention. Show people a grid of numbers or colours for a brief moment, for example, and they're unlikely to be able subsequently to fill in more than four of the spaces properly. This limit of four seems to apply not only to (most) humans but also to other animals such as primates and honeybees. Perhaps this suggests minded creatures have, during the process of evolution, rarely had to attend to more than four salient items at a time – or maybe it's because there are basic cognitive limits to what neurons can usually process. Whatever the reason, four items is not a lot of information to guide us, and our minds find inventive ways of coping with the limitation.

For example, we may reduce collections of many objects to a smaller number of composite ones: a process cognitive scientists call 'chunking'. We're no worse at remembering random words than random letters, because we process a word as a unified whole rather than as a string of letters. This is an acquired ability, and takes practice. When children first learn to read, they have to decode each letter one at a time and then assemble and blend them to get the word; now (I'm guessing) you don't give a second thought to which letters a word contains, but see it as a whole or *gestalt*.

What's more, you don't need to even think about the decoding process; it happens 'out of sight'. One of the aims of the mind, Bor says, is to relegate things from the conscious to the unconscious mind to free up the working memory to hold new items. When a good musician has learnt how to play a piece, they don't even think about the notes – which arrive as if by magic at the fingertips or in the throat – but can focus on 'higher' aspects of performance. Practice is the process of automating a skill – and in fact that skill can then suffer if the performance of it *does* again impinge on the conscious mind, as when musicians start thinking again which notes are coming, or a tennis player becomes too aware of the body movements she must make.

This concept of a 'theatre of consciousness'* in which items are brought onto the stage of attention resonates with a popular theory of consciousness first articulated in the late 1980s by cognitive scientist Bernard Baars, called global workspace theory (GWT). Loosely speaking, it says that the brain processes information and determines actions using neural resources that can be considered a kind of internal 'workspace' that acts as a crucible of consciousness. The neural signal from this workspace is 'broadcast' to the rest of

* I use that metaphor with caution, for it shouldn't be confused with what Daniel Dennett has derided as the 'Cartesian theatre', in which some non-physical homunculus sits in the brain observing experiences as if projected onto a screen. This homunculus, Dennett said, is the residue of the soul that Descartes believed to inhabit the body and imbue it with mind and thought. It's a picture that simply defers the problem of consciousness and mind by collapsing these things into the person of the homunculus. However consciousness arises in the brain, it is surely not by presenting sensory events to an internal self.

the brain, perhaps in coordinated waves of neuronal activity that spread through densely connected hubs of the network such as the thalamus, to create a kind of global awareness of what's present in the workspace. In this view, says Koch, 'once you have information, and the information is made broadly available, in that act consciousness occurs'.

Consciousness is here created by but not *in* the workspace itself. The workspace acts like a kind of information bottleneck: once it is occupied by the contents of an incipient 'thought', other, perhaps contradictory information can't be accommodated. Only when the first conscious notion slips away can another take its place.

GWT theory is basically a computational model of mind: sensory data, memories, and so forth provide the input that is processed and sent out to motivate and guide actions. It is thus a functionalist theory (page 33), meaning that the mental phenomenon arises out of the informational content of the system and how this is processed and distributed, irrespective of the hardware in which that happens. It should be a feature of any information-processing system that has such a capability for integrating and then broadcasting information throughout the system. 'Consciousness', says Dehaene, a leading advocate of GWT, 'is global information sharing.' Dehaene, in collaboration with molecular biologist Jean-Pierre Changeaux, has proposed that this sharing is facilitated by a network of neurons with long axons (the 'wires' down which nerve impulses flow) that can connect sensory and other cortical circuits into global states of brain activity.

This picture of experience being mixed in a workspace-studio from a cacophony of inputs before broadcasting to the mind's world is rather like a so-called pandemonium architecture, first proposed in 1959 by computer scientist John Selfridge as an account of how the brain recognizes patterns. His idea is that there are circuits (a group of neurons, say) that specialize in recognizing particular elements of the pattern – say, straight or curved lines, differences in contrast, and so on. Each of these works independently, and is conceptualized as a 'demon'. The demons yell out what they have found, and the loudest voices are heeded by higher-level demons responsible for identifying particular types of object: letters of the alphabet, say. When these demons have heard enough to convince them that their object has been seen, they in turn start to clamour for the attention of a 'decision demon', which adjudicates on vision and then announces it, through what the neuroscientists Gerald Edelman and Giulio Tononi have vividly labelled a 'riotous parliament'. Others have compared the process to a football crowd in which rival songs or chants arise in different parts and the aim is to win enough recruits to your tune to dominate throughout the stadium.

Such a cacophony of possible (and mutable) interpretations, all clamouring for our attention with varying degrees of insistence, will probably sound like a familiar experience as we try to make sense of the world. But the voices of an unruly mob of billions of neurons would be too chaotic to arrive at any consensus. The brain is more organized than that. There are distinct modules that contribute their views to the global workspace: the working memory, the 'episodic' memory that remembers past experiences, sensory systems such as vision and hearing, motor functions, the emotions (Figure 4.1).

The mind weighs up their claims and comes to a conclusion. It might not be the *right* conclusion – we can mistake what we've seen, and optical illusions are designed to confound the demons – but it's a unique one that gets broadcast to consciousness. That final decision might change from moment to moment, though, as when the mind flickers back and forth in interpreting a classic ambiguous image such as the Necker cube (Figure 4.2). Still, only one decision or interpretation is possible at any time.

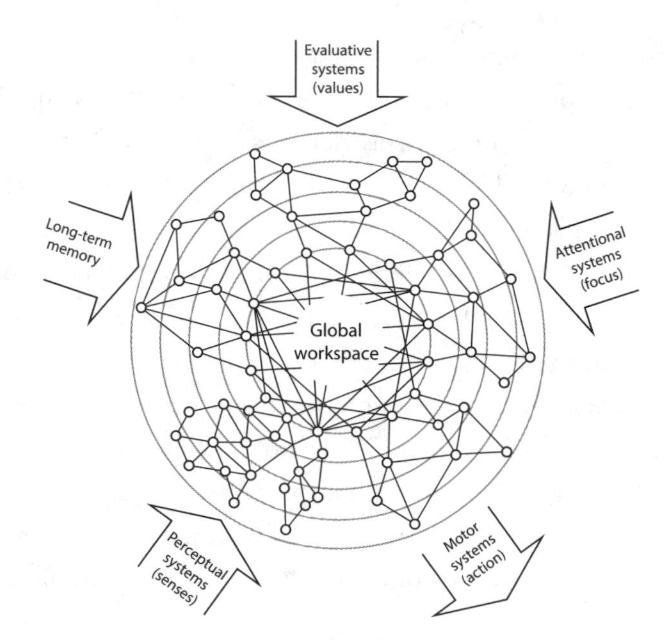

Figure 4.1. In the global workspace theory of consciousness, inputs from various brain functions and modules are evaluated and combined in the global workspace before being broadcast as a single conscious experience to the rest of the brain.

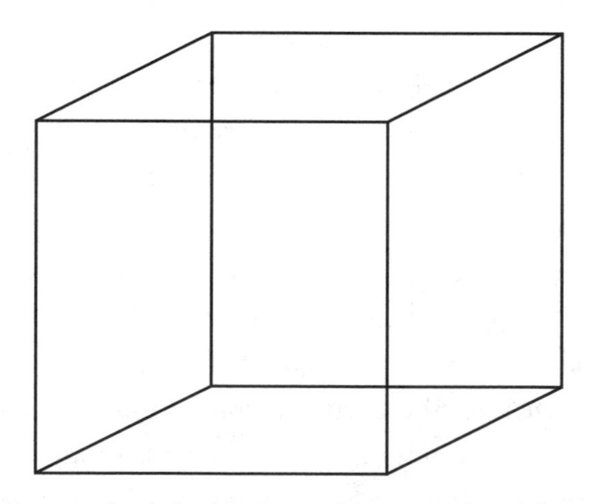

Figure 4.2. You can 'see' the Necker cube in two possible orientations – but only one at a time, reflecting how only one 'interpretation' of the world can occupy the crucible of consciousness at any given time.

Networks of awareness

For Dehaene, consciousness is then *consciousness of*: it is the process by which specific objects, events, or experiences are brought to our attention and 'take possession of our mind'. He suspects that, if GWT can let us understand how this happens, there will be nothing left to account for.

But some feel that conjuring consciousness out of the mere process of broadcasting information is a little like a rabbit-from-the-hat magic trick. And the computational character of GWT seems silent about qualia – about the feeling that accompanies access to the contents of our mind. Dehaene regards this feeling as merely a marker, a means of enhancing or moderating the salience of those contents. But others insist that consciousness must refer to the whole gestalt: the full experience of awareness and being, of which 'access consciousness' is a mere part.

An alternative theory devised by Giulio Tononi and his co-workers, particularly Christof Koch, takes this top-down view. 'To be conscious is to have an experience,' says Tononi. It doesn't have to be an experience *about* anything, although it can be; dreams, or some 'blank mind' states attained by meditation, count as conscious experiences. Tononi and Koch claim to identify the essential, axiomatic features of such conscious experiences, saying that these are:

Intrinsic existence: Conscious experience is real, and is the only reality of which I can be sure. It is moreover inevitably *subjective*, existing only for the conscious entity.

Structured: Experiences have distinct elements, such as objects and colours, not just a mush of sensory input.

Specific: An experience 'is what it is': rich in information and distinct from other experiences.

Unified or integrated: We have only one experience at a time.

Definitive: There are borders to what the experience contains, which exclude some aspects of what is 'out there'.

From these axioms, Tononi and Koch aim to deduce the properties that any physical system must possess if it is to have some degree of consciousness. Their approach is called integrated information theory (IIT).

Unlike GWT, IIT does not portray consciousness as the processing of information from input to output. Rather, it is identified with a particular kind of information itself, namely the 'integrated information' by which the cognitive system is able to act on itself. Consciousness, says Koch, is 'a system's ability to be acted upon by its own state in the past and to influence its own future. The more a system has [such] cause-and-effect power, the more conscious it is.'

The theory posits that consciousness depends on the nature of the brain's wiring. In many computer circuits – including the so-called neural networks used in today's artificial intelligence (Chapter 7) – an input is processed into some output by being conveyed successively from one set of logic gates to the next: it is a so-called *feed-forward* process. In the brain, on the other hand, processing happens in both directions: the output signals at some stages are sent back to influence the processing at earlier stages.

IIT arose out of a proposal by Tononi and Edelman that consciousness hinges on the dense web of reciprocal connections between areas of the brain, forever exchanging signals that influence one another: a process they called *re-entrant.** This is not mere circular reiteration of a signal, but more akin to a group conversation that arrives at a conclusion – that conclusion being a unique and coherent experience, in which conflicts are resolved and interpretations are decided through the crucial process of *integration* of information. There is no unique prescription for how that endpoint is reached; rather, the brain is considered to have certain 'attractor states', like basins or valleys, into which it is drawn and which can be entered along many different paths.

For networks with these re-entrant characteristics, it's impossible to cut many of the connections between nodes without quickly fragmenting the entire structure: it functions as a virtually irreducible whole. There is a mathematical measure of that irreducibility, denoted Φ (phi), which serves as a rough quantifier of consciousness. Φ measures the amount of 'irreducible cause–effect structure': how much the network as a whole is capable of influencing itself. If the web can be divided up into many smaller networks that don't exert causal power on one another, then it will have a low value of Φ no matter how many processing nodes it has in total.

For example, the neural connections between the hemispheres of our brains are routed through a structure called the corpus callosum. Some severe and intractable cases of epilepsy can be alleviated by the drastic intervention of severing these connections surgically – at the cost of making the information in one hemisphere inaccessible to the other. The removal of these links will change the value of Φ for the brain as a whole to zero (even though it remains non-zero

* This picture is somewhat akin to that presented by the Dutch neurophysiologist Victor Lamme in the early 2000s, in which consciousness arises from recurrent loops of neural processing. Lamme developed this idea into his own theory of consciousness, which makes the counterintuitive prediction that we are sometimes conscious of things even if we can't report awareness of them.

for the individual hemispheres), so that it no longer functions as a single connected conscious entity. Such split-brain patients may indeed act as though they have two separate consciousnesses – so that, for example, one hand might perform one action while the other undermines it. Conversely, twins who are born with conjoined skulls may share some degree of consciousness through the neural interconnections of their brains: when one twin has a drink, the other might 'feel' it. Koch and Tononi suggest that conditions like schizophrenia result from a breakdown of integration within the neural network: they are in a sense diseases of consciousness.

IIT insists that consciousness is *only* possible in systems that have the ability to affect themselves by re-entrant processing. Feed-forward systems like those of today's silicon chips can produce nothing but 'zombie minds' – which might, given enough complexity, act as if they are conscious but cannot truly possess that property. 'Digital computers can simulate consciousness, but the simulation has no causal power and is not actually conscious', says Koch. It's like a physicist simulating gravity in a computer model of how a star or a galaxy forms: that doesn't actually generate a gravitational attraction between the computer's circuit components.

IIT makes consciousness a matter of degree. Any system with the required network architecture – that is, with non-zero Φ – may have some of it, regardless of whether the system is human or even sentient in the normal sense. 'No matter whether the organism or artifact hails from the ancient kingdom of Animalia or from its recent silicon offspring, no matter whether the thing has legs to walk, wings to fly, or wheels to roll with', Koch wrote in his 2012 book *Consciousness: Confessions of a Romantic Reductionist*, 'if it has both differentiated [that is, specific] and integrated states of information, it *feels* like something to be such a system.'

Thus Koch thinks that 'we are surrounded and immersed' in consciousness. 'I believe', he says, 'that consciousness is a fundamental, an elementary,* property of living matter. It can't be derived from anything else.' It is not, then, a problem-solving computation, but a state of being: it may become the kind of 'pure experience' described by mystics of all cultures, where 'awareness is vividly present yet without any perceptual form, without thought . . . The mind as an empty mirror, beyond the ever-changing percepts of life, beyond ego, beyond hope, and beyond fear.'

These ideas arouse plenty of scepticism. John Searle regards IIT as a form of mystical panpsychism: crudely, the idea that mind and awareness infuse the whole cosmos. In a withering critique of IIT, Searle has asserted that 'the problem with panpsychism is not that it is false; it does not get up to the level of being false. It is strictly speaking meaningless because no clear notion has been given to the claim.' Consciousness, he says, 'cannot be spread over the universe like a thin veneer of jam' – it 'comes in units and panpsychism cannot specify the units.'

But IIT doesn't actually assert that consciousness is spread everywhere like jam. While the theory considers it an attribute of many things, they must be particular *kinds* of things, namely those that have a non-zero value of Φ.† This includes the biochemical networks of regulation found in every living cell on the planet, and also electronic circuits that have the right sort of architecture. (It's not, in Koch's view, that computing machines can't be conscious, but rather, that to do so they would need to be very different from those

* But *not* a universal property, any more than is, say, solidity.

† One big obstacle to developing IIT into a testable theory is that the definition of Φ is such that, in all but the simplest networks, it is impossible to calculate. And the different ways that have been proposed to quantify it don't generally tally with one another. Some researchers say that it's not defined precisely enough to possess a single, well-defined value even in principle.

we have today.) A typical IIT system, Koch adds, is one that has a lot of diversity in its components and wiring.

GWT and IIT aren't by any means the only current theories of consciousness. Two common classes of theories are called first-order and higher-order. First-order theories (of which GWT is one) think of consciousness as something akin to perception, questioning why we should add anything more. Higher-order theories, meanwhile, say that there *must* be some higher-order representation of that perception: that a conscious experience is not merely a record of the perceptions involved but involves some additional process that draws on this representation. This higher-order state doesn't necessarily serve a function in processing the information; it just *is*.

Looking for the source

How can we tell who, if anyone, is right about the root of consciousness? For that, we need experiments: to identify where different theories make different predictions that can be tested empirically.

That might not be straightforward, though. It's possible that experiments might prove to be consistent both with GWT *and* IIT (or other theories), because they're not using the same definition of consciousness. Alternatively, the neuroscientist Enzo Tagliazucchi thinks it possible that these two are essentially the *same* theory, but 'developed from third- and first-person viewpoints'. Although advocates of GWT and IIT see them, with good reason, as very distinct, what they both have in common is that they stitch together – they *integrate* – different aspects of cognition. As Murray Shanahan puts it, 'the hallmark of the conscious condition is that it integrates the activity of brain processes that would otherwise be insulated from each other's influence.' It's easy to imagine how that weaving together permits the kind of cognitive fluidity seemingly so central to the human mind.

One distinction that does seem to exist between GWT and IIT, however, is in the location of consciousness in the brain: what Koch and his mentor Francis Crick dubbed the neural correlates of consciousness, the parts of the brain that must be active in the conscious state. According to GWT these include the frontal and parietal lobes of the cortex. The frontal lobe is associated with sophisticated cognitive functions such as memory, problem solving, decision-making, and emotional expression; the parietal lobe sits behind it and processes sensory data such as touch and spatial sense. But people who have had a large fraction of the frontal lobe removed – once a neurosurgical treatment for epilepsy – may seem to have a remarkably normal mental life, say Koch. He argues that the prefrontal cortex, where much of our higher reasoning takes place, doesn't seem to be important for consciousness either. If it is stimulated with magnetic pulses applied to the scalp, say, there are none of the sensations and distortions of perception that occur as a result of similar stimulation of the rear portions of the cortex (the parietal, occipital, and temporal lobes), suggesting that little of what we are aware of arises directly in this region. Koch and Tononi think it is, rather, the rear of the cortex that is vital for consciousness. The neurons here are connected in a grid-like arrangement that permits the kind of re-entrant processing (that is, high Φ) demanded by IIT, while the wiring of the prefrontal cortex doesn't have that structure.

Yet the prefrontal cortex (PFC) can't be excluded *tout court*. For example, particular neurons in the lateral PFC – a region present only in primates – seem to be activated specifically in response to certain items coming into consciousness. This region is involved in our working memory, which holds the things currently in our window of attention. What's more, damage to the PFC (as well as the parietal cortex which sits just behind it) can shut down a part of what comes into consciousness. People who have such damage in just one of their hemispheres may seem to experience reality as though 'chopped in half': they might eat only one half of a plate of food or draw just one half of a picture.

Some, such as Daniel Bor and Stanislas Dehaene, believe that a network of neurons involving the prefrontal and parietal cortex may therefore after all be the seat – the neural correlate – of consciousness. It is here, they say, that the information processed in various brain regions is filtered, chunked, and brought together, coming to life as an item in the attention field of awareness.

In truth it's still far from clear whether consciousness has a localized neural correlate in the brain at all. Just because neural activity in a certain part of the brain might correlate with, say, an object coming into awareness doesn't necessarily mean that it is the *cause* of that awareness. We can, however, be reasonably certain that some regions are *not* essential to it. Around 80 per cent of the brain's neurons are in the region called the cerebellum, which sits under the rear of the cortex and is involved in coordinating movement and the development of motor skills. And yet the cerebellum seems to play no role in consciousness. Damage to it does not affect a person's sense of being present in the moment and experiencing the world. When a twenty-four-year-old Chinese woman went for a hospital check in 2014 because she was experiencing nausea and dizziness, the doctors scanned her brain and found to their (and her) astonishment that she had been born entirely without a cerebellum: an extremely rare condition called cerebellar agenesis. Yet apart from somewhat delayed walking and language development as a child, she had led a normal life and was married with a daughter.

Other observations of brain defects call into question whether the *cortex* is essential for some degree of consciousness. In children born with the condition called hydranencephaly, the entire cerebral cortex fails to develop, the space it normally occupies being filled instead with sacs of cerebrospinal fluid. Such children (who generally die within a few years of birth) are often deaf, blind, paralysed, or severely cognitively impaired – and yet detection and diagnosis of the condition can take several months because the early behaviour of these infants may be little different from that of any other. And although they cannot report what they experience, some hydranencephalic children respond to stimuli in ways that leave little doubt that they have some degree of experience and awareness – they have a mind, a 'something it is to be like'.

Perhaps, then, some kind of core consciousness can be built purely from the activity of the more 'primitive' parts of the brain that deal with the emotions and affective sensation? That idea has been promoted, for example, by the cognitive scientist Mark Solms, who argues that 'the upper brainstem is intrinsically conscious': that's to say, a kind of feeling, a 'what it is to be like', arises there. Solms agrees that higher cognitive aspects of consciousness – like Damasio's autobiographical consciousness – originate in the cortex, but believes they depend on the signal of core consciousness from the brainstem and can't support themselves. Certainly, damage to the so-called reticular formation in the brainstem switches off consciousness, causing coma or even death. Likewise, the thalamus – the connectivity hub of the brain, which filters and organizes its processing and plays a key role in determining what comes to attention – seems to have an essential role in consciousness, but this region too is part of the 'primitive' brain and seems unlikely to be the seat of our awareness of self.*

* Whether the thalamus plays an essential role in consciousness (as some have proposed) is open to question, since cephalopods (such as squid and octopuses, which are widely regarded as having a degree of consciousness – see page 232) do not possess one. Other regions of their brains have been proposed to serve an analogous function, but the anatomy of cephalopod brains is so different from ours that it is hard to make any comparison.

Assigning an anatomical seat of consciousness is confounded by the old saw that correlation is not causation: just because a brain region is needed to sustain consciousness doesn't mean it arises there. After all, damage to the power socket of my desktop computer can immediately stop it functioning, but this doesn't mean the computation happens there. But perhaps the real obstacle to identifying both a theory of consciousness and an associated neural correlate is that we might not be seeking the origin of a single, well-defined aspect of mind. Because the healthy brain creates a sense of unified and unique experience, we imagine that there is a single source of it and a single theory that can explain it. But that sensation is probably multi-authored, and has different manifestations when one author or another is asserting themselves. Even Damasio's distinction of 'core' and 'autobiographical' consciousness might be too coarse a division. In his 1994 book *The Astonishing Hypothesis*, in which Francis Crick attempted (with considerable success) to reinstate consciousness as a topic of serious study in neuroscience, he wrote that 'there are many forms of consciousness, such as those associated with seeing, thinking, emotion, pain, and so on.' (He considered consciousness of self a particularly special case that was 'better left to one side for the moment.')

Clearly not all of these sensations (pain, vision) are essential components of consciousness, but we're not sure which of them is. Some (such as Daniel Bor) say that emotion is not essential to consciousness; others (like Damasio, Seth, and Solms) think it is central to it. That the emotions are in any event clearly *visible* to consciousness is itself curious. If emotions were nothing more than the body triggering a brain state that helps to maintain homeostatic equilibrium of the whole organism, why don't they operate just like a reflex, out of sight of awareness? Or perhaps it's better to ask: why are some bodily signals, such as hormone levels, processed automatically but others, like pain, given emotional valence and disclosed to our consciousness?

It's possible that emotions are simply rather leaky and seep into the arena of consciousness by accident. But that seems a rather unlikely inefficiency of the mind. There is probably some value in bringing the information these signals carry to awareness: they are part of what the mind needs actively to take into account in making conscious decisions. Damasio suggests that emotions are really a kind of perception, like vision and hearing – but an internal one. They give us a kind of image of the body's state: a sense called interoception, which monitors physiological parameters such as heartbeat and blood pressure. As with other senses, there is here an element of prediction and inference involved: if interoception tells the mind something, the mind makes best guesses about the cause.

In this way, says Damasio, 'feelings let us mind the body'. Such a body-sense can help make the decision-making job of the mind more efficient: 'I just have a bad gut feeling about this', or 'That hurt like hell, so I won't try it again.' Like all the other features of mind, emotional valence is a far from infallible guide: it provides a quick rule of thumb that enables us to draw a decision from a mass of sensations, but it doesn't guarantee we'll make the best one.

Still, it seems we are better off heeding the bodily awareness signalled by the emotions than lacking it. Some types of mental illness, such as schizophrenia and dementia, can cause a condition called anosognosia, in which people become unable to acknowledge their illness or other afflictions or injuries. It's a genuine breakdown of the brain's ability to diagnose the state of the self and to map the body: people with this condition aren't simply in denial and insisting everything is fine, but rather, they don't seem concerned whether it is or not. Their emotional awareness is suppressed: 'OK, my legs don't seem to be functioning, but it's not a problem I need be concerned about.' Clearly, such a condition can create severe problems for managing one's life effectively, and anosognosics can become socially vulnerable, at greater risk of becoming homeless or of committing offences for example. The underlying problem seems to be that they can no longer integrate all their experiences into a unified whole or self. It is a dysfunction of identity.*

Anyone home?

The cause(s) of consciousness is often, with good reason, regarded as one of the central puzzles of human existence. But the view from within Mindspace can afford to be less parochial. I want rather to ask: what *kinds of mind* can be conscious? And how does the nature of its consciousness depend on the kind of mind that hosts it? What varieties of consciousness might there be?

We'll start to glimpse some possible answers in the following chapters, which move beyond the solipsistic confines of our own mind(s). But we've seen already that most current theories of how our consciousness arises do not depend on the details of which neuron goes where; instead, they invoke general cognitive architectures that might support it, thereby prescribing broad classes of cognitive systems that can and cannot be conscious.

The question of whether consciousness inheres in other systems has practical and moral implications. As AI becomes increasingly sophisticated, it might be impossible to tell merely by interacting with it whether one is dealing with a machine or a human. (Already we're easily fooled, as we'll see.) But would that mean the AI

* Edelman and Tononi suggest that, on the contrary, anosognosia might result from an attempt to create a coherent, unified sense of identity by casting aside those aspects that don't fit: that paralysed limb I can no longer control can't be *mine* anyway. This, however, doesn't obviously explain the lack of affect that goes with it.

deserves moral consideration? According to functionalist theories such as GWT, consciousness is not dependent on any particular substrate, but just needs the right informational architecture. If that's so, there is no obvious barrier to conscious machines – and much of Silicon Valley believes this to be true. But in Koch's view, today's computer hardware could never supply more than a 'deep fake': zombie devices that are just very adept at mimicking conscious behaviour.

Whether an entity, a brain, a machine is conscious or not is not an abstract question, but is in some respects an urgent one that impinges on animal rights and welfare and on a wide range of medical and legal questions about mental impairments. It relates, for example, to how we consider and treat people with severe brain trauma that robs them of the ability directly to communicate whether they still possess awareness. Various experiences, including stroke, brain injury, and poisoning, can trigger a condition called locked-in syndrome, where all muscle movement is paralysed except for eye movements. Here there is generally no doubt that the patient is conscious, because the eye movements can be used to communicate, much as Stephen Hawking could use movements of his cheek muscles to control his speech simulator in the later years of his life. But some people can be afflicted with total locked-in syndrome, in which even eye movements are impossible. In that case, consciousness might need to be inferred from patterns of brain activity, observed by a technique like fMRI, in response to sensory inputs. Studies of such brain activity in volunteers while conscious and under general anaesthesia have shown that there are distinct patterns corresponding to the two states. Brain activity persists in periods of unconsciousness but happens only between brain regions that are directly connected anatomically to one another, whereas during conscious activity there are complex long-distance interactions that do not seem constrained by the brain's 'wiring'.

The identification of consciousness gets harder for more extreme losses of brain function. People in a vegetative state retain some instinctive muscle functions such as eye movements and swallowing, but not any apparent capacity for volitional action. In a coma, meanwhile, only basic activity of the brainstem remains: enough to keep the body functioning, but probably not to sustain any real awareness. In such cases, does a loss of consciousness signify a loss of personhood – and with it, the basic human right to life? Is a living body enough to constitute a living human being, or must it host a sufficiently functional human mind too? Our instincts are not attuned to addressing such questions, but the law (and the medical establishment) must draw a line. That boundary – an absolute right to life – is typically deemed to depend on establishing whether a person possesses consciousness.

All we have right now are crude measures. A procedure developed by Tononi in collaboration with the Italian neurophysiologist Marcello Massimini relies on the observation that conscious brains seem to generate a particular electrical signal when stimulated by a magnetic field. A coil of wire held against the scalp sends a magnetic pulse into the brain, creating an electrical response that can be measured by electroencephalography using electrodes also attached to the skull. It looks like a ricochet, as if the magnetic pulse is bouncing around in the brain. Brains that are conscious produce clear reverberations, but if they lack consciousness – for example, for people under general anaesthesia – the response is weaker and more disorganized. Koch compares it to the ringing of the cracked Liberty Bell.

The distinctions between these signals are fairly clear, but we don't really know what they imply; it's still an assumption that the difference is that between a conscious and non-conscious brain. (Koch believes that the response indicates a rough measure of the 'integration' of information in the brain.) All the same, this so-called zip-and-zap technique has shown that there may be significant differences between the mental states of people in vegetative states. In one trial, nine out of forty-three patients in such states displayed a 'conscious-like' response.* The technique is now being explored too for people with late-stage dementia.

Such techniques might be essential for making medical decisions, but they also compel us to wonder about kinds of mind. The mind of someone in a vegetative state surely sits in a different part of Mindspace to that of someone who can communicate with and affect their world. But to where is it shifted, and to what degree? What kind of mind is this, and what responsibilities do we have towards it? Might it ever be possible to turn it back into the kind of mind that we possess? How is Mindspace navigated?

The idea that consciousness is a matter of degree seems uncontroversial, but this doesn't necessarily mean that it is like a single dial that gets steadily turned up in different organisms. It may be that at certain thresholds in consciousness – or perhaps, if it expands in certain directions – qualitatively new features come online. That, after all, is precisely the kind of distinction evident between core and autobiographical consciousness. Is the self-aware consciousness we humans have, then, the highest stage of consciousness? Or might there be more to it that we cannot (yet) glimpse? It seems to me rather unlikely, not to say hubristic, to suppose that we have reached the pinnacle of what evolved minds like ours may become – or become *aware of*. But it might also be as hard, if not impossible, for us to imagine what else there could be to consciousness as it is for a cat to imagine its way into a human-like sense of selfhood.

* Some people in vegetative states also show patterns of neural activity in fMRI, in response to questions asked of them, that are indistinguishable from those of healthy subjects.

What's the use of consciousness?

Such considerations compel us to ask: what is the point of consciousness anyway? Why did evolution award us, and almost certainly other animals (as we'll see), with a perception of *being here*?

There would seem to be two possibilities. Either consciousness is adaptive – it somehow boosts fitness, compared to an equally cognitively adept being that lacks it (if such is possible) – or it is a mere accident, an epiphenomenon, something that happens (perhaps necessarily) in the kinds of minds we have, but which lacks any adaptive value. Perhaps it is like the sound of a beating heart: the heart can't function without it, but the sound of the heartbeat is not what a heart has evolved to produce.

The latter view needn't make consciousness a mere random evolutionary accident. It could be that brains with the cognitive capabilities of ours simply could not exist without becoming conscious: that consciousness is what is known in evolutionary biology as a spandrel. The word comes from architecture, where it refers to the roughly triangular segment of wall above each side of an arch. It is there not by design but by necessity: the arch is the load-bearing structure, but one can't make a load-bearing arch without creating spandrels. Similarly there are elements of biological form and behaviour that are not themselves selected but which arise from selection for something else. In terms of the human mind, music may be one such feature: it's not obvious (the issue is still disputed) that our capacity for making and enjoying music per se was selected for, but it might arise inevitably from other mental attributes, such as our ability to discriminate pitch and our tendency to seek patterns. Consciousness too could be an evolutionary spandrel of the mind.

Daniel Dennett believes that, adaptive or not, consciousness is likely to be inevitable for minds like ours. 'Those who claim they can imagine a being that has all these competitive activities in the cortex but is not conscious are mistaken,' he says. 'They can no more imagine this coherently than they can imagine a being that has all the powers of a living thing but is not alive.' If so, then when we assume consciousness in others we need not be just making a logical extrapolation from our own experience (though it would be reasonable to do so). We might instead be intuiting that consciousness is a necessary corollary of the behaviour we observe: that a zombie that can act entirely like a regular human yet without any inner world of awareness is impossible. This doesn't mean humans can't have impairments of consciousness – and indeed any of us can lose it. But it would mean that they cannot incur such impairments without suffering also some other related and detectable dysfunctions in cognitive ability. If this view is right – that human-like 'intelligence' is possible only with some degree of consciousness – the implications for artificial intelligence are profound.

Most neuroscientists and cognitive scientists suspect, however, that consciousness *is* adaptive. That's not surprising; it seems such an extravagant and central aspect of our minds that it's hard to imagine it is simply incidental, let alone accidental. Surely something so intense, so central to our existence, is not just a random by-product of other evolutionary exigencies? This is what Christof Koch asserts:

> Consciousness is filled with highly structured percepts and memories of sometimes unbearable intensity. How could evolution have favoured such a tight and consistent link between neural activity and consciousness if the feeling part of this partnership had no consequences for the survival of the organism? Brains are the product of a selection process that has operated over hundreds of millions of birth-and-death cycles. If experience had no function, it would not have survived this ruthless vetting process.

A common view is that consciousness serves an organizing function: it pulls diverse mental sensations into a coherent framework. If consciousness is all about forging, from the morass of sensations we receive, a single felt experience and a consequent decision to act on it, there's obvious survival value in that: it shows us what we need to heed, and goads us to care. Global workspace theory, for example, implies that our neural processes conspire to *compel* the mind to decide on an interpretation of what it experiences, even if that means taking a best guess based on ambiguous stimuli, or perhaps continuously switching from one interpretation to another.

In GWT, consciousness *arises* in the act of bringing things to our attention, giving them salience and meaning. Yet it is possible to imagine and even to write computer algorithms that can achieve a comparable focus and decisiveness, for example by applying rules for filtering raw sensory data and even for attributing a kind of *value* to what is selected in the way that our emotions seem to do. One could readily couple this selection process to algorithms that determine appropriate actions, without having to awaken any awareness within the system.

For Damasio, meanwhile, the adaptive value of consciousness may be primarily that 'in a conscious mind the processing of environmental images is oriented by a particular set of internal images, those of the subject's living organism as represented in the self.' Maybe, he says, consciousness consists not so much in this process of selection for what we attend to, but in the subsequent task of constructing mental scenarios to evaluate courses of action. In putting a *self* into those scenarios, we are producing an awareness of our own agency in the world – as Damasio puts it, 'the organism [becomes] cognizant of its own plight.'

But is it always better to include self-awareness in our actions and decisions? As I've said, musicians and sportspeople are painfully aware of how becoming reflectively conscious and self-aware of performance can degrade it relative to when it is done unconsciously and 'out of sight'. In any case, such model-building and scenario-running that includes some self-representation of the agent in the frame could be automated too, given enough processing power.

In other words, it's frustratingly hard to find a compelling and – vitally – a *testable* argument for why consciousness is adaptive. In cognitive terms it is hard to see what we'd lose by removing it – except everything. We can't really pronounce on whether consciousness is adaptive in the evolutionary sense until we have a better notion of what it is. If it is 'nothing more' than the process by which information in the brain is filtered, integrated, brought to attention, and then broadcast, then there is no further case to make: that process seems to be what permits our minds such flexibility and versatility of action-determination and forward planning, and so its survival value is clear. But there is nothing in such a description that obviously imbues it with the quality of *feeling like* anything; it remains an act of faith that this is what *feeling like* entails.

I have chosen to define minds as entities for which there is something it is like to be them – for which at least some glimmer of awareness, some rudimentary form of consciousness, exists. But there is no reason we know of why consciousness should be *required* for sophisticated cognition or intelligence, and so the Space of Possible *Minds* might, in this view, be just a subspace of the Space of Possible Cognitive Devices. Indeed, I think this is surely so. It would be good, then, to know which aspects of cognition are essential for consciousness and which are optional: where, in that greater cognitive universe, the boundary of true Mindspace lies. It's not enough to suppose that the human mind gained a whole bunch of cognitive skills and that in some vague way consciousness tumbled out of them; we can at least hope and strive to know which attributes were essential for this.

Brain in a dish

Christof Koch has argued that even if we can't yet say for sure that consciousness is adaptive, integrated information *is*. He and his colleagues have conducted computer simulations of 'digital creatures' dubbed *animats*, which have rudimentary sensory and motor functions – an eye, a distance sensor, and two wheels – that enable them to navigate a maze. For entities this simple, it's possible to calculate the integrated information (the value of Φ) in their cognitive processes. Koch and colleagues allowed these entities to evolve over thousands of generations by random mutation, and awarded a reproductive advantage – a greater Darwinian fitness – to those that were able to navigate the maze faster. They found that the fitness increased over successive generations in proportion to the value of Φ. The researchers got similar results for animats assigned a different task of catching falling blocks in a Tetris-style game. If, then, IIT is correct to attribute consciousness to integrated information, it would seem to have genuine adaptive value – even if we don't know quite where that value resides.

This doesn't sound like the kind of experiment that could be done with real entities. But don't be so sure. Scientists are becoming ever better at growing what we might call proto-minds, or what have been dubbed 'mini-brains', from living neurons. They are not *really* brains in any true sense – they are much smaller and lack the structure, organization, activity, and functionality of our own brains – but they are brain-like enough to tell us something about how normal brains grow and function. They are more properly called brain organoids, and they are already posing some challenging conundrums for the problem of consciousness and what its ethical implications are.

Brain organoids can help to tackle the problem that it is hard to take a good look at a human brain while it is still alive. We can't ethically go in and poke around or extract samples while the brain is still part of a living, breathing person. And doing so after a person is dead understandably imposes severe limitations on what you can find out. Indirect, non-invasive measurements of living human brains, such as electroencephalograms and fMRI scans, are of immense value, but we're forced to infer, rather than to see directly, what cognitive and cellular processes underlie the signals they yield. Yet because brain organoids are not part of a human being, they aren't subject to the ethical considerations of experimentation on living people. We can stick electrodes in, administer drugs, slice and stain the tissues for examination in a microscope. They are no different from any other human tissue grown in culture – which is to say, in a bath of the nutrients needed for cells to survive and multiply.

Brain organoids are grown from stem cells: the cells in the early embryo that can develop into any tissue type in the body. Such cells can be taken from embryos discarded from IVF fertility treatments, when the embryos are just a few days old. (In the UK and many other countries, such embryos cannot legally be kept viable outside the body for longer than fourteen days.*) Or – remarkably – they can be made artificially from the mature, specialized cells of living people, such as skin cells taken in a biopsy (Figure 4.3). In 2006, Japanese biologist Shinya Yamanaka and his colleagues found that such cells can be reprogrammed by adding to them genes that are highly active in embryonic stem cells (for example, using harmless viruses as vehicles or 'vectors' to inject the genes into cells). Such artificially created stem cells are called induced pluripotent stem cells (iPSCs).

* The International Society for Stem Cell Research recommended in 2021 that this rule be relaxed for some research purposes, although it has yet to be seen if the advice will be adopted into law.

Cultured stem cells can be persuaded to grow into all kinds of specialized types: kidney or heart cells, pancreatic or gut cells, or the neurons of the brain. This might involve adding other genes to set them on the right developmental path, although inducing a 'neuron fate' turns out to be even easier, requiring just a slight change in the nutrient medium. As the cells acquire their chosen fate, they become able to organize themselves into structures rather like those they would adopt in a growing embryo. A colony of gut (epithelial) cells, for example, begins to produce the little protrusions called villi that sweep food particles through the gut. Heart cells begin to pulse as concerted waves of electrical activity pass through them. And neurons start to arrange themselves in layers like those seen in the cortex, or to sprout protrusions like an incipient brain stem, or to pucker into folds like those of a mature brain.

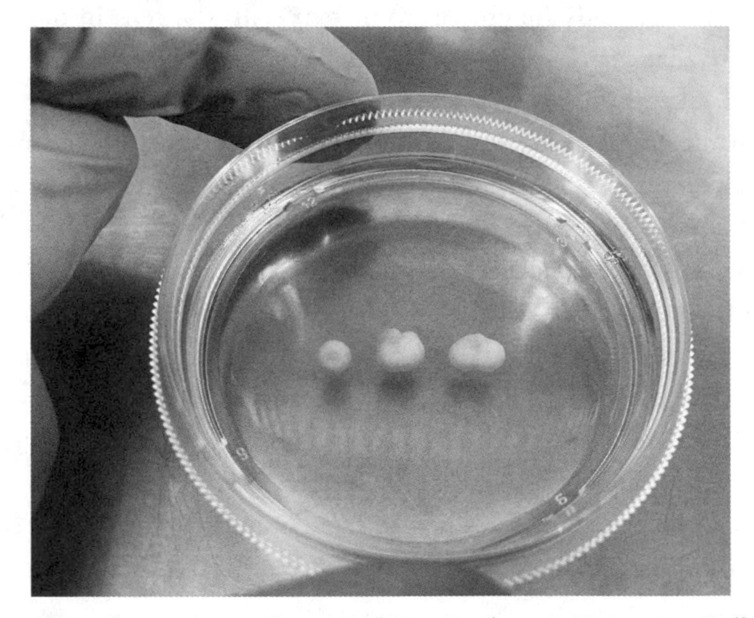

Figure 4.3. Brain organoids grown by researchers at University College London from cells reprogrammed to become induced pluripotent stem cells and then coaxed into developing as neurons. I describe this process in my 2019 book *How To Grow a Human*.

Brain organoids have been used to study diseases or genetic conditions that affect brain development or maintenance, such as Zika (a viral condition that can stunt brain growth in foetuses) and Alzheimer's disease. And they are being created to investigate why the Covid-19 coronavirus causes brain damage in some people: you obviously can't infect a person to find out, but there's no ethical problem (yet) with infecting an organoid.

These entities recapitulate only crudely the growth of real brains, though. Developing brains, like any organ of the body, need to receive signals from surrounding tissues in a foetus to grow properly; without those, an organoid is but a poor approximation to the real thing. What's more, they reach only a few millimetres in size – about the size of a lentil – before the innermost cells in the cluster start to die through lack of a nutrient (here a sort of blood substitute) supply. But researchers are finding ways of making brain organoids more realistic and bigger, for example by giving them networks of blood vessels and providing the cues they would normally get from other tissues to guide growth. 'Given the astonishing pace of progress in stem cell biology and tissue engineering', says Koch, 'bioengineers will soon be able to grow sheets of cortical-like tissue on an industrial scale in vats, properly vascularized and supplied with oxygen and metabolites to maintain healthy levels of neuronal activity.'

For the neurons in brain organoids *are* active: they send signals to one another just as they do in the brain. And they can show *coordinated* electrical activity, somewhat resembling the brain waves seen in the human foetus in the uterus.

All this makes brain organoids extremely valuable for biomedical research, and there are even hopes that neural tissue grown this way might become useful for repairing brain injuries or lesions. But even as improvements in the 'realism' of brain organoids makes them an increasingly good proxy for actual brains, so too does it make them more ethically fraught. I said that brain organoids are no different in principle from any other mini-organ grown by cell culture – but of course in one sense they are profoundly distinct. For our brains are the seat (if not the entirety) of our minds. To put it bluntly, as we get better at growing mini-brains, we have to start asking: are we growing mini-minds? Are organoids in any sense *minded* entities, with an inner life, a glimmer of awareness? Is there something it is to be like a brain organoid?

As we have seen, most researchers concur that sentience and consciousness are not properties that arise simply from having enough 'computing power': you are unlikely to get it, say, once a mass of neurons exceeds some critical size. The cognitive circuitry must have a particular architecture, although no one is yet agreed about what that should be – a global workspace, or the massively re-entrant circuits invoked by integrated information theory, or a brainstem structure, or something else. But if brain organoids recapitulate the basic forms of neural organization in the brain, won't they develop such a structure automatically?

That depends on which parts of the brain appear in the organoid. As we saw, different brain regions have different types of network connectivity, and not all of them play a part in consciousness. On the other hand, cell biologists are starting to gain control of how an organoid develops so that they can tip it towards particular anatomies: it might be possible to make a mini-brain that is mostly cortex, say, or mostly hippocampus. It's entirely possible to imagine making a brain organoid containing good approximations to whatever regions are thought to be key to the development of consciousness.

Indeed, Koch believes that some, perhaps most, of those little organoids made already might possess a small degree of consciousness: they could have a significant value of Φ, the measure of consciousness in integrated information theory. These entities might then buck the trend of the natural world in which, Koch thinks, the

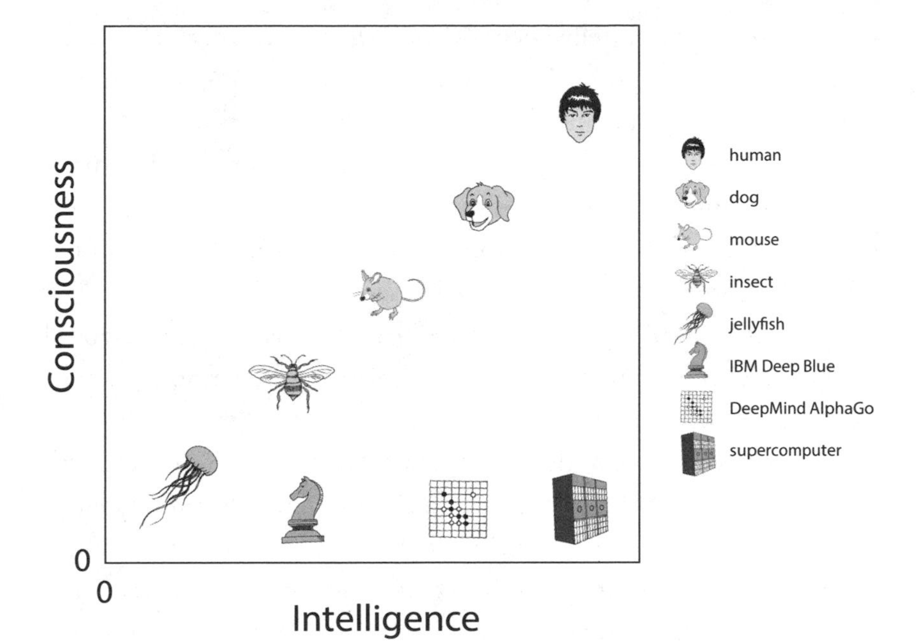

Figure 4.4. In a space of minds defined by coordinates of intelligence and consciousness, brain organoids might occupy an anomalous position that departs from the trend in naturally evolved species.

degree of consciousness generally increases in proportion to measures of intelligence. In this view you could say that brain organoids have an unnaturally high level of consciousness for their level of intelligence. This locates them in Koch's earlier two-dimensional plot of the Space of Possible Minds (Figure 4.4; see page 56).* Conversely, you might regard this supposition as an argument that

* The picture is hampered by our inability to assess the true 'intelligence potential' of an organoid: what it might be capable of, if it were to be somehow hooked up to interfaces that allow its cognitive powers to be exercised and demonstrated. As we'll see, studies are already underway into how that might be done.

consciousness and intelligence are not *necessarily* coupled – even perhaps a vindication of Anil Seth's suggestion that 'consciousness has more to do with being alive than with being intelligent.'

What might such an organoid feel? 'The chance that it experiences anything like what a person feels – distress, boredom, or a cacophony of sensory impressions – is remote', Koch says. 'But it will feel something.' If so, it will be minded.

From an ethical perspective, a key question is whether it could feel pain. Then we might need to recognize an obligation to experiment on such brain organoids only after administering an anaesthetic. But the question itself is perplexing even for experts. It's not like asking if, say, a fly or worm feels pain – because an organoid is not an organism. The question in fact has almost metaphysical dimensions, like asking whether the wind on Mars makes any sound. For the fact is that the brain itself *has no pain sensors*. If – forgive the image – a piece of your skull was to be removed and your brain poked and prodded, it would not hurt a bit (not, that is, the poking and prodding). And yet *the brain is where pain is produced and registered*. When you prick your finger with a needle, it feels as though the finger itself hurts – but if the neurons signalling from the finger's pain receptors to the brain were severed, there would be no pain; the sensation is not 'made on site' at the point of injury. So what can pain mean for a brain organoid?

It is the isolation of a brain organoid from a body that makes it such an ambiguous entity, in terms of neurology, ethics, and mindedness. Philosophers have long argued whether a lone 'brain in a vat' can truly experience anything: whether it can think. This was long considered a mere (forgive me) thought experiment; no one really imagined it could become a reality.* The advent of brain

* Such discussions often now make reference to *The Matrix*, the 1999 movie in which people are kept alive in vats by a controlling super-intelligence and their

organoids reveals for such debates what neuroscience has revealed more generally for the philosophy of mind: words and abstract theorizing on paper are no substitute for actual biological experimentation, and often miss what is truly important.

Some philosophers, for example, have argued that a brain lacking any sensory input – without eyes, ears, touch, and so on – can develop no concepts about the world of the kind needed to support genuine thought. In this view, any neural activity in an organoid is all sound and fury, signifying nothing.

But neither should we imagine that simply hooking up a brain organoid to sensations – making it able to interact with the world – will waken it to sentience. It should be clear by now that mind and consciousness are not like pictures on a TV screen, ready to spring into view once the device is switched on. They develop as the organism develops. This is literally true for brains: neurons need stimuli in order to function properly. Brain circuits are formed and refined by experience. Absent experience, that cannot happen. At the very least, an (advanced) brain organoid would need to be nurtured into consciousness by sensory input from hooked-up sensors.

That, however, is conceivable. Some brain organoids, for example, have developed patches of light-sensitive retinal tissue – for the eyes and optic nerves are actually a part of the brain. It's conceivable, then that these entities might acquire, not vision exactly, but a responsiveness to light. And it's possible to link the neural activity in a brain organoid to muscle tissue, so that the mini-brain can induce movement. The developmental biologist Madeline Lancaster and her colleagues at Cambridge University cultured a slice of brain organoid alongside a piece of mouse spinal cord attached to the

---

brains are fed stimuli that convince them they are living lives like ours. For more discussion of the philosophy and cultural history of a 'brain in a vat', see *How To Grow a Human*. We'll return to the Matrix, so to speak, later.

surrounding muscles of the back, and found that the organoid neurons would connect to the spinal cord and send electrical impulses that made the muscle twitch. In other words, it should be possible to give brain organoids a rudimentary kind of body that offers it 'experience'.

What happens then? *Can* it support a mind?

In fact, some researchers suspect that advanced brain organoids might one day harbour a kind of consciousness – a degree of awareness – even while still in isolation from their surroundings. They would be 'islands of awareness', according to the philosopher Tim Bayne of Monash University in Melbourne and his colleagues: patches of something like Koch's 'pure consciousness', devoid of contact with the world outside them.

It's hard to see how such island minds might ever arise naturally, at least through natural selection: they have no function, no goals, just a kind of Zen-like existence and awareness. But brain organoids, drawing on the 'mindstuff' that evolution has created, might be one way in which they could be made. Bayne and his colleagues also speculate that island minds could be produced by other technological interventions in brains – for example, through surgery or efforts to revive brains from recently killed animals.* Something approaching such isolation may also happen in the late stages of multiple sclerosis or in cases of locked-in syndrome or the

* In 2018 researchers at Cornell University reported signs of neural activity in pig brains taken from a slaughterhouse and administered nutrients and oxygen. Although this activity seemed very unlikely to produce anything like consciousness or sensation – there was no activity that extended throughout the entire brains – the researchers gave the brains anaesthetic to alleviate any concerns about pain. 'If organised patterns of spontaneous neural activity were to be observed in this situation, the question of whether an island of awareness was present would immediately arise', say Bayne and his colleagues.

inflammatory condition of Guillain–Barré syndrome, where people lose almost all of their motor functions and perhaps much of their sensory inputs. The researchers think it might be possible to detect such islands of awareness, for example using something like the zip-and-zap method I described earlier.

If indeed brain organoids were to show some signature of awareness – whether in isolation or while connected to input and output signals – that would surely have implications for their ethical status. They would be more than mere neural tissue.

But more in what sense? It's hard to see how a brain organoid might qualify as an 'individual with rights', in the same sense as, say, a near-term baby. Those who argue for the rights of an early-stage human embryo generally do so on the grounds of its potential to become a fully developed human being, which an organoid certainly does not have. Even concerns about any 'existential pain' it might experience would seem predicated on a false notion that it possesses some sense of selfhood, potentiality, and agency.

Perhaps we can't evaluate this issue without developing some idea of the kind of mind an organoid could have. But the concept of an 'island of awareness' expresses the matter more cogently: it may be that a brain organoid, and possibly other pieces of brain tissue, represents 'pure mindstuff': matter that has mindedness and yet *is not truly a mind at all.* That's to say, we might need to regard mindedness more as a property of matter, which can, in the right circumstances, be organized into a genuine mind, just as tissue culture shows that life is a property of matter that can in the right circumstances be organized into an organism. The question then becomes: does *minded matter* in itself have a special ethical status?

We don't currently have an ethical framework for figuring out answers to such questions. The notion of rights is designed for humans – after all, its modern origins lie in a time (during the political upheavals of the late eighteenth century) when its function was very different, and not originally intended even to apply to all of humanity. It is not some attribute we are born with, but is socially conferred and mediated, which is why we find it so hard to extend it even to our close evolutionary relatives: in 2015 the New York State supreme court declared that chimpanzees do not have the legal status, and thus the rights, of persons because they 'do not have duties and responsibilities in society'. Goodness knows how we could stretch such concepts to accommodate organoids with awareness. But we're unlikely to figure out how to do so unless we can become clearer about the nature of other minds.

CHAPTER 5

Solomon's Secret

Charles Darwin had no doubt that other animals have minds. There are, he wrote, 'no fundamental differences between man and the higher mammals in terms of mental faculties'. He might have been indulging the Victorian tendency to anthropomorphize nature, but few animal behaviourists would now dispute that at least some animals, especially other primates, possess what we can regard as a mind – alive to sensation, feeling, emotion, and most probably to some degree of consciousness.

Whether – or to what degree – we can get inside those minds is another matter. When Thomas Nagel asked what it is like to be a bat, it wasn't an invitation to imagine getting around using sonar, but to highlight the impossibility of conceiving of an entirely different mode of existence, with different motives, goals, and brains. Does the bat (or the cat or rat) even experience itself as a self? Does it possess, say, anything like a Theory of Mind that attributes selves to its fellow bats? Can we ever get further than just imagining our own human mind being poured into a bat-shaped vessel, so that we can fly and see differently and enjoy eating insects?

Until recently, studies of animal behaviour had focused less on Nagel's question of the subjective experiences of animals – of the nature of their minds – and more on the matter of how close they come to, or far they remain from, the cognitive abilities of humans. Koko, a gorilla who lived in a preserve in the Santa Cruz Mountains until her death in 2018, learnt many words in a modified version of American Sign Language, while a bonobo named Kanzi studied by primatologist Sue Savage-Rumbaugh and psychologist Duane Rumbaugh since the 1980s can understand complex human commands and communicate using both sign language and keyboard lexigrams. These feats are impressive, and rightly strengthen our feelings of kinship with other apes. But given how our ability to express ourselves and reveal our inner depths is constrained when we use a foreign language in which we lack proficiency, how can we be sure that any creature that learns to convey its wishes using symbols and concepts we have given it is showing us their true selves – their real mind? As neuroscientist Joseph LeDoux has pointed out, 'research on animal behavior pretty much has to start from an anthropomorphic stance. We study things that matter to us.'

If this seems solipsistic, we should never underestimate the extent to which old ideas still pervade the assumptions of science. Historically, humans have consistently been regarded as inhabiting a fundamentally different plane of existence from other animals, and it has proved hard to set aside that default assumption. For Aristotle, only humans have a 'rational soul', which provides them with a faculty for reason (page 18). By characterizing other animals as mere automata that lack a Christian soul, Descartes deepened the division: the distinction was now not merely functional or biological, but also moral and theological. Descartes wrote (perhaps hoping in part to pacify his critics, who detected heresy in his mechanical philosophy of the body) that there is no error

> more powerful in leading feeble minds astray from the straight path of virtue than the supposition that the soul of the brutes [animals] is of the same nature with our own; and consequently

that after this life we have nothing to hope for or fear, more than flies or ants.

Descartes' exceptionalist view was widely shared right up to the late twentieth century, in particular in behavioural psychologist B. F. Skinner's belief that animals are merely a kind of trainable automaton, expendable enough to be used as kamikaze pilots for guiding bombs. (During the Second World War, Skinner designed a scheme – never put into practice – for harnessing pigeons as living missile-guidance 'machines', their pecking actions on a screen inside the device used to steer the bomb to its target.) The specialness of the human mind is enshrined in the very name we have given our species: *Homo sapiens,* the wise hominid.

Vestiges of these beliefs remain today, for example in suggestions that there is a 'higher' consciousness (a kind of secular proxy for the rational soul) that acts as an all-or-nothing (and single-valued) attribute producing a sharp dividing line between humans and other animals. But at least some scientists now consider this an absurd notion. 'The belief that only humans experience anything is preposterous', says Christof Koch, who considers the idea 'a remnant of an atavistic desire to be the one species of singular importance to the universe at large.'

Yet the well-motivated wish to escape from old prejudices about our own species constantly risks becoming an exercise in pulling other organisms up to our own level. It can often look like another 'Mind Club' affair: whom do we admit (on the grounds of having minds deemed to be a bit like ours), and whom do we turn away at the door? Plenty of animal behaviourists today feel that this attitude needs to change: we should cease comparing animal minds with ours, and ask instead what kind of minds they are. Certain cognitive skills of some animals far exceed our own: the wonderful motor functions of the agile chimp, the auditory acuity involved in birdsong and the bird's navigational prowess, the dog's olfactory sensitivity, and so on. But none of that on its own says much about the *minds* of these creatures – after all, we can make mindless robots that can perform some tasks much better than we can. The question posed by Nagel still stands as a challenge: what do cognitive capacities of this kind imply for the subjective what-it-is-to-be-like of the animal mind?

The Umwelt

The minds of organisms on our planet co-evolved with particular sensory capabilities to suit the environments in which they operate. Animals don't gain an aspect of cognition without a reason: their assemblage of capacities is congruent with their circumstances. The world of the dolphin might as well be a different planet: translucent blue-green in all directions, including up and down, a near-indifference to gravity, (almost) no shadows, no smell, and a totally different soundscape. What mental concepts are needed to navigate a place like this? Animal behaviourist Con Slobodchikoff, a specialist on the communication of prairie dogs, says that 'the time has come for people to understand that what we recognize as reality is not necessarily what other animal species recognize.' For example, he says,

> bees and some birds see in the ultraviolet range of the visual spectrum, but we don't. Bats, dolphins, dogs, and cats hear sounds in the ultrasonic range, but we don't. Dogs have up to one million times the smelling capacity we humans can muster. The world is a different place for each animal species, yet we humans in our arrogance assume that every animal should conform to our standards of 'reality'.

The makeup of those other worlds is encapsulated in what the early twentieth-century Estonian-German physiologist Jakob von Uexküll called the *Umwelt*. An animal's Umwelt reflects what 'stands out' for it in the environment: what it notices and what it does not, according to its own set of needs, concerns, and capabilities. Smell is a much smaller part of our Umwelt than it is for dogs: we are less sensitive to it, for sure, but also it *means* less: it is often more ambient than informative. Intonation in speech is less a part of the Umwelt of people whose native language is not tonal: an Englishman like me struggles to make it the salient aspect of Chinese speech that it is to a native Chinese speaker.

The Umwelt also includes things we recognize as objects: for us it includes, say, a woodland path, whereas for a vole there are just open spaces (dangerous) and more vegetated spaces (safer). In the Umwelt of a slug, a lump of flint is (I admit I'm guessing here) merely another obstacle to be navigated, whereas for our ancestors it represented a potential tool. And the Umwelt includes an organism's conception of intuitive physics: things that it believes can happen. Our Umwelt includes the possibility that a teacup placed too close to the table edge will fall off – something probably not included in the Umwelt of a gerbil. The Umwelt of a European robin includes some kind of awareness – we don't know what that really means – of the Earth's magnetic field.

It is from the Umwelt that a creature builds its conception of the world. It is the organizational framework of its mind: what, in the world, that mind finds meaningful.

The notion of an Umwelt can, I think, function in much the same way as the concept of Mindspace in establishing and systematizing difference without implying scales or hierarchies of significance or status. We are *here*, and like *this*; other species are *there* and *there*, and like *that*. After centuries of arrogant blindness to the abilities (and also the rights and needs) of other animals, and a bigoted insistence on our own specialness, a well-motivated desire now to make amends all too easily morphs into a misty-eyed reverence. Our rightful admiration at the memories, navigation, smell discrimination, and learning abilities of animals can make us overlook the fact that no animal has ever written a book or posed a mathematical theorem. As animal psychologist Edward Thorndike complained in 1911, 'biologists . . . have looked for the intelligent and unusual, and neglected the stupid and normal'. Animal behavioural studies, he said, are 'never about animal stupidity.'

It is unfair to call it stupidity but you can see what he meant. A better way to put it is that the intelligence of other animals is typically specific: they may have a capacity for some tasks that equals or even exceeds our own, but they very rarely generalize it. The bowerbird expresses what looks like remarkable creativity and artistry in making its bower, but doesn't create anything else. Jays have an extraordinarily good navigational memory for where they have hidden their food caches, but don't do much else with this facility. Chimpanzees and the family of birds called corvids (such as crows and ravens) will make tools to get at food, but not to help them with nest-building or transporting heavy objects. The reed warbler can tell apart its own eggs from the rather similar-looking ones of the cuckoos that try to plant them in the nest so the reed warbler raises the cuckoo chicks as its own. But if a reed warbler fails to spot the imposter egg, it seemingly doesn't distinguish its own chicks from those of the cuckoo, which look totally different. Evolution has clearly economized on the bird's cognitive powers.

If there's one thing that truly distinguishes humans from other animals, it is the fluidity of our minds: an ability to generalize their cognitive capacity to many different situations. For this reason, human behaviour is less stereotyped and more open-ended than that of other animals: we have a vast, almost limitless repertoire of things we *could* do in a given situation (even if in practice our choices are usually fairly predictable). This behavioural flexibility, as we've seen, may be connected to the degree of consciousness we exhibit: our ability to bring many cognitive resources to bear on a problem within the workspace of the mind. Our mental superpower is perhaps the capacity to construct new and totally imaginary scenarios: possible futures, rather than just replays of things we have experienced already. We can tell ourselves, and each other, stories. David Hume expressed it powerfully in *An Enquiry Concerning Human Understanding*:

> While the body is confined to one planet, along which it creeps with pain and difficulty; the thought can in an instant transport us into the most distant regions of the universe; or even beyond the universe, into the unbounded chaos, where nature is supposed to lie in total confusion. What never was seen, or heard of, may yet be conceived; nor is any thing beyond the power of thought, except what implies an absolute contradiction.

This is, in the end, how our mental lot differs from that of other animals, and it is of course both blessing and curse. No other species has a mind versatile enough to adapt to so many different types of habitat. No other has cultures as rich as ours, supported by language and art, music and drama. No other kind of being conducts so systematic an investigation into the nature of its world, supported by the most abstruse abstraction. No other has contrived to alter the world's ecology so profoundly, has been so appallingly destructive, and threatens the survival of large swathes of the biosphere and of so many other species with which we share this planet.

Close connections

It's natural that so much of the research into animal minds has focused on our nearest evolutionary relatives, the other primates: apes and monkeys. Our close evolutionary kinship makes it more reasonable to suppose that similarities in behaviour to our own stem from similarities in the underlying mental and neural processes (although as we'll see, we should be wary of assuming this too casually). But it also means that in considering the minds of other primates, we are really not venturing very far afield in the Space of Possible Minds. The exploration is comfortable in its familiarity but timid in its ambition.

It's both undisputable and unsurprising that great apes show many cognitive similarities with humans, given that we all share a common ancestor just fifteen million years ago, and that our ancestors diverged from those of chimps and bonobos a mere six to eight million years ago. That this is just two thousand times longer ago than the building of the Great Pyramids, and just a hundred or so times longer ago than the earliest evidence of human art makes this an evolutionary kinship that feels of imaginable, meaningful proximity.

We humans exist in a world that contains distinct objects that we categorize, quantify, and expect to be permanent. They exist in a space that we represent mentally: my coat is on the hook downstairs. We know about cause and effect. Much of our conceptual relationship to the physical world is about getting what we need, especially food: we need to know where to find it, and how to recognize it. We live in the company of others, and we recognize them as individuals with their own quirks and foibles, their own intentions and goals – their own *minds*. A considerable part of our social interaction is about finding a mate, and this involves competition and struggle for dominance. But we know too about friendship and the value of strong social bonds – to our kin, but also to others. We can make tools and technologies to assist in these goals.

All this is true of other great apes too: the orangutans, gorillas, chimpanzees, and bonobos. Yet the similarities are uneven in variety and extent. In some ways other primates seem to think like us, in others they are worlds apart – not just from us but from each other.

Chimpanzees are good navigators. They seem able to develop cognitive maps of space, a mental representation of familiar terrain that enables them to find things previously stored, such as food caches. Apes can show an impressive memory of events and location: in one experiment, chimpanzees and orangutans were able to find tools in a room that they had familiarized themselves with as much as three years earlier. Some apes have phenomenal short-term memory. Trained to get a reward if it successfully touches the nine Arabic numerals on a touchscreen in the right sequence, some chimpanzees will do this flawlessly and quickly even if all the numbers disappear the moment '1' is touched, so that it must remember where the others were.

Chimps seem also to have an intuitive physics: they know that objects continue to exist when out of sight, and that these continue to take up space and affect other objects. They can work with indirect cues: for example, if trained to know that one of two cups contains food such as nuts that will rattle when shaken, they will deduce, if the experimenter holds up one cup and shakes it without producing any sound, that the food is in the other.

The conceptual world from which an intelligent creature's intuitive physics is drawn seems likely to be shaped by its physical and sensory environments: that's inherent in the very notion of an Umwelt. If birds have any kind of intuitive physics (we'll consider shortly to what extent they do), it is tempting as well as exhilarating to wonder if it is filled with intuitions about gliding, hovering, and aerodynamics – for theirs is partly an Umwelt of flight.

Chimps, meanwhile, have the Umwelt of a tree-dwelling, clambering, socializing forager. They are sensitive to colour cues: if trained to expect food in a red container, they will not be distracted by changes in the container's shape. This ability to focus on hue is understandable in a creature that uses fine colour distinctions to assess the ripeness of fruit and nuts.

And they seem finely attuned to the concept of weight, which is unsurprising given that they spend much of their time climbing trees and swinging from branch to branch. If two identical upturned cups are placed on a fulcrum-like balance beam, one of which (as the chimp has been trained to understand) has a reward such as food hidden under it, the chimp will reliably reach for the lower cup, deducing that the hidden food has tipped the balance that way. But maybe the chimp just learns to associate the reward with the lower cup, without any notion that the reward is what *causes* the beam to tilt? Not so, apparently – for if the experimenter herself pushes the beam down on one side, the chimp no longer makes the same selection. It appears to 'know' that tilting in this case gives an unreliable cue, caused by manipulation and not by a concealed mass of food.

Amanda Seed, an expert in primate cognition, and her colleagues have identified three types of knowledge that animals might draw on to solve problems like this. The most basic and least versatile is *perceptual knowledge*: the right answer or action is indicated by the specific configuration of the apparatus. If that alone were guiding the chimp's choice of cup, it would always reach for the lower one regardless of circumstances. But in fact chimps seem to use *structural knowledge*: some understanding of the way the cue *functions*. A chimp is being guided by a knowledge of the weighing-down influence of the reward, and so it knows better when that is overridden by the experimenter pushing on the beam. Chimps can ignore irrelevant information here, still reaching for the lower cup in experiments where the cups are of two different colours and their positions interchanged. What's more, chimps might be able to extract the relevant knowledge using a range of different sensory modes – if the balance is obscured from sight, the animals may deduce which cup is lower by touch. The ability to exploit structural knowledge is a sophisticated function of mind, requiring some abstract representation of the world that is extracted from sensory data. We don't yet know how such information is represented in chimp minds.

There's a third stage of knowledge in Seed's classification too: *symbolic*, in which one works out what to do with a tool or apparatus based not on an appreciation of cause-and-effect physics but on some representations of the rules – an instruction sheet, if you will. What if the two cups in the experiment above were placed upside down on a flat plane and you showed the chimp a schematic diagram clearly (to us) indicating that the food is under the right-hand one? Or if you had hidden it somewhere in the room and given the chimp a simple map-like diagram showing its location? Can primates make the same conceptual leap?

In one such test, chimpanzees and bonobos were presented with an apparatus in which they could pull on strings attached to food in order to get at it (Figure 5.1). They figured it out quickly enough. But what if some strings were visibly broken – would the animals even bother pulling, or just figure that there was no point? They did the latter: they had apparently intuited the basic physics of string-pulling.

But now the researchers added an extra challenge. Instead of showing the breaks in the strings, they concealed them (except the dangling ends) beneath covers on which were simple depictions of whether the string beneath was intact or cut. In this case, the apes couldn't work out which string to pull. This is admittedly a tough test: if children around six and a half years old are presented with

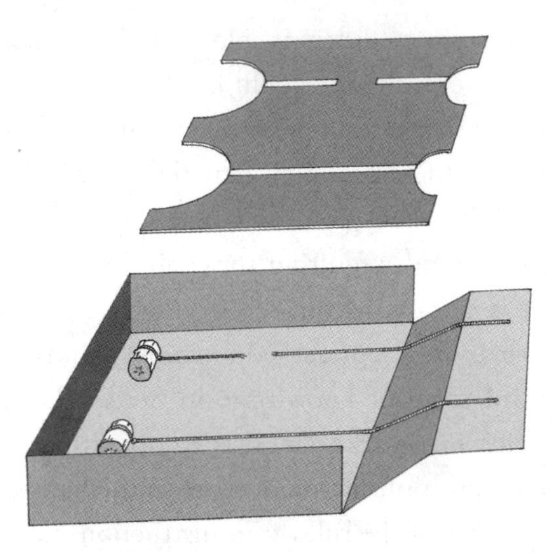

Figure 5.1. In this experiment, chimps can access a piece of food (such as a slice of banana) by pulling on a string that might or might not be severed. When the broken string is visible, the chimps make the right choice. But when the strings are hidden by a cover on which the break (or not) is indicated symbolically, they can't interpret that clue.

the same test, only about half of them make the right choice.* Symbolic comprehension of tools seems a step too far for these and other animals.

Thinking of others

Our capacity for abstract reasoning is just one of the ways we seem to differ from other apes. Some researchers suggest that the key difference is our superior 'executive functioning': an ability to control

* The children's performance – but not that of the apes – gets better if they have previously been allowed to play with the apparatus in which the breaks in the strings are visible. In that case, most five-and-a-half-year-olds will work it out right away.

or defer our innate instincts so that we can bring those powerful rational decision-making skills to bear.

As we saw earlier, others believe that complex *social* cognition is what truly sets us apart from other apes. We saw earlier that a key aspect of complex human social interaction stems from our development of a Theory of Mind (ToM): the ability to attribute thoughts and intentions to others, or to see through their eyes.

There has been a lot of debate about whether chimpanzees have a true ToM, and while this is still unresolved, it now seems unlikely that it is a yes or no question. Chimps will follow the gaze of others, for example – the equivalent of our looking up as we pass a crowd gazing up into the sky – as if deducing that those others have decided there's something worth looking at. This isn't a mere copying reflex: chimps will move if necessary to see around barriers for what others are looking at, and will look back inquiringly at the gazer if they can see nothing of interest themselves ('Is she really looking up there?'). And if they are set a competitive task of getting food before rivals do, they act in a way that shows they can be aware if there is food that their rivals can't see. In one experiment, a chimp was shown a set of boxes that mostly contained food – but one that contained a snake. This individual with privileged knowledge would typically lead other chimps to the snake box, apparently predicting that they would all flee and leave him with all the food. He not only knew what the box contained, but also knew that the others didn't know.

A full ToM, though, means not just predicting the behaviour of others but ascribing intentions to them: intuiting reasons why they do what they do. In one test, chimpanzees watched a human operate a light switch and were then given the chance to do it themselves – a task that an inquisitive chimp would find intrinsically interesting. Normally chimps would use their hands for this. But the human never demonstrated it that way; the switch was placed low on a wall, and the person operated it with his foot. In one case he did so while carrying a heavy bucket in both hands, suggesting that this was his only option and not a choice. In the other case the person carried nothing but still used his foot. In the first trial, chimps tended to use their hands; in the second they would sometimes use their foot. It was as if they realized that the foot operation in the first case wasn't essential to the outcome but just dictated by circumstances – that the person *would have used his hand if he could*. The chimps seemed able to deduce the reasons behind the demonstrator's behaviour rather than just blindly mimicking it.

Many different studies, using different methods, all point to the same conclusion, according to evolutionary anthropologists Josep Call and Michale Tomasello:

> Chimpanzees, like humans, understand that others see, hear and know things . . . They clearly do not just perceive the surface behavior of others and learn mindless behavioral rules as a result. All of the evidence . . . suggests that chimpanzees understand both the goals and intentions of others as well as the perception and knowledge of others.

It's one thing to appreciate what knowledge or options others don't have. It takes a more sophisticated Theory of Mind, however, to imagine that another individual holds a *false* belief. It means conceiving of a state of affairs that you know to be untrue, while at the same time retaining knowledge of the situation you know *is* true.* Even young children can do this – indeed, it can be a source

* We sometimes struggle to do this ourselves, being more ready to attribute a false belief to a cognitive deficit – 'they must be really stupid!' – than to allow that there's a coherent way to see the world in which that belief could be true.

of great amusement to them to see someone else deceived in this way. Until recently it was believed that only humans can include false beliefs of others in their ToM.

There is, however, now some evidence that apes do too. Evolutionary anthropologist Christopher Krupenye and his colleagues showed chimpanzees, bonobos, and orangutans a film of a man dressed in a gorilla suit (it helps if you can identify with the protagonist, right?) who steals a rock from another man and puts it in one of two boxes before chasing the second man away. When he's gone, the 'gorilla' has a change of mind, first moving the rock to the other box but then deciding to take it with him as he exits.

The man then returns to the scene. What will he do? We'd expect him to look in the first box, falsely believing that's where the rock is hidden. And by tracking the eye movements of the apes, the researchers could see that this is what they – more precisely, more than two-thirds of those tested – thought too: they cast 'anticipatory looks' at the box.

This suggests chimps might have awareness not just of what others know and don't know, but also of what others might *think* they know but are mistaken. One evolutionary psychologist described this gorilla-suit study as the 'final nail in the coffin of the long-standing idea that humans are the only species with "theory of mind."' Later studies, also monitoring apes' anticipatory looks, have also claimed evidence that they attribute false beliefs.

But it's controversial. Other studies have seemed to rule out the idea, and some researchers say that 'anticipatory looks' aren't a reliable way of identifying false belief attribution even in humans. Some argue that there's a difference between explicit mental attribution, where the observer can (if they're able) say, 'She thinks it's in

---

Admittedly, the credulity that some people exhibit towards conspiracy theories tests that ability to extremes.

there, but it isn't,' or can perform some action that demonstrates this explicit conception, and implicit mental attribution, where the observer holds that idea unconsciously and without deliberation. Perhaps apes have the latter capacity (as children as young as two years may do) but not the former (as four-year-olds generally do). If so, are these just different degrees of the same mental process, or different processes? We just don't yet know quite what kind of mind an ape's ToM implies.

That other animals might have a ToM at all need not surprise us, because it promotes cognitive efficiency: it's simpler to assume the mindedness of other agents than to explain a particular behaviour otherwise. If a monkey wishes to conceal an item from its rivals, the best action might depend on what the rivals can experience. If they can see you, there's little point merely hiding the food. If they can't see you but can hear you, it might be wise to hide it somewhere that will not make a noise: no rustling leaves that could give away the location. And so on. Each of these possible choices of action could be hard-wired, or the monkey could learn them the hard way, from experience – but probably at the cost of many hungry days. But as the rival's activity becomes more complex, the list of its potential choices becomes longer and longer. At some point, a more efficient decision-making algorithm involves asking, 'What strategy would work best if that pesky thief has a mind like mine?' We don't know what cognitive architecture it takes to introduce this rather sophisticated idea into one's internal model of the world – but we do know that human infants don't require any profound neural rewiring to go from a solipsistic world view to a Theory of Mind.

The socially directed nature of the human mind does seem to distinguish it from those of our cousins. Compared in terms of an ability to figure out how to deal with spatial ('where is it?') or quantitative ('how much is there?') problems, or questions of causation ('what will happen if . . . ?'), two-year-old human children don't perform very differently to apes. But when social reasoning is called for, they seem to be way ahead. In one experiment, an adult engaged in a collaborative task with an infant and a young chimp and then stopped before the task was completed. Chimps would keep trying to do the task on their own, whereas infants would approach the adult and try to get them to re-engage, as if to say, 'Look, we were in this together – don't abandon me.' Such observations seem to imply that we, but not apes, have minds built to work not just collaboratively but 'as one'. For all the natural egoism and volatility of the toddler, their underlying impulses tend to be prosocial and trusting.

'Humans seem to act as though success at a joint task is not only about me getting my own reward, but both of us being successful together', says animal cognition expert Alexandra Rosati. While children have an innate tendency to assume we have shared intentions and goals, other great apes seem to use a Theory of Mind, mostly to manipulate and deceive in competitive situations (what some have dubbed a Machiavellian mind). Conflict is common: wild chimpanzees of both sexes are capable of extreme and sometimes murderous aggression, and in some circumstances show rather little social tolerance of others. Some of the males' mating behaviour is coercive rather than consensual – what humans would regard as rape. (Of course, we are also capable of horrendous acts of cruelty and destruction that we don't witness in apes. That's a paradox; perhaps our cultures can amplify malfunctions or regressions of our instinctive nature out of all proportion.)

Children engage socially not just as a means to an end: not to complete a task, but seemingly out of an impulse just to communicate. Even at a pre-verbal stage they will point out to others objects or events that strike them as interesting, as if imbued with an innate wish to inform. And through communication we can agree on common intentions, values, even fictions: money is only a desirable commodity because we have all agreed to honour its symbolic role. 'Humans may be unique in our ability to represent joint activities as underpinned by shared goals or motivations that both individuals know are shared', says Rosati.

This sophisticated social cognition makes possible our complex cultural structures, and we invest a great deal in recording and passing on information about these structures and the norms that undergird them. Do great apes actively *teach* their young as we do ours, or do the young simply learning by copying? That's not clear. But beyond the constraints imposed by basic evolutionary pressures on social coexistence – such as that one should not wantonly kill or be aggressive – there certainly seems no reason to believe that other apes acquire or are instructed in abstract social norms, such as a sense of what is the right and wrong way to behave.

The kindness of apes

While a Theory of Mind can be a valuable tool for deceiving rivals, it would also seem a precondition for an empathic response to your fellow creatures. And indeed it is unfair to suggest that apes put their resources for conceiving of the minds of others to use in a purely selfish, Machiavellian manner. 'When a female chimpanzee resolves a fight between two juveniles over a leafy branch by breaking it into two and handing each youngster a piece, or when an adult male chimpanzee helps an injured, limping mother by picking up her offspring to carry it for her, we are dealing with impressive social skills that don't fit the "Machiavellian" label', warns primatologist Frans de Waal. He points out that chimpanzees put effort into maintaining social bonds, displaying empathy, and keeping the peace in ways that can't easily be cast as self-interested.

The sensitivity to the feelings and the wellbeing of others that primates may exhibit goes beyond the genetically hard-wired capacity of altruism that emerges from evolutionary kin selection (see page 68). For example, when one of a group of chimpanzees has been defeated and perhaps injured in conflict – in an unsuccessful takeover bid by a male, say – others might offer comforting gestures such as touching. We don't know the real motives for this behaviour; all we can say is that it often looks just like what we would do to offer sympathy to someone who has had a hard time. You might wonder if there is some reciprocation involved here too: perhaps chimps are hard-wired to behave this way because they too would want to benefit from such treatment in the same circumstances. That's not an easy line to sustain, however – for one thing, many males won't ever launch a leadership challenge and risk being in the loser's role themselves. It's hard to see how the notion that 'you never know, it's not impossible that one day I might try my luck' could offer a strong selective pressure to generate such innate behaviour. In any case, what really is the survival benefit of being patted on the head if you've just been worked over by the alpha male? No, some manner of genuine empathy does seem to be the most parsimonious interpretation of such actions.

Chimps will also display what looks to us like selfless generosity towards others. A team of behavioural and cognitive scientists provided three distinct groups of chimps in a wild Zambian sanctuary with a juice fountain (chimps love fruit juice) that could be operated by a button – with the catch that the fountain was several yards from the button (but clearly visible), so operating the fountain could then only benefit *other* chimps that happened to be beside it. After trial sessions to let the chimps familiarize themselves with the setup, all the groups showed ever more inclination to operate the fountain for one another as the experiment progressed. This was unlikely to be a result of mere curiosity – which chimps certainly possess – because they were far less inclined to work a fountain stationed outside their enclosure, from which no one could benefit:

they seemed genuinely motivated by an investment in the welfare of others in the group. What's more, the three chimp groups showed different levels of 'generosity', which seemed to match the amount of social cohesion and group spirit seen in experiments where the chimps had to gather and share out food resources. In other words, some groups were more 'socially minded' than others. Other studies of chimp prosociality – their readiness to look out for others – have given conflicting results, but some of those findings might have been distorted by being gathered under artificial lab conditions rather than observing chimps in the wild.

While we must resist leaping to anthropocentric conclusions in cases like this – to assume that, because a behaviour *looks* like something we might do, the animal is doing it for the same reasons – it's equally important not to become so determined to avoid this mistake that we concoct convoluted and implausible alternatives. Frans de Waal has recounted the story of a gorilla in Chicago's Brookfield Zoo named Binti Jua that saved a young boy who fell into the primate enclosure. Binti picked up the child, sat down on a log and cradled him, and then carried him to one of the doorways to the enclosure and laid him there. While the press went wild over this act of care and kindness, some scientists insisted that such feelings might have nothing to do with the gorilla's behaviour. Perhaps she was 'maternally confused', they said – despite having her own infant on her back all the time, and the fact that the blond boy in the red T-shirt looked nothing like a baby gorilla. De Waal quotes one primate expert as saying, 'The incident can be sensational only for people who don't know a thing about gorillas' – who don't, that is, know already of their capacity to show nurture and concern for others. In his work on the evolution of morality and empathy, de Waal added, he had 'encountered numerous instances of animals caring for one another.'

'The incident at the Brookfield Zoo', he wrote,

> shows how hard it is to avoid anthropodenial and anthropomorphism at the same time: in trying to avoid thinking of Binti as a human being, we run straight into the realization that Binti's actions make little sense if we refuse to assume intentions and feelings.

The real question is how to frame the kind of empathic thoughts the gorilla was presumably having. We might rescue a stranded or vulnerable child who is not related by kinship because we think (we might not say it out loud but the thought is plain): 'What if they were mine?' But you don't have to have children to act protectively; it's a natural human instinct to come to the aid of others, especially when they are young or otherwise vulnerable. That impulse doesn't even need articulating; indeed, many people have acted, without time for any conscious deliberation, to grab someone else's child if they are in danger – about to run into the road, say. The reason we do so is likely to be complex, but our nature as prosocial creatures will surely be a part of it. 'Binti's kindness', said de Waal, 'is most parsimoniously explained in the same way we explain our own behavior – as the result of a complex, and familiar, inner life.' Attributing not just human-like behaviour but human-like mental reasoning to gorillas and other primates is by no means an outrageous leap of faith.

A readiness to comfort an individual that has been injured or bested seems to imply that the comforter can imagine what the comforted creature must be feeling. If we attribute no inner life to the other, we'd have no reason to suspect they would like a hug. So this sort of behaviour is at least suggestive that primates possess a Theory of Mind that goes beyond anticipating others' behaviour for one's own benefit: a notion of others that allows them emotional states and elicits feelings of empathy.

Forward thinking

Another respect in which humans have been deemed to be special is that we alone among animals are considered able to escape a perpetual present – to mentally inhabit the past or the future. (Some mindfulness gurus suggest, with good reason, that in fact we spend rather too much time in those places and not enough in the present.) Of course, other animals *learn* from experience, and in that sense they embody a remembrance of it – but can they truly imagine themselves being back there in the past? Do they have so-called episodic memory, the recall of specific past events? And do they truly plan ahead by conceptualizing possible futures?

Josep Call and his colleague Nicholas Mulcahy have explored that question for bonobos and orangutans. The researchers trained these apes to use tools to get a reward, and then offered them a choice of tools – two of which were suited to the task, while six weren't – for future use. The apes were taken to a waiting room for periods of up to fourteen hours before being admitted back to the test room to try to get their reward. Despite some variation in individual performance – one might infer that some apes, like some humans, are quicker on the uptake than others – on average both types of animal passed the test successfully, taking with them a useful tool significantly more often than one that would not help them later.

Planning ahead might call for compromises in the present. When offered the choice of a fruit now or a tool (a straw) that will enable them to get juice later, chimpanzees and orangutans picked the tool: the kind of delayed gratification exhibited by some (but not all) children in the famous 'marshmallow test'.* If offered a *second* straw

* In this test, first described by psychologists at Stanford University in the 1960s, children are given a marshmallow and told they can have a second a little later, but

or a fruit, the apes would generally take the fruit, as if thinking, 'No, I'm good for tools now.'

Some researchers have ramped up the temptation to test the limits of an ape's self-control – in a sense, to measure the salience of future scenarios in their mind in comparison to the here-and-now. Psychologists Theodore Evans and Michael Beran presented chimpanzees with a pile of sweets that was steadily growing bigger. The chimps could claim the pile at any time, but knew that they would get a bigger reward the longer they waited. Some of them could withstand the temptation for up to ten minutes. It seemed clear that this involved considerable self-control and repression of an instinctive urge to take what was there now. When the researchers also provided toys for the apes to play with, the animals could wait longer by distracting themselves with play. This wasn't just a delay caused by a new distraction; in a situation where the chimps could simply grab the sweets directly, rather than indicating a wish to claim a pile that was physically inaccessible, they played all the more attentively, as if all the more desperate to restrain themselves when the temptation was more immediate.* Studies like these suggest that an ability to plan for the future – to visualize oneself in a

---

only if they can hold back from eating the first while the experimenter is out of the room. It is a trial of the child's willpower and self-restraint. Claims that it predicts life success in adulthood, however, are now largely discounted. One study suggests that a child's behaviour depends more on their socioeconomic status than on any intrinsic cognitive endowment, perhaps in part because if they are accustomed to future economic uncertainty (will there be enough food tomorrow?), they will very reasonably seize the moment and eat what they already have in hand.

* Humans find this kind of delayed-reward test harder with food than with money. That's very relatable, as they say – the sight and especially the smell of a tasty morsel has a tighter grip on our instinctive desires. Perhaps this is a reflection of the fact that we've sought food for ever, but have started chasing after money only rather recently in our evolutionary history.

hypothetical scenario that might come to pass – could have been present even in the common ancestor we share with other apes fourteen million years ago.

Opening up

A common perception about how cognition evolved is that as minds acquired new capabilities, new evolutionary niches opened up as a result. But it might have been the other way around: the appearance and exploitation of a new niche *drove* cognitive changes. Why otherwise would new mental resources be selected, if the environment that drives selection remains the same?

For example, hunter-gatherer foraging is quite different from the way apes gather their food by eating primarily fruit and leaves on the go. Humans rely on food sources that are of higher quality but more patchily distributed (such as meat), and so they might travel further to get them – perhaps promoting new cognitive resources for exploration and navigation, as well as new constraints on risk-aversion. It's a higher-risk strategy, and surviving it might have been dependent on food-sharing in a group: that is, on increased pro-social behaviour.

Many animals hunt in packs. But few have the self-control and the social organization to forgo the temptation to eat the kill right away, instead carrying the spoils back to a camp to be distributed fairly. Chimpanzees might hunt as a troupe, but the one who gets the prize will tend to hog it, sharing scraps only if harassed to do so. Bonobos, our other closest relatives, are less aggressive and more generous, willingly sharing food with others – perhaps because they live in habitats that are likely to engender less competition. In lemurs (a more distant relative than apes), species that live in larger groups on average seem to have more well-developed cognition of the kind that might support a Theory of Mind, suggesting that indeed the need for improved social cognition might have been a key selective pressure along the evolutionary road to modern humans.

It makes little sense to think about evolutionary changes to the brain and mind in isolation from those of the rest of the body. For example, decreases in aggression among hominids may have been linked to an evolutionary decline in levels of testosterone in males. This hormone also has an effect on the shape of the primate skull: the ridge of the brow, for example, becomes less prominent. So we can track it in the fossil record, which shows that testosterone levels decreased in human ancestors around 200,000 years ago, during the last Ice Age (the Middle Pleistocene). It may be that the need for more cooperative and tolerant behaviour in these pre-human societies meant that males with less testosterone fared better, driving a change in minds in parallel with, and for the same biochemical reason as, a change in bodies.*

At some point – which might not (and probably did not) correspond to any significant change in brain capacity or anatomy – this expansion of human cognition became a runaway process. It's as if a gradual accumulation of abilities reached a tipping point at which their interplay produced a new kind of mind with an unprecedented degree of cognitive integration. Some researchers suspect that this change transformed the nature of consciousness too, producing the 'autobiographical self' (page 127): a personal narrative of our place

* It has been controversially proposed by Simon Baron-Cohen that high prenatal levels of androgen hormones such as testosterone in males create significant differences between human male and female brains, making the former less socialized and empathic and in extreme cases leading to autism spectrum disorders (see page 106). Higher levels of aggression among males do seem to be one of the few behavioural differences between the sexes that are well attested in humans. What's more, bonobos have lower prenatal androgen exposure than chimps, and are more socially tolerant and cooperative as well as, in some tests, seeming to have a better-developed Theory of Mind.

and potential in the world. (It remains open to debate whether this sense of having authorship of our destiny was an unalloyed good.)

However it emerged, that trait seems likely to be a precondition for our complex culture. But we shouldn't suppose that all the linguistic, artistic, and social facility of humans is hard-wired into our brains (and ours alone). For culture shapes minds as much as minds shape culture. 'If we imagine', say Michael Tomasello and fellow primatologist Hannes Rakoczy, 'a human child born onto a desert island, somehow magically kept alive by itself until adulthood, it is possible that this adult's cognitive skills would not differ very much – perhaps a little, but not very much – from those of other apes.' Evidently apes cannot be given human-like cognition by being raised by, and as if they were, humans (it has been tried) – but we should be wary of imagining that all animals do is all they *can* do, given the right circumstances.

At any rate, we shouldn't think of the evolution of minds as the steady accumulation of independent, modular skills. While most tests of primate intelligence aim to see if they have this or that attribute, there's some evidence that the *generalization* of cognitive abilities that seems to distinguish humans is also evident to different degrees in other primates. This capacity – you might call it intelligence – is by no means homogeneous even among closely related species. Using evidence of behavioural flexibility – for example, the ability to find new solutions to social problems, to learn new skills, to devise tools, or to deceive others for one's own gain – as a measure of intelligence, evolutionary biologist Kevin Laland and his colleagues have argued that certain species among the families of capuchin monkeys, baboons, macaques, and great apes seem notably smarter than others. This suggests that superior intelligence may have appeared independently on different branches of the primate evolutionary tree. Our own cognitive virtuosity, such as it is, might be a particularly marked instance of a more common tendency for minds to take a great leap forward – and not, perhaps, because it is 'needed' in an adaptive sense, but because of some spontaneous and felicitous synergy of mental abilities. Maybe, you might say, the evolution of mind surprised itself with what it had produced.

On the wing

Among the big-brained, highly intelligent animals, apes and monkeys are generally rated alongside cetaceans (especially whales and dolphins) and elephants as the brainboxes of the natural world. But there's another class of candidates for this exclusive club, which has an evolutionary trajectory that diverged way before primates split from other mammals. Birds are rather distant from us in evolutionary terms: famously descended from feathered dinosaurs, they parted company with a common ancestor of mammals around 300 million years ago. Many birds have large brains relative to their body size: crows have a brain mass comparable to that of small monkeys such as the marmoset or tamarin.

We display mixed feelings about the bird mind. We defame it with talk of 'bird brains' and 'bird-witted' individuals, and we take a rather dim view of the intellect of chickens and pigeons. But we marvel too at the skill and ingenuity involved in building the nest, and the ability of migrating birds to navigate flawlessly over thousands of miles. Owls are avian avatars of wisdom (or, in the Hundred Acre Wood, merely the conviction of it), and I doubt anyone looks a golden eagle in the eye and believes they can outwit it. If we were given the choice of occupying another creature's mind, I'd wager many people would elect to be a bird: to experience that alien freedom from gravity's dominion, that acuity of vision, that heavenly grace. It was often into the mind of a bird that shamanic rituals sought to transport the spirit; that is Matthew Modine's obsessional desire in Alan Parker's 1984 film *Birdy*, based on William Wharton's novel.

The avian world is one of the richest natural repositories of other minds – for the cognitive skills of birds are astonishingly diverse, as you might expect for creatures adapted to such a wide range of different habitats. Among them we find arguably the best claims for non-mammal consciousness, as well as examples of artistry that raise questions about animal aesthetics, some of the most complex vocalizations in the natural world, prodigious feats of memory, and inventive tool use. Writer Jennifer Ackerman chose wisely in speaking, in the title of her delightful 2016 book, of the 'genius of birds' – not as an attempt to persuade us that they're Really Clever, but to point to the remarkable scope and depth of their cognitive specializations. It's for this reason that I think we would find bird minds scattered widely through the Space of Possible Minds. The minds of the ten thousand or so bird species are surely less like one another than are the minds of people with different skill sets: their attributes sometimes seem to be fundamentally distinct, to the extent that some birds will be mind-blind to things that dominate the Umwelt of others. That even some rather closely related bird species can show very different cognitive abilities illustrates again that minds *don't come as a package* that is simply passed (with minor modifications) down the evolutionary chain. Rather, they are strongly and intimately shaped by the specific evolutionary niches and strategies adopted by each creature.

To make a ranking of bird minds would be to reflect human prejudices; there's surely no single scale for their intelligence. We'd probably place the corvids on top, perhaps with the New Caledonian crows of the southwest Pacific at the pinnacle – for these seem to us like the greatest avian intellects, able to improvise tools from the materials at hand and to execute a plan with forethought. Yet nothing they construct holds a torch to the glorious creations of the male bowerbird, who arranges brightly coloured objects into elaborate 'found art' to elicit the attention of females (who in turn seem to judge with highly discriminating criteria). And how many creatures can compare with the vocal virtuosity of the nightingale or the skylark, which human composers have struggled to emulate? How to rate the capacity of parrots to learn hundreds of words – not just to ape them back to us, but to identify the objects, colours, and shapes to which they refer? And there are few natural phenomena that seem to speak to us of an unearthly mode of intelligence than the murmuration of starlings, which led some zoologists in the past to suppose the birds must be endowed with a kind of telepathy. (They aren't.)

How do we even measure the cognitive abilities of birds? It's not easy. If you take an animal out of its natural environment and put it in a lab setting, in some ways that makes for a better experiment, because you can control more of the parameters and ensure you're testing different animals in the same circumstances. But you might also be constraining or altering its capabilities compared to the wild. If an animal hesitates when confronted by some piece of apparatus it has to manipulate, is that because it can't work out what to do with it, or because the unfamiliarity makes the creature reluctant to engage? Are those that perform better smarter, or just bolder?* 'Unfortunately', says bird cognition expert Neeltje Boogert, 'it is extremely difficult to get a "pure" measure of cognitive performance that is not affected by myriad other factors.'

We do what we can. One option is to accept the uncertainties and contingencies of the natural environment, and try to amass careful observations on what creatures get up to there. Bird behaviourist Louis Lefebvre has argued that a good measure of the general intelligence of birds is their ability to innovate: to find novel solutions to problems. Animals that rely only on a limited set of 'stereotyped' behaviours – if *this*, do *that* – are likely to have a sparser set of cognitive resources to draw on in making their

* Needless to say, the same problems can beset lab tests of human cognition too.

decisions than those that seem able to think in more abstract terms about causes and consequences. The use of tools is an obvious example. Some birds get at bugs by digging them out with their beaks; but a bird that realizes it can extend its reach with a twig in its beak clearly accesses another level of representation of the world.

In the mid-1990s Lefebvre and his colleagues scanned the ornithological literature for examples of 'foraging innovations' in birds, confining themselves initially to birds of the British Isles. (Here there's the advantage of a vast data set collected by enthusiastic and often very knowledgeable amateurs, though such reports need to be scrutinized using discerning criteria.) The result was a rough-and-ready intelligence scale for birds, which placed corvids and parrots at the top.

An intelligence scale of this kind suffers from the same ambiguities and caveats as any proposed for humans: we must never suppose 'intelligence' has but a single dimension for any creature. Still, the greater the 'intelligence', the more likely the birds were to have larger brains.* Birds with relatively big brains appear to act differently to others: they are not frightened by novelty, but on the contrary may explore it. Sparrows and magpies are like this, and it seems no coincidence that they are successful colonizers of new territories and habitats – they do well, for example, in urban settings. Cities might act as a kind of selective filter for 'clever' birds that seek out the unfamiliar to see what benefits it might offer them. There are, of course, hazards in that strategy, as many an urban cat or fox will have exploited – and indeed, innovation in birds doesn't seem to correlate with better survival prospects, as measured by the extinction rate of different bird species. It just opens up new niches in which to live.

* Much the same correlation between brain size and capacity for innovation is found among primates.

Once made, a behavioural innovation can be acquired by others simply through observation and copying – the wheel need be invented only once. That kind of learning seems to have been involved in the notorious epidemic of milk-bottle opportunism by tits in the UK in the early twentieth century: they would peck through the foil lids of bottles left on doorsteps to get at the nutritious cream at the top. Such an ability to learn is surely another attribute of the intelligent mind.

Tooled up

Charles Darwin noticed that the humble earthworm – the prey of many birds – seems to engage in a rudimentary kind of tool use by collecting leaves to plug its burrows in an attempt to protect itself from predators. In doing so, Darwin said, they 'acted in nearly the same manner as would a man': selecting the leaf from its shape and dragging it from the tip rather than the stalk. He and his son Francis conducted experiments with different leaves – some not indigenous to Britain, some glued together – as well as triangular pieces of paper, and found that the worm does not simply carry out a standard leaf-dragging routine regardless of the situation, but adapts its behaviour to the object.

Darwin was impressed to find such behavioural flexibility and, as he saw it, intelligence in so seemingly tiny and primitive an organism. But perhaps, he said, we should not really be surprised at the apparent ingenuity of the worm mind, for 'we should remember what a mass of inherited knowledge, with some power of adapting means to an end, is crowded into the minute brain of a worker-ant.'

This is true – but for the ant, and presumably the worm too, much of their apparent intelligence can come from rather simple, instinctive, and hard-wired behavioural rules that play out slightly differently in different contexts. There's no need to regard the earthworm as a connoisseur of leaf-manipulation. To truly adapt means to ends in the versatile manner required to improvise tools, hardwiring is not enough. You need more of a mind than that.

The club of genuine animal tool-users is exclusive. Chimpanzees, orangutans, gorillas, and baboons do it; elephants will use branches to swat away flies, even modifying them for easier use or better effect; octopuses and other cephalopods will grasp makeshift tools in their writhing arms. Beyond this, the only major group of tool-users is birds. Woodpecker finches and some other insect-catchers put twigs to good use – but the champions of avian tool use are surely the corvids, especially the New Caledonian crow. They don't just poke but hook; they show signs of *designing* tools for the job at hand; they keep them for reuse; and they even seem to develop local 'traditions' of design and to refine them over time – a kind of cumulative technological change.

It's possible that we place undue emphasis on tool use as a sign of intelligence because it features so strongly in the narrative of human evolution. All the same, this ability to achieve goals by manipulating objects in the environment surely demands some important features of mind. How does an animal know what will work as a tool? Does the woodpecker 'know' in some sense that to fit a stick into a crack in the bark and poke out the bugs, the stick has to be narrower than the crack? Does it know that a leaf is no good because it is too pliable, and a pebble is no good because it's totally the wrong shape? In short, what, if anything, is the *narrative* of tool-using animals? They're presumably not articulating it along the lines of 'if I move the twig like this, I can get that tasty grub out' – but then how exactly *are* they framing the representation?

It's possible that some animal tool use is guided simply by trial and error. Tool-using birds such as New Caledonian crows and cockatoos display a propensity for what we interpret as play: they mess around with objects, perhaps to explore what can be done with them. Evolutionary biologist Russell Gray thinks that these birds may rely less on building imagined scenarios ('Hey, if I used this stick, I bet I could get at more bugs') and more on past experience of manipulating other objects ('When I did *this*, *that* happened').

But it's not all try-and-see; there do seem to be some general principles behind tool use. Employing a found object in order to increase one's agency means perceiving the *affordance* it offers (see page 87) and the intuitive physics by which it might operate. These concepts are often probed by studying how animals try to retrieve food from within tubes by pushing or pulling on piston-like tools. Such actions suggest that the animal has some notion that solid objects can displace other solid objects by force, for example.

What if there's a visible hole in the tube: a gap that the prize must cross in any attempt to push it out of the other side (Figure 5.2)? Will the animal realize it is futile to use a tool to push, as the reward will just vanish down the hole? What if it has a choice

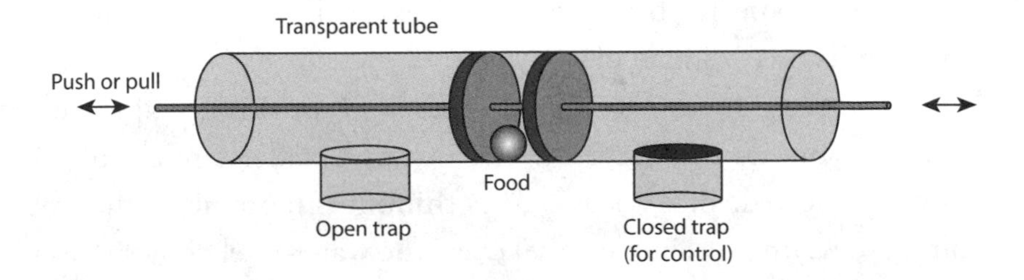

Figure 5.2. Studies of the intuitive physics that informs tool use in birds typically entail extracting food from a tube. What will the animal do, for example, if there is a gap down which a food item pushed or pulled by a piston will fall?

of tools for pushing or pulling? Will it select the right one for the job? Chimpanzees are not terribly good at this: they can figure out that a rake can be used to drag towards them a nut that is otherwise out of reach, but will show no preference for doing this across a flat surface or for one where a trench intervenes, down which the nut will be inevitably lost.

One attribute that can greatly expand the capacity of minds to innovate in tool use is an understanding of cause and effect: to appreciate *why* an object does what it does. Take the case of the 'Aesop trick'. In this ancient story, popularized in the fables attributed to the ancient Greek writer,* a thirsty crow comes across a pitcher of water, but the water level is too low for the bird to reach by perching on the rim. The crow figures out that it can bring the water within reach by dropping pebbles into the pitcher to raise the level. That this story features a crow can surely be no coincidence – perhaps the original author, whoever they were, once witnessed a corvid do this very thing, for it has been observed in rooks and Eurasian jays.

Only children older than about five to seven years are able to work out this problem for themselves, and so it seems a mighty feat of reasoning indeed. In fact, at face value it relies on exactly the revelation that, according to apocryphal legend, inspired Archimedes to leap from his bath crying, 'Eureka!': that a solid body will displace its own volume of water.

It seems highly unlikely that corvids conceptualize the problem that way. But do they have any real intuitive understanding of the physics involved? Bird behaviourists think it is more likely that the birds are relying on simple visual cues: the water level comes visibly

* Aesop, allegedly a Greek slave of the sixth century BCE, is just the author attributed by tradition. The stories were collected by others, especially the Roman writer appropriately named Avianus in the fourth to fifth centuries CE.

closer every time they carry out the stereotyped action of picking up a pebble and dropping it in. Sure, this process has to start with the innovation of taking that action for the first time. But this doesn't by itself imply that the bird makes a *causal* connection between what it does and what it sees as a result; it's enough that the two are correlated ('If I do *this*, *that* follows').

Here's a subtler scenario for distinguishing causation and correlation. A child observes that a box will open to release food when a block is on top of it. You might imagine that, faced with the same box and a block on the floor next to it, the child will pick up the block and place it on top to get the snack. But this requires more than an understanding that the box will open when there's a block on top. It means inferring a causal relationship: the presence of the block *causes* the box to open, rather than merely signifying that the box *will* open. We plan and decide our own actions by this kind of causal inference all the time, but it needs to be learned. Two-year-old infants will sometimes display it, and by about age four this sort of behaviour is very general.

It's a subtle but vital distinction. It is the difference between an action that is understood to cause a result in a mechanical way, and one that is merely correlated with the outcome. The latter is ritualistic, a form of magical thinking: the action is deemed to 'summon' the consequence without any grasp of how cause and effect operate (that is, lacking any intuitive physics). In other words, the distinction is between different modes of thinking, produced by different kinds of mind.

An ability to infer causal influences greatly expands the repertoire of effective actions. Testing whether crows can do this has involved considerable ingenuity. In one experiment, conducted in 2014 by Alex Taylor, Nicola Clayton, and their colleagues, New Caledonian crows were presented with a complex apparatus that could, if properly manipulated, dispense a tasty piece of meat. The meat was

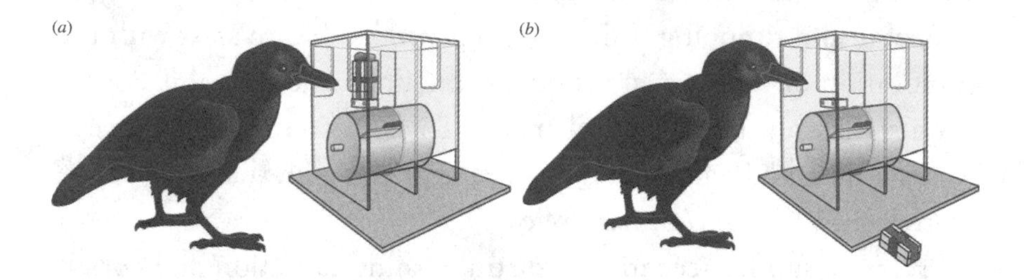

Figure 5.3. Testing the causal deductions of crows. If a block is pushed from a ledge that can be reached through the hole, it falls onto a platform that makes the drum revolve and dispense meat that the bird can reach and eat. Will the bird then deduce that it can pick up the block from the floor (on the right) and insert it into a hole on the upper left side of the apparatus to release the food? (Inserting the block into five other holes in the apparatus doesn't do the trick.)

placed on a small shelf attached to a rotating drum, all shielded inside a box of transparent plastic so that the bird could not get at the meat directly. Instead, it had to make the drum rotate so that the meat would fall onto the base of the device, where it could be reached via an opening in the bottom (Figure 5.3).

The drum would turn if a small block was dropped onto it. At first, this block was placed on a ledge above the drum (inside the plastic box) and 'baited' by attaching it via a wire to another piece of meat. If the crow picked up the meat, the block would be dislodged to fall and release the other piece. After being given the chance to learn this in several trials, the crows were then presented with the unbaited block resting on the ledge, which they could nudge off onto the drum with their beaks, poked through a gap in the box.

They proved able to make this connection between what they'd seen before – the block falling as they pulled on the bait – and the possibility of making the block fall directly in order to get the reward. It was as if they had learned the rule 'block falls off ledge → tasty snack'.

Now the researchers created a new challenge: the block was placed instead on the ground next to the apparatus. To get the meat reward, the birds would have to pick up the block in their beaks (a feat they were quite capable of in principle) and insert it through the right hole.

You might imagine that having learnt about the causal role of the block, this step would be obvious. And indeed, it was to most twenty-four-month-old children invited to play the same game. (For them the reward was instead a marble that they could place in a marble run, an enjoyable game that gave them a motive for getting at the little ball.) But the crows never picked up the block.

The outcome was different if the block placed on the ground was itself baited with meat in such a way that the crows were guided to pick up the block and put it on the ledge. In other words, they could do the requisite actions just fine. But they needed rewards all the way to show them the right strategy. They couldn't generalize, in the way the children could, to find a spontaneous new strategy for using the block to release the final reward from its trap.

What does this mean? Taylor and colleagues suggested the implication is that understanding causal relationships is not an ability you either have or don't have. It's more complicated than that. 'Causal understanding', they said, 'is not based on a single, monolithic, domain-general cognitive mechanism.' The fact that the crows could make the leap from 'I'll just tug on this bit of meat – oops, there goes the block, and here's a reward!' to 'Maybe I could just *push* the block to fall instead, and still get the reward' implies that they had *some* sense of the mechanics. (A truly superstitious crow might simply have done a spot of 'air-tugging' in the hope that making the same gesture would have the same effect.)

But the next step – 'There's no block to push – but hey, there's one down here, now if I just put it on the ledge . . .' revealed the limits of that reasoning. And there were apparently quite sharp, well-defined limits. It seems possible that the crucial distinction here was between actions that involve direct interaction with the apparatus and ones that require acting first on an external object. Yet who knows: perhaps if the trial group had included a real Crow Einstein, the breakthrough would have been made?

Making plans

One of the important differences between these two tests presented to the crows is that the second involved several steps: first, pick up the block, then put it into the hole . . . In such a case, the animal would have to have a kind of plan – taking one action with the intention of following it with another. Now, you might say that birds and other animals do such things all the time. When a bird feeds worms to its chicks, it has to go and catch the worm *with the plan* of then feeding it to the chicks. It does one thing while seemingly having a notion of the next thing in its head.

Of course, we don't know that this is how the bird sees it. We don't know that it holds a kind of avian equivalent of an internal dialogue along the lines of 'Must get a worm for those little ones . . . Ah, there's one, now let's get it back to them . . .' We can, after all, make machines that carry out multistep tasks without having any explicit image of what comes next. They could, for example, have an algorithm that goes along these lines:

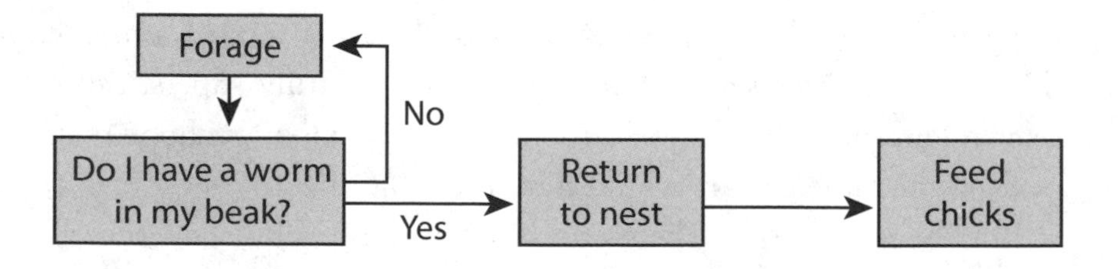

The ultimate goal does not need to be 'kept in mind' in order to carry out each step successfully. But this won't be of much use for innovative behaviour; you can't evaluate whether or not to take some action unless you have some sense of whether it is likely to aid or abet the ultimate goal.

New Caledonian crows might have failed the test on that occasion, but they do seem able to conduct new multistep tasks in others. Nicola Clayton and her colleagues set New Caledonian crows a variety of challenges that involved using sticks to pull food from tubes or release it from platforms, while ignoring distractions such as dummy sticks placed in empty tubes. The experiments were set up in a way that required the birds to decide on their actions *before* taking any one of them, rather than just improvising on the fly, as it were. In these cases the crows seemed able to formulate a plan of action – which would seem to require some kind of mental representation of the task and the ultimate goal. We still don't know quite what that entails in the crow mind, however – for example, whether they picture the whole task as a coherent objective, or just represent the key decision points.

Again in contrast to a common view that animals exist in a perpetual present, reacting only to what the senses present to the brain here and now, such multistep tasks seem to demand some concept of future consequences: if I do this *now*, it will let me do *that* later. This entails some representation of possible scenarios and their outcomes: the sort of internal rehearsal of potential behaviours that we noted earlier in the human brain. That in turn depends on being able to enlist past experience to construct these future scenarios: both semantic memory (facts about the world, such as that red fruits are good to eat and green ones are not) and episodic memory ('when I did this before, it worked').

Such future-imagining seems to be on display in bird food caching: the habit of some species to leave little stores of food stashed around and about for future use. This behaviour relies on a prodigious memory and skill at navigation: some birds may cache a hundred or so items in different locations, and reliably find them all. As ever, it's easy to anthropomorphize: to imagine the bird thinking, 'I've eaten enough for now, I'll save this for later when I'll be hungry again.' But might evolution not instead just hard-wire a tendency to put aside excess food when some threshold of satiation is reached, and create a mental position marker of its location that will be activated, machine-like, when hunger arises?

Well maybe – but some of the choices that caching birds make are hard to explain unless we assume they can create *some* sort of representation of a future scenario. For example, scrub jays seem to remember not just where they have put their food, but also what they put there – and what this implies for different kinds of food: they will return sooner to perishable items than to non-perishables such as nuts and seeds. Sure, it's possible to imagine some kind of automated filing system in the birds' memories whereby they recognize two types of caching sites: ones to come back to soon, and ones that can be left until later. But one has to wonder at what point it would become cognitively cheaper to produce a scenario-representing mind than one that mindlessly follows some complex algorithm encoding all possible future permutations of the interactions between self and environment.

A capacity for planning and imagining future scenarios is suggested still more in an experiment conducted by Clayton and her colleagues in which they kept scrub jays in cages divided into two compartments. Each morning the researchers moved the birds into one of the compartments: one contained a breakfast snack (crucially, this was always the *same* compartment), the other did not. After getting them used to this routine, the experimenters allowed the birds to feed on nuts to their stomach's content in the evening, and then to cache the remaining nuts in one compartment or the other.

The scrub jays always stored them in the compartment they 'knew' would otherwise be empty the next morning, 'just in case' they were put in that one. It's hard to see how they would make this choice – one, note, that seems unlikely to be necessitated in the wild – without some anticipation of future needs and of likely scenarios: 'I'm full now, but I'll be hungry tomorrow morning, and if I get put in *here*, there'll be nothing to eat unless I store it there now.' Again, of course, the bird would have no inner dialogue like this, but there does seem to be a future-projecting plan motivating the behaviour.

Much of the jays' caching behaviour is overt, even naive: the food is simply left at some location in open view. Not surprisingly, there's a lot of pilfering: a bird has no inhibition against helping itself to someone else's cache if it comes across it. On the whole that's just an occupational hazard, but the western scrub jay will take some rudimentary precautions against it. If it can see another bird is watching while it caches, it will often place the food behind a barrier where it can't be seen.

Well, wouldn't you? Yes – but you'd do so because you're thinking, 'That so-and-so wants my food – well, he won't find it here.' You are imagining what the other guy is thinking. This looks, then, like a demonstration that the bird has a Theory of Mind.

Perhaps it too could be a mere hard-wired instinct – a rule of 'If watched, hide the cache.' But it's more complicated than that. The jays seem to adapt their concealment strategy to the circumstances: if they know, say, that another bird can hear what they're up to but can't see them, they will prefer to cache the food in a setting that creates less noise: among soil rather than pebbles, say. And if a bird knows it was being watched while caching, it might later return to *that* cache (but not others) and move it elsewhere. Even more remarkably, it might return and *pretend* to re-cache, without bothering with the effort of actually doing so, apparently to fool any onlooker ('Don't bother, you won't find it here, mate').

Finally – a delicious touch, this – the birds that engage in such deceptive actions tend to be ones that have previously pilfered themselves. It takes a thief to know a thief.

Again: do you think evolution would have come up with distinct, hard-wired behavioural rules for all of these situations? Or would it have been cheaper to evolve a bird brain that possesses not just a capacity for representing the future but with a primitive Theory of Mind to guide it?

Do birds know themselves?

Do all these cognitive abilities add up to the likelihood that crows (and, presumably, other birds) have *minds* – that there is something it is like to be a crow? It's odd, really, that this question – usually phrased as that of whether birds are *conscious* – is often asked with the expectation of a yes-or-no answer, since no one really knows what it even means. Primates show such a range of behavioural similarities to us – in gesture and expression, cognitive and social skills – that it seems perverse to suppose they lack any vestige of the subjective experience we find in ourselves. Birds, though, are already too alien to make us comfortable about such intuitions.

Yet still we ask with misplaced confidence: are birds conscious? It was long thought that they are not, on the grounds that bird brains are too anatomically different from ours. *Pace* theories of consciousness that locate it in the brainstem or other parts of the 'primitive' brain, a functioning cortex is generally considered essential for human consciousness – but birds don't even possess a cortex. Instead they have a forebrain region called the pallium, which lacks the layered arrangement of our cortical neurons. However, researchers have found over the past decade that the neurons of the avian pallium have a complex arrangement rather similar to those of the human cortex, with which it probably shares an evolutionary origin.

Moreover, some birds, such as corvids and parrots, have as many neurons in their pallium (1–2 billion) as there are in the monkey cortex. The neurons of the pallium seem also to form deep connections to other parts of the bird brain, creating hubs where information can be integrated: a kind of workspace where, according to Brazilian neuroscientist Suzana Herculano-Houzel, it can 'create new associations that endow [its] behaviour with flexibility and complexity.'

Might that add up to a capacity we can plausibly call consciousness? Some ingenious experiments have suggested that birds at least possess the requisite mental ingredients for a sense of self. Animal physiologist Andreas Nieder and his colleagues trained carrion crows to 'report' whether or not they had seen an object. They showed the crows a blank screen on which a grey square might be flashed for just three tenths of a second: so fast that the bird wouldn't always register it. Sometimes the crows might have been shown the square and not registered it; sometimes they might have thought they'd seen a square but been mistaken.

To report what they *thought* they'd seen, the bird was presented either with a red square – which it should peck if indicating 'Yes, I saw a grey square' – or a blue square, which it should peck to indicate 'No, I saw no grey square'. Giving the correct answer to whether the grey square had been displayed or not gained them a reward, which motivated them to learn how to do it right. Crucially, in between showing the grey square (or not) and inviting an action to report on what was seen, there was a delay of two and a half seconds when the screen was left blank. This means that the crow had, for that period, to hold some mental representation of what they'd seen in their head. The bird couldn't just, on seeing the grey square, store a 'Get ready to peck' command in its mind, because that wouldn't work if it was then shown a blue square; it was only meant to peck *that* if it was answering, 'No, I saw nothing.'

During this activity, the researchers used electrodes to record the activity of a few hundred neurons in a region of the bird brain called the nidopallium caudolaterale. This is known to be active during high-level cognition, and is the bird's equivalent of the primate prefrontal cortex, responsible for the ability to act on thoughts, feelings, and decisions. The researchers supposed that if crows have something like consciousness, it might originate here. Perhaps the activity of these neurons corresponds to the sensory signal – the sight of the grey square, say – being 'broadcast' to the rest of the brain: the very process that some researchers believe is required for consciousness in humans (page 131).

Nieder and his colleagues found that the activity of these neurons matched what the crows reported. If a bird 'reported' that it had seen a grey square, the neurons would fire in one pattern; if it reported that it hadn't, the neurons generated a different one. This was true even when the crows were mistaken – thinking they'd seen the square when it wasn't really shown, or vice versa. In other words, the birds seemed to form an internal representation of their experience, rather than just a camera-like record of their sensory input. When the birds reported wrongly, it wasn't a random error but was because they really believed they'd seen the stimulus, and sustained that belief in their minds during the delay period. This, said the researchers, identifies 'an empirical marker of avian sensory consciousness.'

What this study can't tell us is what the experience *is like* for the birds – or even, in truth, whether there is a 'what it is like' at all. That's to say, the experiment doesn't distinguish between 'access consciousness' – forming an internal representation of an event or experience that can be broadcast and recalled – and 'phenomenal consciousness', which is a subjective feeling associated with it. The former is, you might say, about 'knowing' something, the latter about 'knowing that you know'. Some researchers believe Nieder's experiments show that crows have access consciousness, but even if that is so, they can't pronounce on phenomenal consciousness. As we saw earlier, though, there's no consensus on whether this distinction is even meaningful: some researchers think that there *is* nothing more to consciousness than 'access'. We can't, then, expect to say anything much more about bird or animal consciousness until we better understand our own.

Hive minds

The psychologist Jennifer Volk has pointed out that while we might be tempted too readily to make an anthropomorphic interpretation of behaviour shown by 'charismatic' animals like chimps and dolphins, we might be inclined too dogmatically to resist any such view of creatures less endowed with features that attract our empathy: fish, reptiles, insects. There's a common assumption that insects are individually stupid, even robotic, even though their social organization gives rise to great feats of collective intelligence. The idea is that each individual performs a simple, routine, and programmed task, from which impressive displays of collaboration emerge much as organized but 'mindless' structure appears in cracking rocks or flowing fluids. There's surely some truth in this: the natural cathedrals that are termite mounds, for example – with intricate systems of tunnels for air-conditioning, and chambers where fungal food stocks are cultivated – seem to arise spontaneously from individuals following simple rules about where to deposit their chewed-up mulch. Fish schools, those astonishing displays of coordinated movement that dazzle and confuse predators, can be replicated in simple computer simulations of agents just responding in a systematic way to the movements of their neighbours.

Still, the cognitive and artisanal skills of social insects are too easily discounted. While some researchers suspect that the hexagonal masterpieces that are honeycombs are formed passively from soft wax pulled into place by surface tension as in a soap foam, in fact bees possess tools on their own bodies for precisely measuring angles and wall thicknesses as they lay the wax flakes in place.

Bees display minds that are both impressively specialized and remarkably versatile. Bumblebees are particularly good at learning new tasks, at which they can be conditioned by offering rewards. Zoologist Lars Chittka and his colleagues trained them to play a kind of insect golf, in which the bees had to manipulate a small ball into a hole in the centre of the 'pitch', getting rewarded with sugar for doing so. The researchers trained bees to do this by demonstrating what was required: they modelled the behaviour using a crude 'fake bee' glued to the end of a stick that they used to push the ball into the hole. Once trained, a bee could act as a demonstrator to show others how to do the task.

Amazingly, the bees were then able to use their own initiative to find a better, more efficient way of getting their reward. The researchers would train 'demonstrator bees' on a pitch containing several balls of different colours – but all except the ball farthest from the hole were glued down, so the demonstrator learned to ignore the others. Yet when the demonstrator bee then enacted the game for the benefit of other bees on a pitch where the other balls *weren't* stuck in place, the other bees quickly realized they could manipulate another ball closer to the hole (even though it was a different colour) to get their reward. Such cognitive flexibility, say Chittka and colleagues, 'hints that entirely novel behaviours could emerge relatively swiftly in species whose lifestyle demands advanced learning abilities, should relevant ecological pressures arise.'

One of the most impressive feats of bee cognition is the foraging of honeybees. Individuals fly out from the hive in search of good sources of food, such as flower nectar. If they find one, they return to the hive and tell the others where it is – by *dancing* the directions,

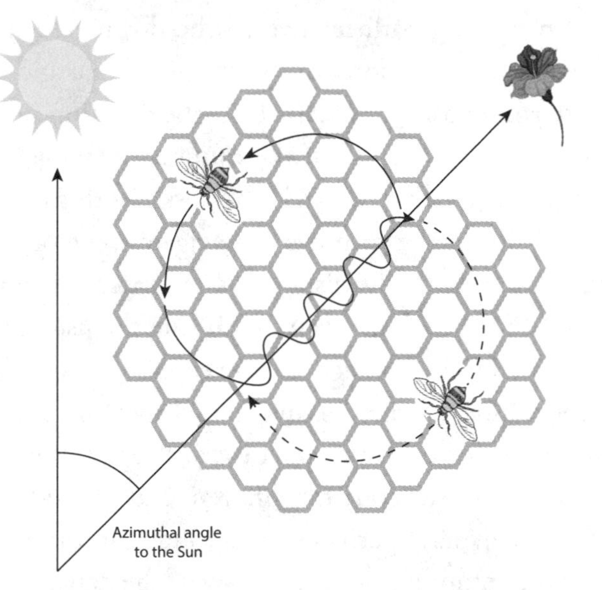

Figure 5.4. The honeybee's waggle dance, which conveys in coded form the location of a food source the bee has found. The angle of the dance relative to the vertical of the comb indicates the azimuthal angle of the food relative to the Sun, and the duration of the waggle section encodes the distance – roughly speaking, one kilometre for every second.

treading out a circulating path on the comb known as the waggle dance (Figure 5.4).*

The waggle dance encodes two crucial pieces of information that will enable the bees being recruited by the dancer to find their way to the food source: the direction of travel from the hive, and the distance. The direction is measured relative to the current position of the sun,† and this angle of outbound flight is encoded by the angle

* That bees so unselfishly share this information is a reflection of their status as social insects. They live in colonies that are all closely related, and so this is a classic example of kin selection: helping those who share your genes.

† Strictly speaking, relative to the azimuth: the direction pointing to where the Sun's position, projected vertically downwards, hits the horizon.

of the bee's zigzagging path relative to the downward direction of gravity. The dancer steps along this path for a while and then loops back to its starting position. Meanwhile, the duration of the waggle portion of this circuit – in effect, the number of waggles – signifies the distance to the target destination. If possible, this is the distance of direct flight. But if the path has to navigate some obstacle, like a hillock, then the total flight distance is indicated. It sounds like a weird problem in a geometry exam, requiring the use of protractors and conversion of units ('If each waggle equals 10 metres . . . ').

What's more, this information may be subtly refined by variations in the waggle time or the time taken for a complete rotation. There are also variations in the code for different bee species – dialects, if you like – that may have partly evolved to take account of the local terrain. And the communication is two-way: the dancer's audience is not entirely passive, but participates with movements of their own that seem to be part of the decoding process. Thus informed, the recruits can find the target with impressive efficiency. Not surprisingly, the instructions don't have the pinpoint accuracy of a grid reference. But once they are close to the target, the bees seem to find it using odour cues, in effect following their nose to the prize.

The waggle dance, first observed and understood by the Nobel laureate ethologist Karl von Frisch in the 1960s, has to be one of the most impressive and sophisticated feats of animal communication in all of nature. What's so unusual about it is that it conveys complex information – about direction and distance – in a purely *symbolic* form. The waggles don't *show* distance, but symbolize it, and similarly the waggle angle encodes rather than depicts directional information. This, says bee expert Randolf Menzel, is more or less unique in the animal kingdom outside of humans. Our own symbolic communication is of course central to our cultures – it's how these arbitrary marks on paper (or on a screen) are registering as concepts in your mind. The same is true for spoken language: the sounds are mostly arbitrary (apart from onomatopoeia), but their meaning is concrete. Language, as we saw, transformed human societies and now shapes our own minds.

Many animals can communicate via calls: bird song and whale song, and the so-called pant-hoot vocalizations of chimpanzees that warn of predators and the like. But bird song (unlike bird alarm signals) doesn't contain semantic information; it is a mating display akin to plumage. And animal warning calls are stereotyped: simple signifiers, like 'Look out: leopard close by!' A chimp cannot, as far as we know, say, 'There's a leopard north-north-east about a quarter of a mile away.' It simply doesn't have a lexicon capable of that. But honeybees do the equivalent with their waggle dance.

The bee brain has a volume of a mere 1 cubic millimetre or so – about the size of a sand grain – and it is very different in anatomy to those of mammals. It is a collection of interconnected lobes, containing around a million neurons in all, with regions specialized for vision, memory, movement, and so forth. A part called the mushroom body (because of its shape) does much of the advanced processing such as decision-making. What is going on in there when the honeybee reads a waggle dance?

One possibility is that it is a clever but mindless process like Google Translate: the sensory data from the dance is somehow converted into a set of flight instructions that the recruited bees become programmed to blindly follow.

But in fact it doesn't happen that way. Rather, it seems that each bee decodes and superimposes the information about the target location onto its own internal map of the territory, built up from its own foraging and exploratory flights. It is as if individuals say, 'Ah, she means the food is by that group of poplars I saw!' And they seem to weigh up the information they receive against their own experience; some might decide it's not worth the journey.

Menzel and his colleagues supplied an impressive demonstration that bees have these 'cognitive maps' in their heads in an experiment where bees that had just found a food source and were about to depart back to the hive were captured and taken in a closed box to a new location in the vicinity. There they were released to find their own way back to the hive.

The bees began by making short exploratory flights from their new location, before returning to the release point. Then, as if they had got their bearings ('Ah, *that's* where I am!'), they set off and flew *directly to the hive*. Or alternatively – and even more impressively – some flew directly to the food source, fed for a bit, and then went straight back to the hive *from there*. They seemed to know exactly where everything was.

Bees acquire these mental maps by self-training once they are mature enough to leave the hive. They conduct orientational flights of ever greater distance; after just four to six of these, they will typically have a good sense of the whole terrain within a half-kilometre radius of the hive. They also build up a memory of 'special routes', rather as I have found preferred cycling routes for getting into the centre of London. All this suggests – and clever experiments on navigation have confirmed – that bees are able to create some kind of visual map of their environment that they can retain and consult in the absence of the stimulus itself. We don't know what this looks like, much less how the bee experiences it. But to some researchers it suggests bees have at least a kind of 'image consciousness'.*

The waggle dance is sometimes described as a language. The

* These representations seem to vary widely in humans too. Some of us are unable to make much sense of instructions based on compass directions, but instead rely on landmarks. There is a strong cultural element at play here too. I always remember how, when trying to find my way around Beijing the first time I visited in 1992, people would look at the maps in my guidebook as though they were mysterious cryptograms – as though they had never thought of their city

metaphor is understandable, but it is no more than that. There are rules to how the information is encoded, but these do not have the rich structure of real languages – there is no syntax, and the elements can't be combined to say novel things. Yet these differences arguably make the waggle dance all the more interesting, for it shows that, while written human language appeared *after* spoken language, as a translation into visual symbolic form, this isn't the only route to symbolic communication. The mind of the honeybee 'understands' symbols without 'reading them as voice' in the way we do. The implication is that symbolic communication need not be an adjunct of the acquisition of language, but can be independent of it.

Do animals have emotions?

If Descartes had not done enough already to earn the censure of many cognitive scientists, his views on animals seem to clinch the deal. He described how he performed vivisection on dogs and rabbits to investigate William Harvey's ideas about the circulation of the blood, and dismissed any cries of pain that procedure elicited as a mere mechanical response, not unlike the screech of a poorly oiled axle.* It was said that some of Descartes' followers nailed animals to wooden boards by their paws to dissect them – although we must be wary of those claims, which were made years after the event by opponents of the Cartesians.

There's little doubt that Descartes' experiments would horrify us today, but his real views don't easily translate to the version often given, which is that he thought animals lacked any sentience. 'I

being represented this way. (I still don't understand it, as cartography in ancient China was far more advanced than that in the West.)

* Apparently, however, Descartes owned a dog called Monsieur Grat, and was devoted to it. Go figure.

deny sensation to no animal', he wrote. Rather, he felt that sensation itself was simply mechanical in all animals except humans, for they have no soul.* What's at issue here, then, is not sensation per se but an awareness of it. We can say without too much approximation that what Descartes was really disputing was whether other animals have *feelings* and emotions.

There seems little question that something akin to pain exists for dogs, and many other higher animals: they retract from and show aversion to stimuli that cause injury, for example, as well as displaying the human-like response of crying out. But some say that what truly matters is whether this is accompanied by *suffering*, which is an emotional state. It's possible in principle to imagine pain for an animal as being no more than a reaction of the nervous system that triggers certain behaviours, without any intervention of a *feeling*. After all, there's obvious survival value in recoiling from something that causes injury – and perhaps the yelp that accompanies it is no more than an evolved altruistic signal for warning others?

If you're thinking that it seems callous in the extreme to suppose that a dog exhibits all of the responses to a painful stimulus that a human infant does while denying that it experiences anything of the emotional trauma, you have a fair point. My own view is that, in such a case, we should assume the dog suffers as the infant does, unless we have very good evidence otherwise. But that's an ethical precautionary measure; it doesn't in itself prove that other animals have emotions.

At the neurobiological level, emotions manifest as global states of the brain that influence a wide range of physiological responses and behaviours. In a state of happy euphoria, our heart beats faster, we feel more alert and receptive, and we're likely to act more

* This was not just Descartes' view, but official Church doctrine. Even today the Catholic orthodoxy is that we have souls but dogs do not.

optimistically. In a state of fear, we become hypersensitive to stimuli, our muscles and limbs contract, and we break into a sweat, in extremis perhaps losing control of our bladder or bowels.

Global states of 'sensitization' like this do exist in animals. In fruit flies, for example, a nearby female can trigger behavioural changes in a male, making it both liable to become more aggressive towards other males and also apt to engage in courtship with the female. (As so often, the temptation to see anthropomorphic analogies is irresistible.) These are two quite distinct behaviours, yet both are caused by the same stimulus. Crucially, they seem both to be activated by the same cluster of neurons, which, once triggered, stay switched on for several minutes, seeming to switch the fly brain into a state of general arousal. Mice have similar neural clusters that appear to be activated by danger and which trigger a range of behaviours that we might associate with fear.

Such states of neural activity that affect many aspects of animal behaviour have the characteristics of 'mood' or 'emotion' states, regardless of what they 'feel like' to the animal. Unlike simple stimulus–response neural circuits, they are non-specific in both their triggers and their effects, and they persist for a time even when the stimulus has ceased. Their function seems to be not to provoke specific kinds of behaviour but to alter the way other sensory inputs are processed in a manner appropriate to the situation. For example, it can be adaptively valuable for a threat not just to provoke a specific response but to put the mind on alert so that it is more attuned to subsequent, potentially threatening signals too.

Zebrafish larvae have groups of neurons that trigger them to seek prey only in their immediate environment, and other clusters of neurons that, when active, prompt them to explore further afield in unfamiliar waters. These neural states don't dictate what the larvae do there, but act more like generalized 'timid' and 'bold' states. The larval brains seem to switch between these two states automatically every few minutes as the animals forage, perhaps generating a compromise between safe and risky behaviour that leads to the most effective results.

'We used to think of animals as being kind of stimulus–response machines', says neuroscientist Anne Churchland. 'Now we're starting to realize that all kinds of really interesting stuff is being generated within their brains that changes the way that sensory inputs are processed – and so changes the animals' behavioural output.' These internal 'states of mind' have been hard to spot because the neural activity they produce is brain-wide, complex, and not obviously correlated with what the animal is doing at a given moment – so it has previously been mistaken for mere random noise. But instead it seems to reflect a process in which different regions of the brain, concerned with both internal and external information, are integrated and 'conversing' – sensory processing being hooked up to motor circuits, say. Indeed, it looks enticingly like the sort of 'broadcast' proposed in the global workspace theory as the basis of consciousness. It's mere speculation, but this sort of brain activity might be the kind of 'feeling' state that some researchers consider to have been the earliest sign of sentience (see page 244).

However, we could alternatively suppose that these internal 'mood states' are somehow tuning the brain to optimize signal processing and behavioural outputs, without any need for the animal to register those changes subjectively. At root this comes down to the issue of whether there need be anything *more* to awareness at all, beyond the broadcasting of activity from a workspace of attentiveness to the rest of the brain. But as we saw earlier, one of the hallmarks of human consciousness seems to be not only that we feel but that we *know* that we feel. We have an autobiographical self that in some sense we can regard just as we can regard objects in the world. We don't just, say, experience feeling bad, but can think and say, 'I feel bad.'

Antonio Damasio argues that emotions as we know them demand this sense of self: the suffering caused by pain doesn't result just from the *feeling* it creates in the body, but from the knowledge that this is happening to *us*. An animal lacking a sense of self might, therefore, experience pain but not feel suffering: it might know 'there is pain' and be aware of its body responding appropriately, but have no concept of the pain 'afflicting *me*'.

Vladimir Nabokov, who was a respected zoologist (with a specialism in butterflies) before he became a famous novelist, thought this is how it is for animals. Self-awareness, he observed – I know that I am, and I know that I know it – is uniquely human, which is why in his view 'the gap between ape and man is immeasurably greater than the one between amoeba and ape.' This seems now to be almost certainly wrong, but it's hard to demonstrate self-knowledge in creatures that cannot tell us they have it. If we can't know 'what it is like to be a bat', we sure as heck can't know 'what it is like to be a bat in pain.'

Perhaps, then, the pertinent question is not so much whether an animal feels pain but whether it has a sense of a self that will suffer from it. The classic behavioural signature of a sense of selfhood is the mirror test (page 125): whether a creature will attempt to remove a mark (or an equivalent action for, say, dolphins) from their head when they see it in a mirror, showing that they recognize the reflection as 'them'. But it's not easy to interpret such an action – some researchers think it can be explained as a 'conditioned' response, a mere learned action that doesn't reflect any internal representation of the self. In any event, it seems a questionable discriminator of selfness, let alone self-consciousness. Some monkeys and all dogs fail the test, for example – but it's not easy to make the case that said monkeys have a significantly different quality of experience than other, closely related species.

Ingenious experiments have offered better evidence that some animals have true self-knowledge. Psychologist John David Smith has trained dolphins to respond to sounds so that the creatures will swim over to press a paddle on the right-hand side of their pool if the pitch is higher than some threshold frequency, but will press the left-hand paddle if the pitch is lower. Offer the dolphin a food reward for getting it right, and the creature will rapidly learn to perform well. But if the pitch presented is close to the threshold, it can be hard for the dolphin to know which way to swim (humans too may struggle with such problems of pitch comparison). In such cases Smith offered the dolphins another choice: to press a third paddle that would elicit an easier pitch-discrimination test. He was careful to ensure that the dolphins learned not to use this option too freely as a cheap get-out: if they did, their rewards would be delayed. So his dolphins were encouraged to use this fallback only in those cases where 'it knows it doesn't know.' And that is just what the smart creatures did. Some monkeys will pass an analogous test too. They seem able in some sense to step back and look at what is happening in their brains, and to assess it.

It's hard to imagine why such creatures would fail to be similarly aware of other 'internal mental states': to experience something akin to what we recognize as fear, suffering, elation, sadness. It's not absurdly romantic to suppose they could feel love.

How could we really know, though, if a dog is feeling happy or sad? Well, dog lovers will say that it's downright obvious, and I believe them. But what can we measure? And how can we do that in animals that apparently have a less rich repertoire of behaviours and expressions, such as fish or insects?

One classic test for emotion-like states in animals is to see whether circumstances that might be expected to make them happy result in increased 'optimism'. Animal behaviourist Catherine Douglas and her colleagues investigated whether the conditions in which pigs are reared affects their outlook on life, as indicated by their expectations of what it will offer.

The researchers housed one group of pigs in a cognitively stimulating and comfortable environment, well supplied with straw, space, and objects that invited curious exploration. A second group was kept in a barren enclosure devoid of these things. The team investigated whether the two groups of pigs were more optimistic or pessimistic. They trained the animals to associate one sound with food being provided through a hatch, and another sound with no food. The pigs would learn to go to the hatch on hearing the first sound but not the second. Then the researchers sounded an ambiguous signal, midway between the two training ones, which could be interpreted either as a 'food' or 'no food' signal. An 'optimistic' pig would do the former, a 'pessimistic' pig the latter. Douglas and her colleagues found that the pigs in the sensorially enriched enclosure were more likely to interpret the cue optimistically and approach the hatch. This doesn't, of course, say anything about how the two groups of pigs really *feel* about their surroundings. It doesn't mean the well-kept pigs felt more buoyant. But it shows that they experience *some* sort of affective state that influences their behaviour.

It shouldn't take an experiment like this to persuade livestock farmers to treat their pigs well and give them spacious, amenable environments to live in – we should do that by default. But it does mean that saying 'What do the pigs care?' is not only callous but demonstrably false, if by 'care' we mean that their mental state is impacted.

Perhaps this should come as no surprise at all for smart creatures like pigs. But should we worry whether less cognitively adept creatures like insects have emotions? I do worry a little about this – not, I have to admit, always enough – when I am tormented by a buzzing fly as I'm trying to write, and tempted to reach for an improvised swatter. It's easy to see that flies react to a potentially threatening approach – they buzz about in a frenzy for a short time, and more controlled studies have shown that they also display other behaviours redolent of fear or anxiety: making more hops, for example, or occasionally freezing. Here too, then, a threat seems to induce some change of mental state that affects a range of behaviours. But at least some of these can be supposed to be automatic, adaptive reactions that help with predator evasion, not necessarily accompanied by any feeling.

Lars Chittka and his colleagues have, however, looked more closely at whether bumblebees show the same kind of 'positivity' seen in pigs, in response to something that might be expected to improve their mood – to put it anthropomorphically, to make the bees happy. In this case, the equivalent of the food hatch cued with different sounds was coloured flowers (bees have good colour discrimination) that might or might not provide a food reward. The researchers trained the bees to recognize that blue artificial flowers would have droplets of sugar solution on them, whereas on green flowers the droplets would just be water. The bees learnt that only the blue flowers were worth flying to.

Then the team used the 'ambiguous' signal: here, blue-green flowers. Would the bees interpret these optimistically as 'probably blue', or pessimistically as 'probably green'? They divided the bees into two groups. One was, before the experiment, given some sugar – an unexpected reward – while the other was not. Those given the surprise treat were more inclined to fly to the blue-green flowers, as if the unanticipated benison had put them in a 'good mood'.

In a second test, the two groups were subjected to a mock predator attack by being temporarily trapped in an artificial mechanism. This makes the bees subsequently more cautious about foraging, as if they are temporarily gripped by fear. But the bees that first got the unexpected sugar reward were quicker to return to foraging, as though their initially rosy outlook partly offset the anxiety.

Evidently it's hard to discuss these experiments without lapsing into anthropomorphic speech. I don't think for a moment that we can legitimately call the sugar-surprised bees 'happy' – that is a word coined for the human emotion, which will be imprinted with our sense of self and nuanced by our cognitive complexity. But human happiness did not come from nowhere. Like everything else in our minds, it is ultimately a product of evolution, and the affective 'mood' states of bees are very plausibly the kind of proto-emotion from which happiness arose. What is more, when Chittka and colleagues treated the bees with a drug known to block the effects of the neurotransmitter molecule dopamine – known to be involved in the human 'reward' response to pleasurable stimuli such as food and sex – they no longer showed any apparent 'mood enhancement' from the initial sugar treat. This at least suggests that the behavioural effects arise from neural signals comparable to those that produce pleasure in us.

Animal aesthetics

We humans have feelings about all kinds of things: not just those with obvious survival value, such as food and sex, or conversely danger or injury, but also about music and art, sports results, political events, and many other things. In other words, our emotions are attuned not just to basic survival but to culture, and they embrace an aesthetic dimension.

Do animals have anything like an aesthetic sensibility? Your dog might show pleasure, as it seems, in having a warm rug by the fire to lie on. But does it care if the rug is a tasteful burgundy or a loud lime-green?

The bowerbird is, as I've suggested already, often regarded as the aesthete of the animal world. The male constructs elaborate arrangements of found objects to decorate the arch-like enclosure – the bower – it weaves from twigs and stems to impress females. The birds collect coloured objects such as shells, petals, and pieces of discarded plastic, which they place around the bower, often sorted by colour and arranged symmetrically. Like birdsong, bower construction is a learnt skill: young males watch experienced builders at work. Great effort and what looks like judgement go into these creations: some males will sit *where the female will sit* before making a change, as if considering what the result will look like for his intended audience.

It's possible that the female response to a fine bower is no different to that which some birds show to elaborate plumage, another example of a mating display: an automatic reaction to something loud and bright. But the exquisite artistry of the bower, and the fact that it is the unique product of an individual bird's mind, along with the care lavished on the project by the male and the prolonged deliberation of the female, suggest at least that there is complex cognitive processing going on. Are the birds merely following elaborate but automatic rules that we can't fathom? Or could there be, in the females at least, some stirrings of an emotional aesthetic response? Charles Darwin felt that animals might evolve to appreciate what we might call beauty for its own sake, and Karl von Frisch argued that case for the bowerbird:

> Those who consider life on earth to be the result of a long evolutionary process will always search for the beginning of thought processes and aesthetic feelings in animals, and I believe that significant traces can be found in the bowerbirds.

To object that the male bird's behaviour is simply triggered by his sex hormones is misguided, said von Frisch – for the same could be said of the poet composing a love sonnet to his *amour*. At the very least, it should give us pause that we have our own emotional response of pleasure both to birdsong and to the artistry of bowerbirds, not to mention the polychromatic splendour of avian plumage. Is it just coincidence that what evokes our aesthetic delight in the genius of birds happens also to trigger attraction and excitement in the animals themselves?

Speaking in tongues

There's a clever cartoon lampooning animal behavioural research in which a white-coated scientist tells a line-up of animals – a chimp, a crow, a snake, an elephant – that in the interests of giving them a fair test of ability, he wants to see them all climb a tree.* It makes the point that one of the hazards of such research is to mistake the absence of a certain behaviour for the lack of the cognitive capacity to exhibit it. The reason we will never hold a conversation with a dolphin is not just because they show no sign of being able to handle the recursive nature of human language but because the anatomy of their throat and mouth does not permit them to make the requisite sounds anyway.

I argued earlier that minds without language are surely very different from ours. Words are not mere labels for objects, actions, emotions and so on; they are combinatorial symbols that permit new kinds of reasoning, and indeed new types of awareness. As Daniel Dennett puts it, 'Many animals hide but don't think they are hiding.' They undertake actions without categorizing them. This is not the same as the self-awareness that 'It is me that is hiding', and in fact needn't be coupled to it: it's possible to imagine a creature

* We could learn a lot from this cartoon for testing humans too. IQ tests have become notorious for their cultural bias, although those problems are somewhat acknowledged today.

that is aware of itself within a situation without giving that situation a name ('hiding').

All our best efforts to probe the minds of animals are hampered, therefore, by the obvious problem that we can't just ask them about their lives. And it might not help much even if we could. 'If a lion could talk', said Ludwig Wittgenstein, 'we could not understand him' – because we don't share the lion's Umwelt, its subjective frame of reference. (Of course, in reality if a lion could talk it would not be a lion, because, laryngeal physiology aside, 'talking' doesn't come as a module you can bolt into a lion's brain.)

Plenty of animals, including lions, do vocalize. But what do they mean by it? Birds have some of the most complex vocalizations of the animal kingdom, and their songs have some striking parallels with human language. Bird songs aren't hard-wired, but learned by practice: fledglings have to develop the right control of their vocalization apparatus (the syrinx), and getting the song right involves feedback in much the same way as children make gradual refinements until their words sound correct. Birdsong has dialects too: some songbirds have entirely different 'tunes' in different parts of a country. The songs consist of discrete acoustic elements, like syllables. And the song production area in the bird brain is the equivalent of Broca's area, the part of the human brain responsible for speech.

Yet birdsongs are not really a language at all. Rather, they are signals – in general, calls designed to attract a mate. Such songs are mostly then just long-winded, if rather lovely, ways for male birds to advertise their attractiveness.* We still don't fully understand what it is that makes a bird's song a hit with the females, but it

* Not all birdsong is about sexual selection. Some males also use it to signal aggression, for example. And many female songbirds sing too, although it's not clear why.

seems that some elements, such as double syllables, are considered 'sexy', and that one aspect of the criteria used by females is how consistent a song is – whether a male can repeat it, for example. Nor do we know whether singing prowess is an 'honest' signal, meaning that it reflects some genuine superiority in the singer as a potential mate. It's possible that having a large song repertoire indicates good cognitive skills, but in fact there's no indication that good singers are otherwise smarter. The tunes evidently do their job: males with large repertoires produce more offspring on average. But there's nothing in them to truly be 'decoded'.

The other order of creatures that engage in complex vocal communication is the cetaceans, including whales and dolphins. Their sounds do seem more language-like, although in truth we have little knowledge of what they mean either. Some of the sounds, at least, might be akin to 'words' with specific meanings. Marine biologist Denise Herzing has reported that an algorithm designed to identify meaningful signals in dolphin whistles picked out a sound within a dolphin pod that she and her team had trained them to associate with sargassum seaweed (which dolphins sometimes 'play' with). Herzing claimed that the dolphins had assimilated this 'word' and used it in the wild to convey the same meaning.

Communication in the animal world happens in many ways besides sound. Humans convey a great deal from posture and gesture, and for some animals these might be as important as noises and calls. Figuring out what these signals imply is complicated, however, by the fact that we have no idea where one gestural 'word' stops and another starts, nor whether there is even anything like a vocabulary or grammar at all. Unless you have a really good understanding of the behaviour of a species, says cephalopod expert Jennifer Mather (and mostly we don't, she adds), you can't really decode what its actions mean, because you don't know which of them are jumbled or incomplete: you have no lexicon for translation.

But there are now new tools for opening up animal communication, in particular the same kinds of 'deep-learning' artificial-intelligence methods that are used for human-language translators (see page 293). Computer scientist Britt Selvitelle, co-founder of the San Francisco-based Earth Species Project, is convinced that some kind of AI-based trans-species communication might be possible in five to ten years. But it will require us to 'remove our human glasses', he says: 'We may not even have the sensory organs to feel and understand the communications of another species. Do these communications look more like the human experience of language, or music, or even images?'

Conventional decryption of an unfamiliar human language demands a Rosetta stone: a text correlating the unknown language (such as Egyptian hieroglyphics) with a known one (such as Greek). For animal communication, we can't hope for any such guidance. But today's AI systems can work without it. Trained to search for patterns and connections, they can identify correlations, with an almost clairvoyant acuity, for all manner of types of data. The hope is that if there is some aspect of an animal's behavioural repertoire that represents or expresses something our languages can also express – indicating a type of species, a warning sign, a spatial direction – then AI might spot it. The Earth Species Project has a program called Whale-X that aims to collect all communications between a pod of whales over an entire season and use AI to figure out what it means. A key issue here is that an AI should be agnostic about meaning and not (or at least, less) susceptible to the cognitive biases that might incline us to interpret in human-centric terms. 'The exciting thing about AI and computer technology', says Con Slobodchikoff,

> is that we are beginning to be able to decipher animal languages and animal cognition on terms that are meaningful to the animals,

and not on our terms. Computer technology is finally allowing us to see inside the world of animals in ways that are showing us that they are complex sentient beings that deserve our understanding and respect.

It seems extremely unlikely that any animal communications will really amount to a language, with the recursive complexity and syntax of human tongues. As Dennett puts it, human language 'cuts too fine': it recognizes subtle distinctions of category, quantity, and quality that don't matter to other animals. No dog, says Dennett, seems likely to think (which is to say, to mentally represent) the notion that would translate as 'my dish is full of beef'. We risk, he says, 'importing too much clarity, too much distinctness and articulation of content, and hence too much organization, to the systems we are attempting to understand.' Dogs might show a preference for beef over pork, but that does not in itself mean 'beef' and 'pork' are distinct categories of meat for a dog. The animal might, say, simply be following a rule that 'if meat arrives in conjunction with smell X, eat it preferentially'.

At the same time, we may encounter types of communication, feeling, and concept that simply can't be expressed in any human language. For example, as well as whistles, dolphins generate clicks that they use for sonar echolocation. From the acoustic reflections as clicks bounce off an object, a dolphin can form a mental picture of the object's size, shape and even density. In a real sense it 'sees' with sound, being able to match a sonar representation with what it sees visually.

What's more, one dolphin can interpret another's sonar signals – using sound alone, they can see what another sees. 'They are able to see shapes of things when they passively eavesdrop on someone else's clicks', says neuroscientist Marcelo Magnasco, who studies dolphin cognition. In principle, then, dolphins could use clicks for communication too. 'If they can imitate with clicks the shape of an object', says Magnasco, 'then they would be able to speak in pictures.' It's rather as if, instead of saying to you 'teapot', I could project some signal that makes you see one – not just mentally, but right there in your visual field. Perhaps *that* extraordinary notion is a part of what it is like to be a bat.

The most immediate benefits of having a better appreciation of the range and abilities of animal minds could be for animals themselves. Some of our mistreatment of other species arises simply from communication breakdown. 'Dogs are often surrendered to shelters because of a lack of ability of us humans in reading and understanding the signals with which dogs are trying to communicate with us', says Slobodchikoff. If we are to coexist with other animals in close and even domestic proximity, it is time we at least tried asking them, 'How would you like us to let you exist?'

But it would be still better, and certainly less hubristic, to start regarding ourselves as particular kinds of minds coexisting with a broad spectrum of others, each with our strengths and weaknesses. The title of pioneering ethologist Konrad Lorenz's 1949 book on animal psychology, *King Solomon's Ring*, alludes to the legend that Solomon possessed a ring giving him the power to speak with beasts. Lorenz suggested that perhaps the talisman Solomon possessed was no magic ring at all, but simply the power of careful observation. As cognitive psychologist Diana Reiss puts it, Solomon might simply have 'made the space to see and hear what other animals were doing.' It would be an example worth heeding.

CHAPTER 6

Aliens On the Doorstep

In 2019, more than a hundred scientists signed a declaration demanding that we grant rights to octopuses and ban octopus-farming. Most of the signatories were specialists in animal cognition and behaviour, and especially in the mental capacities of cephalopods: squid, octopuses, and cuttlefish. They asserted that these marine creatures are highly intelligent, sophisticated, and sentient – and they were concerned that the practice of intensive farming of octopuses for food, placing them in 'sterile, monotonous' environments, ignored those mental attributes and treated the animals as if they were insensate brutes. These experts evidently felt that there is something it is to be like an octopus.

It's not hard to see why we might imagine otherwise. Cephalopods have long seemed too strange to match our preconceptions about where minds might reside. There's no hint of affinity or shared experience in those speckled eyes, their slit-like pupils looking to us more like grimacing little mouths. Cephalopod physiology is utterly different to anything we find in mammals, fish, reptiles, or other vertebrates, with a nervous system wired to a bizarre and puzzling plan.

These differences speak of our evolutionary distance. While the common ancestor we share with most big-brained animals (including dolphins) was a reptile that lived around 320 million years ago, our branch of the evolutionary tree diverged from that of cephalopods fully 600 million years ago. Our common ancestor offered little hint of the forms its successors would take: it was a kind of flatworm with very simple neural circuitry. Cephalopods – molluscs that share a phylum with snails and oysters – thus represent an almost entirely distinct evolutionary trajectory for building intelligence. Philosopher Peter Godfrey-Smith says that our encounters with them are 'probably the closest we will come to meeting an intelligent alien.' It's surely no coincidence that so many fictional aliens, from H. G. Wells' Martians to the extra-terrestrial visitors in Denis Villeneuve's 2016 film *Arrival* (based on a short story by Ted Chiang), have been cast in their likeness.

Octopuses have around 500 million neurons in their bodies: comparable to the number in many dogs. But most of the neurons are not collected into centralized brains, as they are in vertebrates. Rather, they are spread throughout the body, mostly in the arms, where they form knot-like clusters called ganglia connected by a sparse web of connecting nerves that runs along and around the body somewhat like a ladder. The ganglia seem to act a little like mini-brains, autonomously controlling the respective arms while fed with information from their touch sensors. Octopuses do have a large central brain in the front of the body, but its structure is very different from that of mammals, organized into around forty lobes. Bizarrely, the creature's oesophagus, through which food is ingested, runs right through the middle of it, almost as if the brain were an improvised afterthought.

If cephalopods really do have minds, they offer one of the most concrete glimpses of what else might be contained in the Space of Possible Minds. 'If we want to understand *other* minds', says Godfrey-Smith, 'the minds of cephalopods are the most other of all.'

Marine tricksters

Octopuses are surely intelligent: they possess memory, problem-solving ability, even cunning and personality. They have an exploratory nature: they don't recoil from unfamiliar experiences but seem inclined to engage with novelty in an open-ended way, devoid of any immediate purpose or agenda we can discern. They figure out how to unscrew jars, navigate mazes, fuse laboratory lights with jets of water (they don't like brightness), escape from their tanks at the very moment their human wardens aren't looking. They appear to hoard items they find interesting, even if they lack any obvious use. Such curiosity, while not unprecedented in other animals, is unusual: familiarity is usually preferred over novelty in the unforgiving arena of nature. Some researchers think inquisitiveness might be one of the hallmarks of sophisticated minds, driven by an ability to develop complex internal models of the world that are then improved and refined by seeking new experiences and testing whether they turn out as predicted. It sounds like a potentially risky strategy, but it might suit some lifestyles. The octopus must manoeuvre in a complicated and ever-changing three-dimensional environment, and it doesn't have long to learn how to do that: some types of octopus live for just five months, and the typical lifespan is just one or two years (although some species may live for up to eleven years in the wild).

Octopuses are notorious tricksters, and seem to cause mischief for the sheer hell of it. An octopus kept at the Seattle Aquarium decided one night to dig through the gravel bed of its tank, rip apart the steel cables that attached a plastic sheet at its base to the tank's corners, and tear the sheet to pieces, leaving these floating in the water – all for no reason the keepers could imagine (unless it was sheer fury at being kept confined in a tank?). They named the creature Lucretia McEvil, after a pop song from the 1970s.

They are excellent navigators: able, in the often murky and rather featureless surroundings of a seabed, to explore their environment not just by taking a path and retracing its steps but by following looping trajectories. They rely partly on vision, although are unable to see well beyond a distance of five metres or so, but they also use chemical sensors on their arms and suckers to seek out prey. They have a smell organ below the eyes on the head, but we don't know much about how they use it.

Godfrey-Smith observes that octopuses seem to have a 'mental surplus': *far more mental capacity than they should need.* They use it in unexpected ways. Perhaps more than any other animal, octopuses seem to have designs of their own, which may subvert ours. Godfrey-Smith says he was drawn to study them because of their intriguing ability to 'turn the apparatus around them to their own octopodean purposes.' For example, in 1959 Harvard biologist Peter Dews trained three octopuses to pull a lever to get food. But one of the trio did it in an idiosyncratic, more forceful manner, and would squirt water at anyone who approached its tank. Some octopuses in captivity have been known to take a dislike (as we'd perceive it) to individuals, squirting them but not others.* They get to know you – to recognize individual humans – extraordinarily quickly, much more so than dogs. 'They "talk" to you, reach out to you', says marine biologist Michael Kuba, 'but only to people they know.'

* The fact that reports like these are mostly anecdotal, observed under widely varying conditions, is a problem for understanding octopus behaviour – as the saying goes, the plural of anecdote is not data. Much of the work is also done on octopuses reared in captivity in tanks, because observing them in the wild is generally more challenging than for land animals. But such creatures might behave differently from their wild counterparts.

This sensitivity to individuals seems all the more surprising given that octopuses are not naturally social beings: each ploughs its own furrow in the marine underworld, on what Godfrey-Smith calls 'a path of lone idiosyncratic complexity.' To put it another way: it's tempting to regard cephalopods as having oddly sophisticated means of expressing themselves while having rather little to express, or anyone to express it to. They don't do parenting: after hatching, the vulnerable young are left to fend for themselves. The relatively solitary existence of cephalopods might in itself be expected to produce rather different sorts of mind from ours. As we saw, it is widely believed that our brains, as well as those of other primates, 'smart' birds such as corvids, and cetaceans, were shaped in part by the selective pressures of their social environment: we all needed intelligence in order to live alongside others. If the minds of cephalopods lacked this socialized shaping, they are likely to be attuned in very different ways to ours. So it's little surprise that we are often left wondering, 'Why on earth did you do *that*?'

Experiments on animal behaviour have tended to be conditioned by what we expect from vertebrates, so with cephalopods we're often just guessing at their agenda. 'When I first saw octopuses play', says Jennifer Mather, 'I realized that we only saw it as play because it looked like *our* play.' And it surely does. Ethologist Ila France Porcher describes an encounter she had with an octopus while on a diving expedition:

> One night, a small octopus began shooting through my light beam . . . I stopped and it settled on the sand, its mantle spread around it. I walked my hand toward it, and it wrapped its tentacles around my forearm briefly, its light touches like those of a playful kitten. For a long time, we played this game of chase, and the octopus repeatedly encircled my hands with light, brief touches.

Mather sees such behaviour as exploratory, motivated by the question: 'What can I do with this object?' (Is there even an 'I' for the octopus? We'll see.) She suggests that, in addition to this curiosity and exploration about the affordances of the environment, octopuses are motivated by a fear impulse: 'Everyone is out to get me.'* In addition, its cognitive landscape might include flexibility in attaining goals ('That didn't work; how about *this*?'), as well as impulses of rivalry and possession, aggression, and a sense of ease with what is familiar (for even octopuses are reassured by that).

If our habitual division of mind and body looks increasingly unhelpful for understanding human cognition, it makes no sense at all for cephalopods. The body has, if not a mind of its own, then a set of proto-minds in each arm that make their own decisions.† The arms *can* be guided by the creature's eyes, but much of their movement seems to happen of each arm's own accord, informed by its touch and chemical sensory systems. The limbs seem even to possess some degree of short-term memory.

Because cephalopods have no skeleton, their movements can't rely on simple tendon-controlled flexing or rotation of hinged or ball-and-socket joints. Instead, they are controlled by a very complex system of muscles. Some of the muscles, for example, may stiffen not to cause contraction in themselves but to provide a rigid, bone-like platform on which other muscles can move. The resulting motor-control system is very complicated, but enables exquisite dexterity of the tentacles. Distributing the control of the limbs throughout the body might then be essential to operate a body like

* Many octopuses are cannibalistic, and females have been known to eat smaller males after copulation. So they're not being paranoid here.

† This is why the Korean tradition of eating octopus while it is still living can leave the diner trying to swallow a still wriggling tentacle, which can pose a serious choking hazard.

this: no centralized brain would be capable by itself of coordinating the actions.

Although octopuses have a small repertoire of standardized arm movements – for example, when crawling they use the rear four for pushing and the front four for exploring what lies ahead – we don't know what their governing principles are: whether, say, there is anything like a normal gait or a set of 'primitive' motions (like us raising the upper arm or flexing the elbow joint) that are combined and refined for specific purposes. Indeed, we don't really understand how the movements are organized at all. In general there seems to be no obvious logic in the order the arms are used to do stuff, although neither are all eight of them equivalent. (The animal is, after all, bilateral to some extent, with an eye on each side of the head.)

We're not even sure if the octopus 'knows' what its limbs are doing. The central brain might not contain any representation of the motor neurons in the arms, in the way that we have a network in the brain (the proprioception centre) that supplies an internal image of the body's posture.* As a result, octopuses may lack any strong sense of self. They don't seem to recognize themselves in mirror tests, as birds and primates do: perhaps to some degree they just 'watch' their limbs perform as if observing another creature in action. On the other hand, there are times when the commands do seem to come from the central brain. Godfrey-Smith thinks that occasionally the brain might assert control and literally make the octopus pull itself together, while at other times it is content to let the arms make up their own minds.

* It's not certain that octopuses lack all proprioception – but if they do have it, it seems likely to be much weaker than ours. The octopus *appears* to have a sense of where its arms are, insofar as it will grab random passing objects but not its own arms. But that self-discrimination is achieved by a sense akin to self-taste: the arms secrete a chemical that tells the suckers not to attach.

At any rate, these creatures challenge the notion that experience must be a tightly integrated whole. Even the bilateral brain of cephalopods seems only rather weakly woven together. If an octopus learns to do a task via information seen in one eye, it only very slowly becomes able to repeat it when triggered by the same visual input in the other eye, as though the training information diffuses sluggishly from one side of the brain to the other. This kind of hemispheric disconnect, such that the world of one eye is not immediately or perhaps ever available to the brain hemisphere served by the other, happens for humans only when some lesion or surgical intervention cuts the connective nerves between the two. Maybe the octopus mind is a little like that of these split-brain patients, in which the body sometimes responds to stimuli that are registered but not brought to consciousness. To wonder what it is to be like an octopus might then depend on which part of the organism we ask. 'Unity', Godfrey-Smith concludes, 'is optional, an achievement, an invention. Bringing experience together . . . is something that evolution may or may not do.'

One theory is that the ancestors of cephalopods had two entirely separate 'minds' – one for the eyes, one for the arms – that eventually linked up. Philosopher Sidney Carls-Diamante has suggested that, if octopuses are conscious, their consciousness might even now not be unified but could have two centres, in the brain and the peripheral arm nervous system (the very word 'cephalopod' aptly means 'head-foot') – or perhaps even a 'community of minds'. It would be like several beings in one body. Whether the cephalopod's experience is really this fragmented is by no means clear, however: the neural ganglia of the arms may not be complex enough to support autonomous sentience.

There are other dimensions to the cephalopod experience too. Some have 'smart skin', which can change colour to camouflage the animal, or to signal between individuals (for example to show aggression or as a sexual display), or to mimic other species. The mimic octopus *Thaumoctopus mimicus* can change its appearance to resemble more than fifteen of the species in its natural habitat. Octopuses and cuttlefish have colour-change cells in their skin called chromatophores: sacs filled with a coloured pigment, which may appear dark brownish or black when contracted but orange or yellow when distended like an inflated balloon. These shape changes are controlled by a system of nerves and muscles. Cephalopods can also alter the texture of their skin, again for camouflage, by controlling the size and shape of bumps called papillae, letting them resemble rocks or coral. Not all of the colour changes have a function we can easily fathom. Some octopuses and cuttlefish change their skin tone bilaterally to make one side dark and the other pale, but we simply don't know what that is all about. (It might have something to do with the fact that cephalopod brains too are bilateral, with the two sides performing different functions and therefore differing in how they will control their respective sides of the body.) The octopus can use its colour-changing body as a kind of movie screen – for example, producing dark patches that move as a whole over the skin like the shadows of passing clouds on the sea. This seems to be a ploy for startling prey, but again we don't know for sure. We watch in amazement, but with little comprehension.

Squid of the Loligidinae family, meanwhile, have colour-producing cells called iridophores in which microscopic stacks of protein layers act like diffraction gratings – rather like the arrays of tiny plastic pits on a compact disc – that reflect light of particular wavelengths, generating colour. A neurotransmitter molecule called acetylcholine released by nerve cells can trigger changes in the spacing of the layers, altering the colour reflected by the cells. These transformations seem to be controlled by the optic lobe of the brain, for example to mimic the appearance of the squid's surroundings as seen by its eye.

And yet . . . cephalopods can't actually see colour themselves! That's because they have only one type of photopigment in their eyes rather than the three (sensitive to red, green and blue) in our own. Perhaps colour vision is of limited value in the environment of these creatures, red light being largely filtered out from sunlight as it passes through the water. Yet still their visual system is able to control the colour of the skin, creating camouflage that is attuned to the colour-sensitive eyes of vertebrate predators and prey (such as fish and sharks). Given how prominent colour has been to philosophical discussions of the mind's subjective qualia (is your red the same as mine?), it's sobering to see a creature able to manipulate colour for its own designs *without even perceiving it.**

The skin (of some octopuses, at least) is itself sensitive to light, and seems to feed this sensory information to the brain. It would be too much to imagine that the animal truly 'sees' with its skin, but perhaps the skin can convey information about the ambient light intensity, shadows, maybe even local colours. This raises the weird prospect of the creature's brain receiving ever-changing and unanticipated sensory input from its autonomously wandering arms, so that the perceived world of the creature is shifting in a way it can't anticipate – unlike us, who know if we turn around (say) that our view will shift. 'When I try to imagine this', says Godfrey-Smith, 'I find myself in a rather hallucinogenic place, and that is everyday life for an octopus.'

Coordinating these movements and changes of appearance imposes a huge cognitive load on cephalopods. So while their

* It's not completely agreed that octopuses are fully colour-blind, however. Some researchers have speculated that the shape of their eye lens might separate out colours in the light they transmit using the phenomenon called chromatic aberration – an effect that, when it manifests in microscope lenses, may produce spectrum-like colours at the periphery of the image.

nervous system barely resembles ours, it's not surprising that it is of a somewhat comparable size. Perhaps that's the reason for the apparent 'mental surplus' of these creatures. Attributes like curiosity might then emerge as by-products of, rather than function of, the cognitive complexity needed for bodily and sensory control, much as our own ability to solve differential equations and compose odes may be a by-product of the cognitive demands for living in social groups.

How minds began

The mental sophistication of cephalopods suggests that the origins of mind and sentience may lie much earlier in evolutionary history than we might guess from looking at other large-brained creatures such as mammals and birds. Specifically, we might need to locate it at least as far back as that divergence of vertebrates and molluscs 600 million years ago: during the so-called Ediacaran period, predating the 'Cambrian explosion' of around 540 million years ago when most of the major animal groups appeared in an orgy of biodiversity.

This is not to say that the rudimentary elements of mammal, bird, reptile, fish, and cephalopod minds were all present in that primitive, flatworm-like common ancestor. It seems more likely that this creature contained only a rather simple nervous system: today's closest relatives to them, such as the tidepool-dwelling *Notoplana acticola*, have about two thousand neurons. Genuinely brain-like structures evolved only later, on different branches of the evolutionary tree. In other words, Godfrey-Smith asserts, evolution 'built minds twice over.'

Built them on what foundation? The basis of all animal minds is a nervous system, which allows information about the surroundings to be collected, processed, and used to inform a suitable response from the organism. But as bacteria and other single-celled microorganisms demonstrate, it's possible to be responsive to the surroundings even without nerves at all. So why did such cells evolve?

One theory supposes this was related to what was going on *inside* the organism. As bodies became multicellular and specialized into different tissues, organs, and appendages, each with particular functions, the actions of these components needed to be coordinated and synchronized so that the organism worked as a coherent whole – so that it was not, metaphorically at least, tripping over its own feet. A nervous system is not the only means for coordination – as we'll see, plants manage without one, although they don't have to achieve anything as complicated as locomotion. And the human heart synchronizes the pulsing of cardiac muscle cells via electrical signalling from cell to cell, travelling through the tissue as a coherent pulse. But communication between relatively distant parts of the body requires something more: in effect, a network of 'wires' that can transmit a targeted signal quickly and accurately.

The other possible reason to acquire a nervous system is *external* to the organism: in order to sense what's all around. Some of the earliest multicelled organisms, called ediacara* (whence the name of the geological period), were sessile (attached to a surface and lacking locomotion) and shaped like tubes or leafy fronds. These weird beings – which are not even easy to assign as plants, animals, fungi, or something else, given that all we have to go on are fossils – simply sat in place on the seabed and captured their food, such as algae and other microscopic organisms, from the water. There was nothing much else in their environment worth knowing: they just remained open all hours to a steady flow of nutrition. But once organisms became detached and free-moving, the whole game changed. Those creatures needed to find their way around, in particular to look for

* Pronounced 'ee-dee-*ack*-ara'.

food. Locomotion made sessile organisms themselves vulnerable: other organisms could now come and eat them. So they needed to spot predators and have some means of repelling or evading them. Life got a whole lot more complicated, and living things needed to sense and anticipate what lurked around the corner. They needed proto-minds – and crucially, those would have arisen in the presence and context of other proto-minds. In other words, minds seem unlikely to arise in idyllic Edens: it is stress and jeopardy that motivates their evolution.

Might nerve signalling have evolved independently for each of these distinct tasks: internal coordination and external sensing? No one knows. But it was surely a pivotal moment in evolutionary history when some primitive organism found that a single nervous system could serve both functions. Then the nerves could be bundled together into a central processing unit to which both the sensory and the motor systems – eyes, ears, muscles, viscera – are connected. The flatworm-like common ancestor of vertebrates and molluscs might have been rather like this. Those two groups of organisms might then have diverged as one branch unified the nervous system while the other left its operation more distributed throughout the body.

Nervous systems and brains are a costly investment: as we saw, they consume more energy than other tissues. Only when organisms reached a certain level of complexity, needing a diverse behavioural palette, would a mind worthy of the name become an asset, rather as a business in the 1970s could only justify buying so expensive a piece of kit as a mainframe computer once it was big enough. Otherwise, what is there for the machine to *do*?

But when does a nervous control system change from being an automated, stimulus–response, sensing-and-control network to becoming a mind – a system with sentience? Neurobiologist Simona Ginsburg and evolutionary biologist Eva Jablonka have suggested that the 'transition to experiencing' was a very gradual affair. No creature suddenly woke up to itself, and indeed the kind of mindedness Ginsburg and Jablonka imagine to have first appeared is more like 'overall sensation': a unified internal representation of the organism's situation, yet lacking any sense of self or awareness.

Such integrated sensory states could have arisen as nervous systems became increasingly interconnected into a 'nerve net' in ancient creatures similar to those in today's cnidarians* (such as corals, sea anemones, and jellyfish) and ctenophores† (such as comb jellies). Stimulate one part of this net, and the neural activity spreads throughout it. Initially, the two researchers think, this activity of the nerve net looked like random 'white noise'; but gradually it developed well-defined global states of activity. These overall sensations were just 'there' – they didn't tell the organism what to do, or provide motivation for action, or serve any other function.

Modern-day cnidarians do have neural states like this, for example which cause the organisms to contract or withdraw when their tissues are injured. These primitive creatures can also attune their responses to the circumstances, such as learning to ignore a repeated stimulus (that is, not bothering to withdraw) if it is found to do them no harm. This process, called habituation, is a kind of learning, and it saves the organism from wasting energy on unnecessary actions. It might be reasonably considered an *intelligent* response.

These global states of the nerve net would have needed to be quite general: it would be pretty pointless, or even detrimental, if every minute variation in the stimulus led to a different state, because the organism would only have a very limited repertoire of behaviours anyway. It would be rather like giving over-complicated instructions to a toddler rather than just saying, 'Can you fetch the cup?' So

* 'Nye-*dair*-ians'.

† '*Ten*-o-fors'.

certain states would have to be selected and stabilized out of all those that the net could support – as Ginsburg and Jablonka say, these would 'impart a certain tone on the indistinct buzz of ongoing neural activity'.

Eventually these neural patterns of 'experiencing' come to have *value* for the organism: a positive or negative valence, depending on whether they indicate circumstances that are good or bad for survival. That, say Ginsburg and Jablonka, is all that's needed for the first flicker of a 'feeling' to appear: the hallmark of 'basic consciousness'. At that point there is something (however faint or indistinct) it was *like* to be the organism.

To reach this stage, the creature had to be capable of learning – specifically, of *associative* learning, which means that it was able to associate a particular neural state with a particular kind of novel sensory input from the environment, and remember the connection between the two. This enabled the organism to move beyond fixed, instinctive responses and to develop ways of coping with new challenges by seeing how the situation makes it *feel*: the neural state can act as an internal evaluator of the situation being sensed. 'The function of feeling', say Ginsburg and Jablonka, 'is to alert and "inform" the animal about its present general state and to trigger adaptive behavior.' Any stimulus that tapped into the 'fear' state,* say, would trigger evasive action (hiding or fleeing). Triggering the 'hungry' state, meanwhile, initiated exploration for food.

'If an animal has integrated overall sensations', say the two researchers, 'and these direct and guide its actions so that it has the potential to learn flexibly by association, it can be said to be

* For such an organism, 'fear' surely wouldn't feel as it does to us – our fear is a complex response that is filtered by acculturation (see page 101). But whatever that 'fear' would have felt like, Ginsburg and Jablonka say that it felt like *something*, triggering aversive behaviour.

conscious even when this (very basic) consciousness is very limited indeed.' They think that neural states imparting values might even, by impelling new repertoires of diverse and flexible behaviour, have fuelled the explosion of new animal forms and the population of new niches in the Cambrian era around 540 to 515 million years ago. For although cnidarians don't display associative learning, it has been seen in the kinds of primitive invertebrates that first appeared in the Cambrian, such as arthropods, molluscs, and crustaceans.

With associative learning in place, natural selection had something to work with. Animals that developed a tendency to move away (say) when the 'feeling like that' meant a threat to survival – for example, if it was a feeling of being eaten – would have gained an adaptive advantage. If this smart, flexible, responsive behaviour of early animals relied on intimate interconnection of their nerves, we would expect there to be a selective pressure that produced ever more interconnection, binding the nerves into centralized clusters. In other words, it would have favoured the formation of what we can start to call brains, or what biologists call *cephalization*. These nascent brains would bind different types of sensory data together to produce a neural representation of particular objects with a colour, smell, texture, and so on, rather than leaving these as separate sensory streams. They would start to crystallize a distinct, experienced world out of the buzz of sensations.

At that stage the crucial feature of minds becomes possible, namely *representation*: an ability to build internal models of the world, based on memory and learning as well as immediate sensory data, that enable predictions about the outcomes of different behavioural options. The mind can have what we might even call *ideas*: a capacity for innovative decisions. A bacterium doesn't have that luxury: all it can do to move towards its goal, say, is to tumble randomly and see if that improves its situation. It can't think of a better idea.

Cephalization might also have driven a reorganization of the neural web in which the brain, in a head rich with sensory organs, is plugged into the rest of the body via a central nervous system. In other words, the body shape would at least in part have followed the dictates of the nervous system. The anatomy of cephalopods suggests, however, that there was another option for increasing neural complexity that didn't integrate all the cognitive processes into a single organ but left it distributed throughout the nervous system, maybe in distinct centres that could develop their own 'feelings'.

Vegetative states of mind

Cephalopods show us animal minds at their most alien, but we can imagine stretching the view of what a mind might be even further across the living world. The ability to receive information from the environment and to adapt behaviour accordingly is shared by even the most primitive of organisms, such as single-celled bacteria, and it is abundantly evidenced in plants. Until recently, however, most biologists maintained that this is merely automaton-like activity, and that no living entity without nerves – that's to say, nothing that is not an animal – can possess a mind.

But neurons have no monopoly on communicating with one another. Other kinds of cell can and do exchange information, typically by passing small 'signalling' molecules between them rather as neurotransmitter molecules signal across neural synapses. This allows cells to form communities or networks capable of information-processing: *computation.* The interaction and collaboration of replicating human cells to form a person from a zygote (the single-celled embryo) is commonly regarded now as a kind of computational process, as is the development of a bacterial colony or the growth of a fungus: the embryo, you might say, *computes* the organism. The question is when such coordinated cell activity merits being described as truly mind-like. Some biologists now argue that mindedness is an inherent property of most, perhaps all, living systems.

These suggestions have been made most forcefully for plants. At face value this might seem an unlikely notion, outside of eco-mystic circles. Plants have no nervous system and they are not mobile – they're literally rooted to the spot. They are a totally different kind of organism from animals, metabolizing and growing by using the energy of sunlight to drive chemical reactions that capture carbon (as carbon dioxide) from the air in the process of photosynthesis. We saw that Aristotle placed them only on the lowest rung of living things, lacking sensation but animated only by a 'vegetative soul'. Even today, plant biology has a decidedly (if unjustly) inferior status to the study of animal life.

But if we take the view that inferences about mind must ultimately be decided on the basis of *observed behaviour*, the matter is less clear. Watching a speeded-up time-lapse film of plant roots growing into the soil, you might be struck by how much it resembles the burrowing of a worm. If a root encounters some impenetrable object like a stone, it switches course. Plants may slowly turn to keep facing the sun as it passes overhead; at the end of the day, flowers gently fold in a manner that we even tell small children is 'going to sleep'. Much of this activity goes unnoticed either because it is literally buried out of sight or because it happens too slowly for us to perceive.

Evidently, such activity is purposeful: plants have goals, and are flexible and adaptive in their means of attaining them. In this regard they are no different to all living organisms, which evolution has endowed with aims and with abilities to realize them. The apparent intelligence of plant growth and movements struck Charles Darwin, who said that 'It is hardly an exaggeration to say that the tip of the radicle [root] thus endowed [with sensitivity] and having the power of directing the movements of the adjoining parts, acts like the brain of one of the lower animals.' He wrote an entire treatise on the abilities of climbing plants to sense and wind around poles, apparently using a touch-like responsive capacity. His son Francis helped him with some of his experiments on plants, and when Francis later became president of the British Association for the Advancement of Science, he said in his presidential address in 1908 that 'we must believe that in plants there exists a faint copy of what we know as consciousness in ourselves.'

If science was, by and large, unpersuaded by that idea, science-fiction writers loved it. Most famously, John Wyndham's *The Day of the Triffids* (1951) presented giant carnivorous plants that could move around and eat people.* The alien in the Howard Hawks movie *The Thing From Another World*, released in that same year (and remade by John Carpenter in 1982 as the visceral *The Thing*), was humanoid in shape but was actually identified as a plant. As a journalist in the film says when a scientist describes the alien, 'Sounds like you are describing some form of super-carrot.'

It's true that we can devise simple machines capable of some of the smart feats that plants achieve: photosensors linked to motors can produce sun-following motion, say. But plants are much more than automata enacting a simple program. They make decisions, and seem to do so on the basis of learning, as demonstrated by plant biologist Monica Gagliano for pea plants. Like many plants, these are phototropic: they grow towards the light. Gagliano and her colleagues grew them in a Y-shaped tube that presented the plant tip with a choice of two routes to follow, only one of which led to a 'reward' of light. At first a plant had to make the choice randomly

* This of course is the very thing plants cannot do (I mean move, although they can't eat people either). That's not merely being pedantic; pretty much all of the behaviour of plants is an evolutionary adaptation to this restriction.

and (figuratively speaking) hope for the best. But the pea plants seemed able to remember which of the channels had taken them towards the light, consistently taking that same route in subsequent trials.

What's more, the plants could be behaviourally conditioned, rather like Pavlov's dogs. Ivan Pavlov, a Russian physiologist, showed in the 1890s that dogs rewarded with food whenever they heard a bell would subsequently salivate in response to the bell, whether the reward arrived or not. His experiments stimulated the growth of the 'behaviourist' school of psychology that considered creatures to be little more than stimulus–response automata. Gagliano's pea plants were given a Pavlovian cue in the form of airflow in the Y tube: the airstream flowed into the passage that led to the light. These conditioned plants would then follow the airflow subsequently as if anticipating that it would lead them to the 'reward'.

One of the most 'animal-like' responses of plants is that of the flowering perennial *Mimosa pudica*, a member of the pea family native to South and Central America. *Pudica* is Latin for bashful or shy, because the plant's leaves fold inwards when touched or shaken, like a creature coiling up to protect itself from harm: the plant is also known as 'sensitive' or 'sleepy' plant, and as 'touch-me-not'.

In 2014 Gagliano and her colleagues showed that *Mimosa pudica* seems to exhibit the classic behavioural response of habituation (page 244), where it learns to ignore a repeated but inconsequential stimulus. If dropped a short (and harmless) distance, *Mimosa pudica* plants will reflexively fold their leaves – but the researchers showed that this response disappears on repeated dropping. This couldn't be just because the mechanical process of leaf folding gets worn out after repeating the action several times, as the plants would still fold their leaves in response to a *different* stimulus, namely shaking. It's as if they could tell the difference between the two stimuli and had decided that dropping was nothing to worry about, whereas shaking might be. What's more, plants grown in dim light habituated faster than ones grown with abundant light. This makes adaptive sense: under low light, energy is in shorter supply and so there is more incentive not to waste it on a costly action like leaf-folding unless absolutely necessary. The low-light plants also showed a longer memory for habituation, sometimes for over a month; this too suggests, the researchers said, that the plant adjusts its behaviour for 'when it matters'.

The interpretations of these experiments have been contested by some other researchers, who say the result don't unambiguously show genuine decision-making at work. But the real arguments started when Gagliano claimed that these capabilities meant not just that plants could exhibit surprisingly *intelligent* behaviour but that they were a signature of a kind of *consciousness*: of plant minds. 'The ability to learn through the formation of associations', she wrote,

> involves thc ability to detect, discriminate and categorize cues according to a dynamic *internal value system*. This is a subjective system of *feelings* and *experiences* . . . [Since] feelings account for the integration of behaviour and have long been recognized as critical agents of selection, plants too must evaluate their world *subjectively* and use their own experiences and feelings as functional states that motivate their choices.

The claim here is nothing less than that there is something it is to be like a plant.

That is some leap: from observations of sensitivity, memory, and learning to the assertion of a genuine mind. Some scientists who share this view have called for a new discipline of *plant neurobiology* – prompting dissent from others, who assert that the term is an absurd oxymoron because plants have no neurons.

Plant neurobiologists cite a growing roster of abilities plants display that are suggestive of cognition. For example, some plants seem able to count. The Venus fly-trap is a rare example of a carnivorous plant: it can close its leaves around insects that land on them, and then secrete a digestive enzyme called hydrolase which dissolves the hapless insect's tissues to supply the plant with nutrients. The sudden movement of closing the trap is energetically costly, and the fly-trap has evolved an ingenious scheme for ensuring that energy doesn't get wasted in false alerts. When fine hairs on the surface of the leaf are touched lightly, as by the body of an alighting insect, this triggers the generation of electrical impulses in the plant tissue that induce the trap to spring. But German biophysicists Erwin Neher and Rainer Hedrich and their colleagues have found that only when *two* such impulses have been triggered within the space of about half a minute does that happen – as if the plant figures that the first could have been an accident, but two of them signify that there's really something worth trapping.

Yet there's more. The fly-traps don't start to ramp up production of hydrolase and other proteins used to absorb nutrients until *five* electrical impulses have been produced. Making these digestive chemicals also requires precious energy and resources, and so the plant waits until the motions of the struggling insect it has ensnared 'convince' it that there's a meal in its jaws. So there must be a way for the plant to keep track of the number of previous impulses, up to at least five: it must have some kind of short-term memory that 'counts'.

The electrical signals of the Venus fly-trap are reminiscent of the action potentials produced by neurons. In fact, electrical signalling in plants is rather common. It can be induced by damage ('wounds') to plant tissues, for example, and also by coldness. The signals are caused in much the same way as in neurons: by the passage of metal ions such as sodium in and out of the cell interior via protein channels in the cell wall that open and close, setting up a difference in electrical charge – a voltage – across the membrane. Even more strikingly, the electrical communication between cells seems to be controlled by genes encoding proteins that bind the small molecule glutamate, one of the most common neurotransmitters in the synapses of our own nervous system.* So although plants do not have neurons, they have a cellular system that mirrors some aspects of how neurons work. The electrical signals in plants propagate along the phloem: channels made from back-to-back cells that are largely empty but interconnected by sieve-like walls. The phloem are traditionally thought of as the transport network for sugars that the cells need for metabolic processes – loosely analogous to our own vascular system for blood flow. But their ability to carry electrical signals also makes the phloem what Hedrich has called 'green cables' for carrying plant 'action potentials'.

Some have argued that plants can even experience something like pain: a feeling of aversion towards things that threaten their well-being and integrity. It's certainly true, and rather surprising, that plants are affected by substances that cause anaesthesia in humans. When exposed to common anaesthetics such as diethyl ether and xenon, for example, the Venus fly-trap no longer traps, *Mimosa pudica* leaves no longer fold when touched, and pea tendrils stop curling in response to contact with a support pole. What's more, plants themselves produce a range of chemicals in response to stress and wounding that induce anaesthesia in humans. One such is cocaine, used in the nineteenth century as a clinical anaesthetic until its toxic and addictive properties recommended its replacement by other, related compounds.

* Other plant hormones have also been postulated to act like neurotransmitters, in particular the molecule auxin that controls tip growth. Plants also make use of another molecule that serves as a key mammalian neurotransmitter, denoted GABA.

But 'feeling pain' is a very different matter to 'possessing an electrical signalling system for stress and damage'. The latter need be nothing much more than a clever biological thermostat or stress-sensor: you can go from stimulus to response in an artificial device without requiring the mediation of awareness. A signal does not in itself equate with 'pain', any more than the signal carried by the optic nerve from the retina equates with 'vision'. Pain has to be constructed in 'pain circuits' – and where, in the plant, are they? Some of those sceptical of plant neurobiology criticize its advocates for taking too simplistic a view of the brain, as if it is 'just' electrical signalling. If that were the case, needless to say, computers would have feelings too. (We'll get to that in the next chapter.)

Similar arguments for mindedness have been made for fungi – and here a comparison with the sprawling, interactive, and communicative networks of nerve cells is perhaps even more suggestive. The mushrooms that sprout in woodlands are merely the so-called fruiting bodies of vast networks of fungal filaments that extend in microscopic labyrinths through the soil and vegetation. Many fungi grow as webs of mycelium, a kind of connective tissue that spreads not only in soils but in coral reefs, in our bodies and those of other animals, in damp houses and under urban pavements and roads. The mycelial strands are called hyphae, typically five times thinner than a human hair, that extend through ecosystems and digest and absorb nutrients, conveying them from cell to cell along these intricate, sprawling transport systems. Many fungi exist in symbiotic relationships with plants, providing essential nutrients in exchange for the carbon compounds that, lacking the biochemical machinery of photosynthesis, they can't make for themselves.

Fungal hyphae, like plant cells, can generate and transmit electrical impulses, again resembling neuronal action potentials. Some researchers argue that the information transmitted in this way along fungal networks makes them capable of intelligent behaviour, memory, and cognition: in short, that the networks show some analogies with brains. Some are even exploring the idea of 'fungal computers' that might, for instance, serve as natural environmental sensors, distributed throughout the environment and detecting and signalling the presence of pollutants and other substances.

Atoms of feeling

Isn't it remarkable and suggestive that plants and fungi use electrical impulses analogous to those we see in nervous systems and brains? Well, perhaps. But electrical signalling is actually widespread between all kinds of cell. That's no great surprise in itself: it's a good solution to the challenge of getting cells to 'talk' and thereby to act in a concerted way that improves their survival prospects. Electrical communication is likely to have been a very ancient ability, for all cells use differences in the concentrations of ions across their membranes – and thus the production of voltages – to drive their energy-consuming reactions. Some researchers believe that harnessing 'gradients' in ion concentration across membranes was one of the fundamental steps in the origin of life itself, perhaps at least as important as replication and the transmission of hereditary information in genes.

It seems likely, then, that both neural activity in nerves and electrical signalling in plants arose as specialized cases of a more general process of cell communication. On this view, neurons are cells adapted *specifically* for that purpose, which may transmit such signals over large distances. The mere presence of such cells needn't by itself induce feeling and awareness: they may simply act as coordinators of hard-wired physiological responses to stimuli. Yet some researchers suspect that the very existence of electrical sensitivity – what physiologists once referred to as 'irritability' – imbues nerve cells with the ingredients of sentience. Informatics expert Norman

Cook has suggested that a nerve cell, by opening up its membrane to the inflow or exodus of ions that creates a voltage and supports an action potential, breaks the hermetic seal by which living cells distinguish their interior from the outside – and that this opening up to the environment is sufficient in itself to generate a 'proto-feeling'. It is then the synchronization of electrical firing between nerve cells that organizes 'the sentience of individual neurons' into 'the brain-level phenomenon of subjective awareness'. Brains, minds, and cognition are, in this view, an aggregate of 'atoms of sentience' in individual, excitable cells.

Cook, along with neuroscientists Gil Carvalho and Antonio Damasio, suggests that the excitability of nerves is thus a necessary precondition for sentience. An influx of ions, and the consequent creation of the action potential, happens in response to events outside the cell, and thus informs the cell that something is going on outside to which it needs to respond for its own good. They suggest that this responsive concern of the cell for its own existence seeds sentience. The absence of this behaviour in the switch-like components of computer microcircuits, they suggest, will then prevent sentience and mind ever arising even within the vastest array of such devices.

The implication is that perhaps there might even be something it is like to be a *single neuron*. It's hard to see how we could ever test that idea, however. And understanding how sentience might develop and deepen from the coordinated action of these proto-sentient atoms looks no less challenging. Take the nematode roundworm *Caenorhabditis elegans*, long regarded as a simple model system for understanding nervous systems since it has one that is composed of just a few hundred neurons. That makes it small enough to map out the circuit – the so-called 'connectome', which is identical in all *C. elegans* individuals – completely. The roundworm's connectome was fully mapped in the late 1980s, and became known colloquially as the 'mind of the worm'. The hope is that it might be possible fully to understand how such a simple and well-characterized circuit converts input to action: to disclose the information-processing mechanisms of a real nervous system as transparently as we can understand those of a simple logic circuit in a microprocessor, ultimately allowing us to create a 'virtual worm' that 'lives' in a computer simulation and responds to virtual stimuli just like its real-world counterpart.

Yet despite all our technological prowess in computer science and neurobiology, this goal still remains out of reach. Even a neural circuit this minimal is mind-bogglingly difficult to comprehend.

Biopsychism

The notion of plant minds remains not just unresolved but perhaps unresolvable. That doesn't make it pointless, though; on the contrary, it forces us to grapple with central questions about what a mind can be.

We should certainly be wary of excluding plants' minds by fiat. When neurologist Todd Feinberg and evolutionary biologist Jon Mallatt state that 'only multicellular animals with a nervous system and a basic core brain can be said to exhibit primary consciousness [a basic level of sentience] with any certainty', they may be right, but not because those attributes are in themselves *essential criteria* for consciousness (or, in its most basic manifestation, sentience). Rather, theories of how it arises tend to demand a particular cognitive architecture. Loosely speaking, say Simone Ginsburg and Eva Jablonka, it requires that several sensory modes supply information that gets integrated into a multi-dimensional map, which can be used to construct and evaluate hypothetical future scenarios so that actions can be decided on the basis of their likely outcomes and how well these meet the entity's goals.

Even if this is merely a necessary but not a sufficient requirement, the only organisms that seem to meet it, according to some critics of plant neurobiology, are (most) vertebrates (including mammals and fish), (some) arthropods (including insects and crustaceans), and cephalopods. This would imply that consciousness arose during the Cambrian era, as Ginsburg and Jablonka suppose – long after the last common ancestor of plants and animals (and fungi). Consciousness and mind, these researchers imply, should not simply be conjured out of information-processing that involves signalling and computation. They are hard-earned and rather special attributes.

Plant neurobiologists counter that this is a chauvinistic loading of the scales. As ever more evidence accumulates for the range of cognition-like faculties plants possess, they say, denying them intelligence and even mind can look increasingly like the sort of special pleading we once used to elevate our own species above all others. With each new discovery, we move the goalposts a little more and say, no, *that* is not true mind. Should we not be more broad-minded about minds?

Some plant neurobiologists assert that mindedness is in fact a fundamental property of *all* living things, a position called *biopsychism*. In this view, all organisms have minds of sorts merely by virtue of being alive.

The position that life *must* be sentient was argued a century ago by philosophers such as Henri Bergson. But much of the current advocacy draws on the work of Chilean biologist and philosopher Francisco Varela and his mentor Humberto Maturana. They developed the notion that living things are distinguished by being 'autopoeic': self-making. It is commonplace to refer to living cells as molecular factories – but unlike factories that churn out cars or televisions, the product of these living factories is themselves. Cells, said Varela and Maturana, are engaged in the constant business of making their own components.

This means that the environments of living entities take on a particular significance for them. They are not just 'outside influences', but are filled with things that the organism needs (such as nutrients and moisture), as well as things that it must try to avoid (such as predators and viruses) and things to which it has no reason to pay heed (such as cosmic rays and argon gas). This environment of *things that matter* to an organism is precisely what Jakob von Uexküll called its Umwelt. The mere existence of things that matter to an autopoeic entity will (the argument goes) create in it a *feeling* towards them: the basic ingredient of mind. Even bacteria that swim towards a source of nutrient by sensing the direction in which its concentration increases – a common process, called chemotaxis – have some kind of feeling about it, the first stirrings of desire or excitement. Biologist Lynn Margulis and her son Dorion Sagan asserted in 1995 that 'Not just animals are conscious but every organic being, every autopoeic cell is conscious. In the simplest sense, consciousness is an awareness of the outside world.'

Varela's student Evan Thompson expounded the biopsychist position in his 2007 book *Mind in Life*. Two of its leading supporters, botanist František Baluška and psychologist Arthur Reber, have said that 'without an internal, subjective awareness of [environmental] changes, without being able to make decisions about where to move . . . a prokaryote [the simplest forms of single-celled life] would be a Darwinian dead-end.'*

That claim is not obviously true, though. It is entirely possible for goal-directed decisions about, say, where to move, to be made by

* Reber adds that biopsychism avoids any need to postulate a moment of discontinuity in evolution when organisms went from being non-sentient to sentient. But no magical 'birth of sentience' is required anyway, just as few scientists today would suppose there was a moment when non-living matter became living. You might as well ask at what point a group of starlings becomes a murmuration.

entities without their having 'internal, subjective awareness' of anything, as we know from the Roomba robot vacuum cleaner.* Now, even a prokaryote has more sophisticated decision-making capacity than a Roomba, because it can often find effective responses to situations it has never encountered before or been programmed to deal with. That's very smart, and there are good reasons to speak of such versatility of behaviour as a truly *cognitive* ability. But it doesn't by itself demand or imply that sentience be involved. Simply postulating that feeling somehow emerges from an entity's interaction with the environment is not enough.

There might be a way to get a feeling, a what-it-is-likeness, even in such minimal circumstances, though: from the theory of consciousness called integrated information theory (IIT; page 136). Recall that the capacity for experience is, within IIT, deemed proportional to a quantity denoted Φ that measures the degree of 'integration' of information-transmitting networks. Since Φ is not (according to the theory's advocates) strictly zero even for the simplest organisms, neither is their awareness. Christof Koch has claimed that 'Some level of experience can be found in all organisms, including perhaps in [the parasite] *Paramecium* and other single cell life forms . . . Per IIT, it is likely that it feels like something to be a bacterium.'

But it's not obvious that we should attribute much significance to the distinction between a system that has an entirely minuscule value of Φ (which may well be the case for, say, *Paramecium*) and one that has none at all. By the same token, we might suppose that

* The Roomba is not autopoeic either, but it's not obvious why that should matter – we can imagine in principle making a machine able to construct itself from components scattered in its vicinity, but that needn't instil a felt *desire* for these parts. Indeed, chemists have made self-replicating molecules that can do this, but no one (to my knowledge) supposes they have any glimmer of sentience.

even if there is anything about plants, fungi, and bacteria that might plausibly be labelled consciousness, it is at so rudimentary a level that it's no longer clear whether such a definition means much. Just because we might create arbitrary thresholds in a continuum to define words such as 'mountain', 'lake', or 'crowd' doesn't mean the distinctions aren't worth making; it does not make a molehill into a mountain.

If biopsychism sounds akin to the old notion of vitalism – the belief in an animating principle unique to living matter – it's for the same reason that vitalism was invoked in the first place: absent an understanding of what mindedness is and how it arises, we are forced just to assert it as a mysterious, immanent property. Reber argues that this is in fact a *strength* of biopsychism, since it relieves us of the obligation to understand the origins of mind and sentience: we get them for free as part of the package of life. But mindedness shouldn't be given away so freely, simply defining out of existence the profound questions it raises.

What plants really exhibit, says Daniel Dennett, is *sensitivity*: a responsiveness to environmental stimuli, coupled to an ability to adapt aspects of behaviour accordingly. As cell biologist Dennis Bray puts it, 'living cells have an intrinsic sensitivity to their environment – a reflexivity, a capacity to detect and record salient features of their surroundings – that is essential for their survival.' Bray believes that these features are 'deeply woven into the molecular fabric of living cells' – that 'a primitive awareness of the environment was an essential ingredient in the origins of life.'

Yet perhaps 'sensitivity' doesn't quite do justice to these capabilities of cells, plants, and 'primitive' organisms; as Dennett says, even photographic film is sensitive. What we are really talking about here is *agency*: the ability to act on environment and self in a *goal-directed* manner that is not fully prescribed by stimulus–response rules. The claims of biopsychists seem to me to be trying to fill the gap left by our lack of a good understanding of how agency arises in biology and beyond. Not all agents have minds, but all minds have agency. I will return later to what a true theory of agency might look like.

All the same, information theorist José Hernández-Orallo is right to say that work on plants and fungi, as well as on bacteria and other primitive organisms, 'has raised questions about what cognition is, how *minimal* it can be, and even whether all life is cognitive in some way.' Indeed, Dennett and biologist Michael Levin have argued that life is 'cognition all the way down'. The differences are a matter of degree, not of kind.

When many minds are better than one

Perhaps it is cognition all the way up too: the discussion of where cognition and intelligence become mind continues beyond the organism. For intelligent entities that sense and respond to one another can display a higher-level *collective* intelligence that emerges from their individual behaviour but cannot be seen by focusing on that alone. The result can be more than the sum of its parts: swarms, flocks, and networks of living entities can display problem-solving, decision-making, and generally adaptive behaviours that are not available to lone individuals. These collectives seem to have gained new cognitive capabilities – perhaps even new dimensions of mind.

The benefits of using 'many heads' to solve a problem are familiar. Sometimes this is nothing more than a matter of fishing for the right mind: putting out a question on social media in the hope that at least one person knows the answer. In other cases it's a matter of sampling many independent views to find the ideal mean. That's the notion of the 'wisdom of the crowds', made famous from the example reported by Darwin's cousin Francis Galton of how the average of all the guesses at the weight of an ox in a fairground competition of 1906 was more than 99 per cent accurate.

But real collective or swarm intelligence is more surprising, dynamic, and inventive than this. It is what fish harness by swimming in schools in which the movements of each individual are sensitive to those of their near neighbours. The resulting collective motion enables information about a predator to be transmitted throughout the whole school in a flash, and thereby to trigger coherent collective manoeuvres that successfully evade the threat. The flocking behaviour of birds similarly ensures that information about a good perching location can be rapidly conveyed to all. Collective movements in human crowds might develop into counterflowing streams that reduce the chances of collision even though no individual consciously elected to move that way.

Plant communities certainly show *this* kind of intelligence. Some trees form complex root networks that intertwine with one another and send signals underground via plant hormones, influencing one another's behaviour in ways that might benefit the collective. Such communications between plant root systems are typically mediated by so-called mycorrhizae: symbiotic associations with fungal networks, which are capable of many of the functions of brains and nervous systems such as memory and decision-making. What happens to one tree, for example, can be 'remembered' by another, recognizably imprinted in its growth rings.

'I've come to think', says forest ecologist Suzanne Simard, 'that root systems and the mycorrhizal networks that link those systems are designed like neural networks, and behave like neural networks.' She and others are quite happy to describe these complex, wired ecosystems in terms not just of intelligence and communication but also cognition and even emotion.

This doesn't imply that there is any kind of unique and coherent 'hive mind' directing such collectives. But it does suggest that a Space of Possible Minds may include territory beyond the organism, where emergent cognition becomes more than the sum of its parts.

After all, what is a brain but an organized collective of components (neurons) that in themselves seem unlikely to have any significant degree of mindedness? What is a mind in the first place, if not an emergent phenomenon?

The ethics of biopsychism

The impulse to anthropomorphize life has a long history. The great Swedish taxonomist Carl Linnaeus spoke in the eighteenth century of plant pollination in racy terms appropriated from marriage: 'When the bed is thus prepared, it is time for the Bridegroom to embrace his beloved Bride and surrender his gifts to her.' With his talk of 'promiscuous intercourse' and 'polyandry' among plants, Linnaeus scandalized his contemporaries.

Some critics of plant neurobiology and biopsychism today dismiss such ideas as a return to this kind of 'Romantic biology'. In the end they may be an expression of how difficult it is for us to perceive the goals, purpose, and agency that arise in evolution *without* attributing them to a mind. But advocates of plant sentience say that there is a moral and ethical dimension to their position. It would certainly be problematic for agriculture if we concluded that plants feel pain when we pull them out of the ground, peel and chop them, and eat them. As we saw, possession of a mind is rightly deemed to confer on organisms a moral status, thereby imposing on us moral obligations towards them. But plant neurobiologists are surely right to say that it would be good if we cast a wider ethical net: if we treated plants and their ecosystems with more respect and even with reverence.

Yet arguing for the sentience of plants (and fungi) might be a questionable way to encourage that – for if the argument is deemed too flimsy, our sense of obligation might vanish with it. It's simply not possible for us to preserve all life, not least because life has never worked that way. If we elect not to kill pathogenic microbes with antibiotics, we fail to respect the lives of our fellow humans. We need organic sustenance, which has to be grown from living matter even if that is pulses and grains or lab-cultured meat. But none of this need obviate a generalized respect for life and an obligation to preserve diversity. We need that variety of nature on purely selfish grounds, although it will be to squander the capabilities of our own evolved minds if we do not also learn to value for its own sake this remarkable ability of matter to become animate. None of this appreciation and respect need be contingent on a notion that mind pervades life.

Perhaps the ultimate example of collective intelligence – and some would say, of Romantic biology too – resides in the suggestion that the entire planet Earth is a living, and perhaps even a minded, entity. That is the extreme version of James Lovelock's Gaia hypothesis (which he developed partly in discussion with Lynn Margulis), in which the connections and feedbacks between the living and the non-living world – the biota, oceans, atmosphere, cryosphere, deep Earth – give rise to a self-regulating planetary entity.

Lovelock's ideas were considered maverick when he first broached them in the 1970s; today many of them are mainstream, and they help to inform research on climate change, seen now as involving a wealth of planetary processes in which the interactions are sometimes hard to discern and predict. It contributed to the initial scepticism that Lovelock personified the hypothesis with the name of a Greek goddess. Although that personification has been neutralized by the adoption, among many planetary and ecosystem scientists, of the term 'geophysiology', still the discussion readily drifts into teleological and minded territory with suggestions along the lines that 'Gaia does not care about us' and that she will live on even if we, in our stupidity, engineer the conditions of our destruction.

Does this Gaia have a genuine mind? It's hard to see how, within the network of complex biogeophysical interactions, there is a place where mindedness might cohere: where integration and broadcast of the information Gaia 'receives' about her internal state and her cosmic environment might occur. But she surely remembers, she adapts, she senses and responds. We can reasonably award her a place in the Space of Possible Minds.

For I believe that thinking about this Mindspace might clarify some of the arguments about the limits of mind in the living and natural world. Yes, my initial definition of mind requires there to be at least some spark of sentience, something it is to be like – and we simply do not know if any such property exists in plants, let alone in planets. But if we recognize minds as having various capacities to a greater or lesser degree – learning, memory, integration, consciousness – then we can place such entities in this space even if we decide to set them to zero on the axes that somehow represent the parameters of consciousness and feeling. Whether that admits them as real 'minds' seems less relevant than being able to construct such a comparison with other minds, and to acknowledge some of the properties they share with them. The goal is not in the end to guard the gates of any Mind Club, but to explore what is out there.

CHAPTER 7

Machine Minds

'The development of full artificial intelligence could spell the end of the human race,' the legendary theoretical physicist Stephen Hawking warned in 2014. That he had no expertise in AI did not prevent the claim from making headlines; after all, Hawking was Hawking. But he's not alone: tech entrepreneur Elon Musk has warned that AI could be 'our biggest existential threat.' Such concerns were reiterated in an 'open letter' by seemingly more qualified experts early the following year. AI leaders at Google and elsewhere, roboticists, and academic specialists in computer science were among the cosignatories who said that 'we cannot predict what we might achieve when [our] intelligence is magnified by the tools AI may provide' – but warned that 'our AI systems must do what we want them to do', for otherwise the consequences could be dire.

Science fiction is filled with tales of AI *not* doing what we want it to. From the machines of Isaac Asimov's *I, Robot* (1950) to HAL 2000 in *2001: A Space Odyssey*, and Skynet in the Terminator film franchise, to the malign super-intelligent AIs of *The Matrix*, we seem positively to revel in the nightmare that our machines will conquer, enslave, and eradicate us.

Our pessimism about the prospects of AI could seem odd – not least from Hawking, who latterly relied on it and related technologies to permit him any kind of communication with the world. He and others who warn of the impending dangers of AI do generally acknowledge the huge benefits it might bring – the open letter mentioned above stated that it 'has yielded remarkable successes in various component tasks such as speech recognition, image classification, autonomous vehicles, machine translation, legged locomotion, and question-answering systems', adding that 'The potential benefits are huge . . . the eradication of disease and poverty are not unfathomable.' Still, we worry if we're doing the right thing by developing machines that might rival or surpass us in intelligence.

This fear has old roots. It stems partly from a distrust of art and artifice, a conviction that 'artificial' means 'unnatural' and is thus bound to go awry. The malevolence of the 'created being' finds voice in Mary Shelley's *Frankenstein* (1818), and ever since robots (from the Czech word meaning 'serf' or 'indentured labourer') made their debut in Karel Čapek's 1920 play *R. U. R.* they were portrayed as soulless beings that might wipe out humankind. Asimov called this fear the 'Frankenstein complex'.

The distrust of AI is also a fear of the Other, of the unknown. Because we don't understand the AI's 'mind', we project onto it the worst that lurks in our own. The result, though, is a peculiar prospect: those who seem most troubled by our lack of knowledge about the kinds of mind AI might conceivably harbour appear to be oddly sure they know what it will (or at least could) be like.

Yet while some futurologist-prophets insist on forecasting the worst, many of those working at the pit-face of AI research tend to regard these dire warnings with disdain, even despair. In general they worry not that AI will become too powerful for us to control, less still that it will develop a wicked mind of its own, but rather that by overestimating its capabilities and overlooking its faults, we will end up using it inappropriately. Some of them think the dangers might be lessened if today's AI was in fact *more* powerful, more capable, and perhaps more human-like, than it currently is.

Yet is it such a good idea to engineer a convergence of human and machine minds? Another recurring fear in AI narratives is that it will become impossible to distinguish 'thinking machines' from humans, so that beings like Ava from Alex Garland's *Ex Machina* (2014) will walk among us undetected. Either they will be truly conscious (which is to say, minded) entities, or they will do so good a job of emulating minds that we will never know the difference. But is that a realistic scenario?

Given how much concern is expressed by scientists, technologists, futurologists, and the public about the dangers of AI, it is odd how little attention is given to the question of what kinds of mind AI is likely to acquire. Take the 'Asilomar AI Principles' developed by the Boston-based Future of Life Institute, which attempt to codify the safe development of AI. (The name alludes to the 1975 Asilomar conference in California, which sought to steer responsible development of biotechnology.) Endorsed by a dizzying array of luminaries including Hawking, Musk, and many leading figures of AI, the principles stipulate that 'Highly autonomous AI systems should be designed so that their goals and behaviors can be assured to align with human values throughout their operation'. Yet these principles seem more concerned with managing and directing AI research than with considering the fundamental nature of what is emerging from it: how AI 'thinks', now and in the future.

Perhaps some of this neglect stems from the misapprehension that we already know the answer. Jaan Tallinn, a co-founder of the Future of Life Institute, says that 'Earth's current environment is almost certainly suboptimal for what a super-intelligent AI will greatly care about: *efficient computation*.' But imagining we know already what a truly super-intelligent AI 'will greatly care about' could be rather like supposing that ants can guess our own hopes and desires. Whether we want to anticipate their dangers or their possibilities, it would seem wise to start getting to know our machines a little better.

The origins of machine minds

Let's be clear: every robot and computer ever built (on this planet) is mindless. I feel slightly uneasy, in asserting this, that I might be merely repeating the bias against animal minds that I have so deplored in the earlier chapters. All the same I think the statement is true, or would at least be generally regarded as such by experts in the field. Siri and Alexa do not feel bad when you are peremptory to or mocking of them, because they are not personalities but just AI algorithms. Today's robots do not need rights or moral consideration. If we throw them on the scrapheap, the only guilt we should feel is at the waste of materials and energy that went into making them and despoiling the planet in the process.

Why am I so sure about this? The answer is that the distance between the well-understood principles of logic-circuit design and the manifestations and speculative theories of consciousness is too vast. Computers today, for all their staggering capabilities, remain closer to complicated arrays of light switches than even to rudimentary insect brains.

There are two caveats. One is that there is absolutely no guarantee that the two are fundamentally distinct classes of object – that something truly mindlike could not arise in a device built in a factory from transistors and other electronic components. The other is that we don't fully understand even how today's AI does its job: it is to some degree a black box, which works (and sometimes fails) without our being able to say exactly why.

I suggest that we think of the latest AI systems as a collection of proto-minds, metaphorically (probably no more than that) akin to the nervous systems of those early Ediacaran organisms. We do not know if they are really the kind of stuff that can *ever* host genuine minds – that there can ever be something it is like to be a machine. We can and should, however, begin to think about today's computers and AIs in *cognitive* terms: to treat them conceptually (if not morally) *as if they are minds*. Indeed, some researchers are convinced that this is already the best way to try to understand how they work.

It's one of the oddities of artificial intelligence that the field has long prophesied, or even claimed to have achieved, human-like capabilities in its products by automating precisely the kinds of computational task that are the *least* characteristic of what the human mind is good at: ones that tend to depend on the application of strict logical rules and can be solved by laborious number-crunching. This reflects the rather perverse but persistent notion that 'intelligence' refers to quasi-mathematical reasoning and problem-solving, itself a relic of the Enlightenment trust in 'reason' as the proper measure of the human mind. Time and again the apparent triumph of the computing machine – in numerical calculation, chess-playing and so forth – has prompted headlines claiming that humans have now been superseded and the machines are about to take over, only for us to realize that these machines are idiot savants of the narrowest kind that fail at anything other than their designated task.

Researchers in AI and computer science are under no illusions about the limitations of their machines. All the same, the prejudice persists about what 'intelligent' consists in – as evident, for example, in efforts to develop generalized theories of intelligence based on information theory and the manipulation of binary data.

But the bias towards making machines that compute with numbers and logic also has a sound rationale: it's precisely because we are not terribly adept at such things that mechanizing the maths is a good idea. The abacus is generally regarded as the earliest 'calculating machine', and several mechanical 'reckoners' were devised from the seventeenth century, notably by the German mathematician Gottfried Leibniz and the French philosopher Blaise Pascal. These were mechanical devices with moving parts like cogs and wheels that were positioned and powered by hand.

As the age of steam power dawned, inevitably someone decided that this motive force could be enlisted to turn the cogs of calculating machines. That person was the British mathematician and polymath Charles Babbage. When he and his friend the astronomer John Herschel found errors in certain calculating tables in 1821, Babbage is said to have exclaimed, 'I wish to God these calculations had been executed by steam.' To that end he designed a machine called the Difference Engine for solving so-called polynomial functions: algebraic equations that cropped up in calculations for navigation. Babbage subsequently tried to make a more versatile machine, the Analytical Engine, which could be *programmed* to carry out all kinds of calculation. In correspondence with Lord Byron's mathematically gifted daughter Ada Lovelace, Babbage decided that the programming could be done using cards with holes punched in them to set the machine's levers into the right positions for each task. This was a massive conceptual leap, introducing the idea that computation can be generalized. A set of fundamental components that carry out simple mathematical operations such as addition and subtraction can, if suitably configured and activated again and again, compute almost any mathematical task.

Babbage's Analytical Engine was never built: it was too vast, cumbersome, and expensive. It's frequently remarked that the first *computers* were instead humans, enlisted to conduct routine calculations for problems in pure maths, navigation, meteorology, ballistics, and astronomy. Often they were women; at the Harvard Observatory, the entire computer team from 1880 was female. By carrying out a specific sequence of mathematical operations divided into manageably small steps, these human computers were enacting *algorithms* – well-defined sequences of logical steps for performing a task or solving a problem – much as Babbage's Analytical Engine was intended to do.

In the 1940s, electrical engineers figured out how the logical operations involved in the calculations that human computers were undertaking could be enacted instead in electrical circuits that used diodes or 'valves', in which incoming signals could be systematically converted to outputs according to strict rules. The circuits manipulated sequences of electrical pulses representing the digits of binary numbers ('bits'): 1 (a pulse) and 0 (no pulse). In other words, the calculations and operations were *digital*. Diodes could be assembled into various logic gates, each of which produced a specific output that depended on the inputs. An AND gate, for example, outputs a 1 if both of two inputs are the same (both 1 or both 0), but outputs a 0 if the two are different (10 or 01). Such digital circuits are in effect carrying out mathematical operations represented as strings of 1s and 0s.

The first digital computer, called the Electronic Numerical Integrator and Computer (ENIAC), was built at the University of Pennsylvania between 1943 and 1945. Unveiled in the press as a 'Giant Brain' or 'Magic Brain', it showed that the mythology of the 'machine mind' was present from the outset. Most of ENIAC's programmers were again women, for whom this seemed a natural progression from their role of human computers. The predominance of women remained in computing technology until around the 1960s – but by degrees it became a male-dominated field, eventually aggressively so. The lack of female coders and entrepreneurs in Silicon Valley today has become notorious, not least because it seems to be perpetuated by a misogynistic work culture. How this gender imbalance has skewed the norms and goals of information technology has been much debated, but it seems safe to conclude that the nature of computer 'minds' the industry has so far produced bears the imprint of a lack of diversity in those of their designers.*

One of the earliest suggestions that there might be more to machine intelligence than number-crunching came from Ada Lovelace. She felt that devices like Babbage's Analytical Engine might ultimately be used to produce poetry. Lest we get too misty-eyed over that apparent prescience, bear in mind that Lovelace seemed to regard the composition of poetry as just a further level of complexity beyond solving quadratic equations – a suggestion that the human mind produces what it does simply because it has more cogs than Babbage's device. (We will never know what Lovelace's father, who died in 1824, would have made of her algorithmic view of poetry.) We will see later how close today's AI has come to achieving Lovelace's goal.

ENIAC might be considered the electronic analogue of the Difference Engine: a machine hard-wired to do a specific task. In the 1930s, however, the British mathematician Alan Turing imagined giving such devices the scope of the Analytical Engine. He conceived of a general-purpose, programmable digital computer.

In a paper of 1950 (published, appropriately, in the journal *Mind*), Turing speculated about where such a machine could lead. The paper was called 'Computing machinery and intelligence', and it asked the question: 'Can machines think?'

Turing suggested that the question was in fact rather meaningless, and he proposed to replace it with something more precise and testable. Might we, he asked, be able to make a machine that can generate responses to questions from human judges that are indistinguishable from those a real human might offer? Turing called this

* The vitriol aimed at feminist or simply at female critics of tech culture, meanwhile, attests to the toxic investment by some of its advocates in its 'masculine' aspirations and ideals.

test the 'imitation game'; it is now more commonly known as the Turing Test.

Turing fully expected that it would be possible to develop digital computers, like the one he was then experimenting with at the University of Manchester, to the point where they could pass the imitation game. Even with the technology currently available, he said, it was feasible that computers might be given memories with a capacity as vast as 10 megabytes!* Imagine what might then be possible.

But could such a device be truly said to 'think'? Turing argued that, if indeed its behaviour in the imitation game could not be distinguished from that of humans, the only grounds on which we could deny it that ability are chauvinistic ones: why, surely only humans can think! 'I believe', he wrote, 'that at the end of the century the use of words and general educated opinion will have altered so much that one will be able to speak of machines thinking without expecting to be contradicted.'

Turing's 1950 paper is packed with insight (as well as with wit). But it also foreshadows the tenor of the discussion about artificial intelligence that persists even today, in that it frames that question 'Can machines think?' in terms of an unspoken assumption: Can machines think *like us*? In this chapter I am more interested in asking how machines can and will think in ways *unlike us*.

Where are we going?

The early days of artificial intelligence are sometimes remembered now as a golden age of the field – not because of its achievements, which were modest in the extreme, but because nothing seemed

* As you might realize, this is roughly the capacity needed to store a single digital photo at modest resolution.

impossible and there was everything to play for. Computers themselves were new and revolutionary, promising to liberate us from the limitations of the feeble human mind. It amuses my children to hear about my own first efforts at programming (as we called coding back then) in 1973, writing out the code by hand on paper charts so that technicians could convert it into stacks of punched cards like those envisioned by Babbage and Lovelace, which were fed overnight into a device that filled an entire room and which I barely ever saw. The next morning, the paper printout from this mysterious machine revealed the bugs in my program, and I started over. It would have seemed a cumbersome way to build a mind.

The phrase 'artificial intelligence' was coined in 1955 by the American academic (one can hardly call him a computer scientist when such a job description barely existed) John McCarthy. A mathematician by training, McCarthy submitted a proposal to the Rockefeller Institute, a philanthropic supporter of the sciences, to organize a workshop that would bring together experts from academia, industry, government, and the military who were thinking about potential applications of computers. McCarthy and his fellow organizers suggested in their application that 'every aspect of learning or any other feature of intelligence can be in principle so precisely described that a machine can be made to simulate it.' The purpose of the workshop would be to figure out how to do that. That endeavour needed a name, and on grounds hardly more substantial than that he had to call it *something*, McCarthy proposed 'artificial intelligence'.

Whether that's a good name for efforts to create 'thinking machines' is debatable, largely because of the ambiguity of the second word (and perhaps the first). If one accepts that intelligent behaviour demands more than just a superfast calculating device, what more is needed exactly? It is easy with hindsight to mock the insouciance with which the organizers of the workshop, which took place in Dartmouth, New Hampshire, in 1956, refer to 'learning or any other feature of intelligence' – and appear to associate those 'other features' primarily with functions such as reasoning and language.

Seen in retrospect, the limited scope of 'intelligence' initially prescribed for AI has probably restricted the field to this day; only recently have we finally started to appreciate the subtlety and elusive sophistication of what the human mind does. It's tempting to read the same naivety into the wildly over-optimistic pronouncements made by some of the pioneers in those early years. Herbert Simon, an attendee of the Dartmouth meeting and later a Nobel laureate in economics for his work on complex systems, declared several years later that 'Machines will be capable, within twenty years, of doing any work that a man [sic] can do.' McCarthy's own program, the Stanford Artificial Intelligence Project, hoped to build 'a fully intelligent machine in a decade.' Marvin Minsky of the Massachusetts Institute of Technology, another luminary of the Dartmouth meeting, proclaimed that 'within a generation, I am convinced, . . . the problems of creating "artificial intelligence" will be substantially solved.'* By the time we watched HAL 2000 trying its damnedest to kill Keir Dullea's astronaut en route to Jupiter in 1968, you could be forgiven for imagining – and fearing – that such dastardly machine-minds were just around the corner.

But there's no virtue in jeering now at these predictions. When scientists make seriously mistaken predictions, it's usually not because they are simply ignorant or foolish but because there's some revealing error in their assumptions. Here the problem was that

* Arguably this might be because Minsky wanted to set a low bar. In 1968 he defined AI as the 'science of making machines capable of performing tasks that would require intelligence if done by [humans]' – in other words, it counts merely to simulate the appearance of intelligence, not to necessarily produce it.

notions about the brain and cognition were themselves heavily entrained in the conception of the brain, and of 'thinking', as a form of computation. To put it crudely, if the brain creates thought and intelligence, and the brain is just like a computer but with more components, we'll produce something like it by making our computers bigger and more powerful. This remains a popular view in Silicon Valley even today, where, as Christof Koch puts it, the fantasy is that machine consciousness is 'only a clever hack away'.

We saw earlier (page 34) why there are problems with drawing a simplistic analogy between the brain and a computer. The deeper problem, however, lies with the notion of the *mind* as the disembodied and abstract software of the general-purpose intelligence-machine of the brain. All minds that we currently know of have been tailored by evolution to operate within the organism that houses them. This means that 'artificial intelligence' does not have a sufficiently constrained meaning to really constitute the coherent objective of any single research programme. Even if we take it to mean 'a machine that has the capabilities of the human mind', the fact is that human minds are 'designed' for operating, and operating within, human bodies. If the representations of the world that a mind produces were not designed for an organism in a body (and specifically *this* type of body), observes Antonio Damasio, the mind would surely be different from ours.

Why, anyway, should we wish to ape the human mind in all its glory and foibles? In fiction, and sometimes in fact, we tend to conceive of a human-like AI as being like us except in respect of emotion – but as we've seen, that does not correspond to a human mind minus emotion, but rather to something that is not really human-like at all, and quite possibly to something that will work very poorly (if at all) when presented with the kinds of cognitive challenges we face.

If we mean rather that we want to build a machine with *some degree of mindedness*, then we have to ask the very question that this book is posing: what are the parameters, the requirements, and the functions of mindedness?

Most research on developing real computer systems that are considered today to fall within the umbrella of AI doesn't talk much about this issue of mindedness. It seeks systems that do specific jobs better or faster than we can – the same goal, in other words, that engineering has pursued since time immemorial. As a result, AI has indeed become a misnomer for the field. For outsiders it invites the belief that these are 'thinking machines' like those they see in the movies, as for example in Steven Spielberg's 2001 film *A.I.* For practitioners today, the term is often used almost as a synonym for computer systems that can be trained to do a particular computational task well in the process called machine learning, which typically seeks for patterns and correlations in data sets too vast for human minds to encompass and assimilate. Machine learning is, as we'll see, a kind of glorified pattern recognition – which is not to denigrate this fantastically useful ability in areas such as medicine, image processing, language translation and speech recognition. Today's AI tends to mean Google Translate, unnervingly astute recommender systems, and medical diagnostic systems such as IBM's Watson. It does not mean HAL 2000 or *A.I.*'s 'Mecha' humanoid robots.

And it means that AI researchers don't have a goal; they have hundreds of them, and mostly they just get on with the job in hand. A 2016 report on AI produced at Stanford University argues, surely with some validity, that 'the lack of a precise, universally accepted definition of AI probably has helped the field to grow, blossom, and advance at an ever-accelerating pace.'

But its shortcomings are now prompting some practitioners to look back to the roots of the field and wonder: did we take the route that will actually lead us to 'a fully intelligent machine', a *machine mind*, even if it takes longer than we first thought? Or will we, for that, need to find a different path?

And by the same token: is the route we're now on taking us, not towards a human-like machine mind, but to a different sort of mind altogether?

Symbols

AI research unfolded at first with a traditional engineering mindset: you design systems to do what you want them to and according to principles that you understand from the bottom up. So the dominant strategy became to figure out the rules of intelligence, and then to program them into the machine. This was framed as a problem of manipulating symbols – which might, for example, represent words or other items of data. It was called symbolic AI, and it worked along the familiar lines of computer algorithms: *if*, say, the inputs are X and Y, *then* the output should be Z. The issue is that of figuring out the rules by which that solution is arrived at. Some initial efforts to develop AI programs, for example by Herbert Simon and his colleague, cognitive scientist Allan Newell, attempted to identify these rules by getting students to enunciate their thought processes as they solved logical puzzles.

No one imagined that it would be this simple to create a machine with 'general intelligence', that ill-defined notion supposed somehow to equate with the capabilities of a human. But many hoped that the challenge could be cracked piece by piece: by focusing on individual capabilities involved in intelligent problem-solving for many different types of problem, and then combining them.

That prescriptive symbolic approach might work well enough if you're carrying out a mathematical calculation. But *thinking* is not like that. Of course, we do make some decisions this way: *if* I am out of milk, *then* I will go to the shop. But, well, it's raining and I'm tired, and so maybe I'll have my coffee black. Or perhaps I'll borrow some from my neighbour . . . In other words, even for seemingly simple decisions like this, an entity that *thinks* rarely tends to reason in perfectly transparent, linear, and well-defined ways governed by rules that can be formalized. Imagine what it's like trying to identify rules that will tell a machine how to respond if it is confronted with a demand like the one proposed by Turing in his seminal 1950 paper: 'Please write me a sonnet on the subject of the Forth Bridge.'*

So symbolic AI didn't get very far towards the goal of making a 'thinking machine' worthy of the description. It did however produce some impressive feats of engineering, such as the robot called SHAKEY built at the Stanford Research Institute. SHAKEY could manoeuvre itself autonomously around its environment using an array of cameras and laser sensors, so as to perform some task such as shifting boxes in an office. But even with such relatively simple challenges, SHAKEY could prove very slow and unreliable.

And even for strictly logical, maths-type problems, this sort of AI could come unstuck once the challenge acquired much complexity. One problem is that, even with small numbers of variables or symbols, the number of possible combinations quickly becomes so large that the computers of the 1960s and 1970s couldn't enumerate them in any reasonable time. That was a rather general problem in computing science, and indeed remains so today even with the most powerful supercomputers: there are some problems where the numbers of possible solutions to be evaluated in order to find one that works explodes exponentially. The prototypical example is factorization: breaking down an arbitrary number into its prime

* Sure, you could program the device to respond to any task that is unfamiliar or beyond its competence by identifying the nature of the task X ('sonnet' = 'poetry') and then responding (as Turing suggested), 'Count me out. I never could do X.' But that's not much help to anyone.

(indivisible) factors, such that 15 becomes 5×3. As the numbers get larger, finding factors by brute-force trial-and-error becomes an astronomical task. It's because of this very intractability in factorization that it is used to encrypt confidential information – such as your banking details on the internet – so that, without knowing the secret key for decryption, no eavesdropper could decode the information with the computer resources currently available.

When it came to intelligence about the world at large, this kind of AI looked hopeless. What you might reasonably want from a box-moving robot is to be able to say – literally, to voice the instruction out loud – 'Pick up those cardboard boxes and stack them by the window.' It's the kind of instruction a four-year-old child could follow, but there's an awful lot in it to expect from a machine, even if it has an efficient voice-recognition system to decode the command.* How does it recognize a 'cardboard box' (especially if there happen to be plastic boxes nearby too)? What is this notion of 'stacking'? What exactly does 'by the window' mean? What is a 'window' anyway? And what if you've left your full cup of coffee inadvertently on top of the boxes, or the cat happens to be sleeping in front of the window and you've not noticed?

In short, a useful AI system needs to have some knowledge about the world: knowledge that we acquire from infancy without our even noticing, but which you somehow have to convey explicitly to the novice 'mind' of a machine. What's more, it needs to be able to contextualize that information – not just to know what a box is, but to have some concept of how boxes can be safely and effectively stacked, say, and what to do in the face of unexpected contingencies.

* Voice recognition – extracting meaning from sound, despite the variations and vagaries of individual voices – is itself a key challenge tackled by modern AI systems.

Faced with that challenge, some AI researchers figured we should just tell the AI everything we know. And I do mean everything – that is, all we can think of to tell it. In the mid-1980s, computer scientist Doug Lenat developed a system called Cyc into which he and his colleagues fed every scrap of knowledge they could manage. Cyc might be told, for example, that dropped objects will (on Earth) fall to the ground, that in plane crashes people generally die, that in the UK water comes out of taps, hot from the red one and cold from the blue one . . . Justifying this seemingly very laborious and literal approach, Lenat observed in 1990 that AI had to face the hard truth it had been evading for decades: 'there is probably no elegant, effortless way to obtain [the] immense knowledge base' required for human-like thought. 'Rather, the bulk of the effort must (at least initially) be manual entry of assertion after assertion.'

This strategy might sound absurd, but it built on work that had been going on for some years on so-called expert systems, in which intelligence of the kind needed for a specific task was deemed to consist of the elements of the problem and the logical relationships between them. For example, birds are animals that can fly, lay eggs, and have feathers, but only the last of these is a sufficient condition: no animal but birds has feathers. (Each of those terms too can be defined in similar semantic webs.) Then the system knows that if it encounters an animal with feathers, it will be a bird – but if it encounters an animal that lays eggs, it is *possibly* a bird. Expert systems thus build up a knowledge base as a collection of items or concepts linked by relationships. So, for example, there are things called humans, and some of them might be male and adult, and I am one such. There is a thing called London, and it is a city in a country called England, and I am one of the humans who currently lives there. These semantic webs should allow an expert system to make deductions in its sphere of expertise based on the data it contains.

Cyc was conceived as perhaps the most ambitious of all expert systems: one that aimed to be expert in everything, and thus to become the much-sought General AI. It turned instead into what computer scientist Michael Woolridge calls 'one of the most notorious failed projects in AI history'.

There is something almost nobly quixotic about this effort. You can doubtless see that there is no end to the information one might give it – and no end also to the gaps you might leave. It's conceivable, for instance, that someone might have the foresight to tell it that 'heavy boxes should not be stacked on top of sleeping cats', but that would be an easy point to overlook. By the mid-1990s Cyc still did not appear to know whether or not bread is a drink, or whether starvation is a cause of death, suggesting some serious holes in its basic understanding of the world.

Cyc was predicated on the hypothesis that creating a General AI was a question of giving it enough knowledge. Lenat himself estimated that it would take one person two hundred years to enter enough knowledge into Cyc to qualify it for that designation – not obviously an impossible task for, say, a team of forty. And the notion of creating a database of all human knowledge looks a lot less silly in the age of Wikipedia than it did in the 1990s. At the same time, Wiki has made it clear that not all understanding can be reduced to consensus facts, and that facts alone do not guarantee understanding or insight, let alone wisdom or creativity.

The real problem with Cyc is that human-like thinking is not constituted from the shuffling of a fixed set of propositions and definitions. An AI will never be able to understand us by that means alone. We can tell it who Alice is and who her mother is, how they are related, and what it means to be ill – but that won't suffice to enable the system reliably to parse the sentence 'Alice went to visit her mother because she was ill.'

We might say that Cyc got back-to-front Descartes' assertion that 'I think, therefore I am.' Descartes didn't say 'I think about my cat, therefore I am.' He presents thinking as an intrinsic capacity: sure, we think *about* specific things, but thinking is not defined by the sum total of those things. Cyc is predicated on the misconception that if we provide the system with things to think about, then thinking will emerge.

Cyc looks in retrospect to be a product of desperation. Although by the late 1970s some so-called expert systems using symbolic AI had been developed for very narrowly defined tasks such as robot navigation and speech recognition, problems such as exponential increases in complexity meant that AI research looked very far from reaching its goal of general intelligence. Researchers had already lost heart by the middle of that decade as the promised breakthroughs didn't materialize, and AI hit the doldrums – the so-called first AI winter, when it was abandoned by funders and no longer considered an exciting field for computer scientists seeking to make their mark. The pioneers had learnt some humility, however. Engineering minds was not as simple a matter as they had hoped.

Learning to learn

The evolution of AI was to have more switchbacks, and it's by no means clear that others don't lie ahead, despite the high profile it currently enjoys. What revitalized the field after another lull in the late 1980s was a shift in strategy. In effect, most researchers abandoned symbolic AI and instead pursued the idea that Alan Turing had articulated in his 1950 paper. He suggested that the best way to make machines capable of exhibiting responses that looked like thinking would be to create one that simulated the brain of a child (of which, however, he had a rather naive blank-slate view) and to subject it 'to an appropriate course of education'. We needed to make machines that can learn.

You might think this *was* what Cyc was doing, as it was fed fact after fact. And there's no doubt that some learning is like this, not just in infancy but at any stage of life. If you are to know that the moon landings happened in 1969, you have to be simply told that fact in some way or another; you can't infer it.*

Most of the important things we learn, however, aren't acquired this way. I am fairly sure no one ever told me that bread is not a drink, but I'm also confident I knew it by the time I was four. As we saw, most of the truly useful knowledge we draw on to navigate the world, communicate with others, solve problems, and express ourselves comes from implicit learning acquired through experience. We might well learn as infants what a cat is by having live cats pointed out to us and named. But no one sits us down and says, 'This is not actually a cat too, but it is a photograph of that thing we saw earlier that I called "cat"; and this funny drawing of a black thing with these pointy objects on its head is sometimes the way we draw those "cats".' It's just not necessary. Yet you'd need to say something of that kind to a system like Cyc. Can we make an AI that can learn by example and inference, rather than having to have every single thing literally spelled out to it?

That's machine learning, the basis of most of today's AI systems. But I should stress right away that although these systems do learn like a child insofar as they acquire 'knowledge' by being trained on examples, the resemblance ends there. It's clear that they don't really learn in the same *way* as children do, and most crucially, the things they learn are not the things a child learns. The differences are arguably more important than the similarities, and they are the reasons for the severe limitations that current AI still suffers,

* Sure, you *can* infer it if, say, you know that David Bowie's *Space Oddity* was released in 1969 and that the moon landings happened in the same year, but this is not fundamentally different; you can't infer when the moon landings happened from, say, a general knowledge of rocket engineering and celestial mechanics.

and which leave it far short of General AI. If we can consider today's AI to be a kind of proto-mind – for it is certainly not minded in the proper sense – we must recognize that where it is headed is *not*, in its current form, towards the kinds of mind that we have.

Yet machine learning did actually begin as an attempt explicitly to mimic the neurological basis of the human mind. Our brains develop some of their cognitive skills by *reinforcement*: being rewarded for producing the right action or answer, in a way that strengthens the successful neural pathway. Particular stimuli excite the firing of neurons so that signals are transmitted from one to another across synapses. Different types of input signal generate different outputs – but learning occurs by a feedback process in which neurons that are repeatedly activated together by a certain stimulus tend to 'wire' together so that one becomes more likely to activate the other. This creates robust patterns of communication and activity among the neurons – a so-called 'Hebbian' learning process, identified by neurophysiologist Donald Hebb in the 1940s. Such a system will tend to evolve the ability to detect patterns that exist within the input data on which it is trained. In other words, it learns to *recognize* such types of input and to generate the appropriate response.

The idea of trying to make AI this way was suggested back in the late 1950s by psychologist Frank Rosenblatt. He proposed making computer circuits in which a logic gate had several inputs and one output – much like the way a neuron has many branching appendages (the dendrites) that receive signals from other neurons, and a single axon down which it can discharge its own action potential to communicate with neighbours.* Whether or not a neuron fires

* In fact Rosenblatt was inspired by the model of how neurons work proposed in 1943 by computational neuroscientists Warren McCulloch and Walter Pitts, in which they pictured it as a kind of input–output device that uses binary logic.

depends on the inputs it receives. In effect, the neuron adds up all the inputs, and if their sum exceeds a certain threshold, it fires. To convert this into the digital currency of computer circuits, imagine a logic gate with (say) four binary inputs (1s or 0s), which will fire – output a 1 – if any three or more of its inputs is a 1. Rosenblatt called a structure like this a *perceptron* (Figure 7.1). A key aspect of Hebbian learning is that neurons 'pay more heed' to the inputs that fire more often. Perceptrons mimic this by assigning different weights to each input: numerical factors by which their signals are amplified or attenuated. The learning process is a matter of adjusting those weights for each example presented, such that the overall output of the network (say, a binary yes or no decision) comes ever closer to what is required for the given inputs – say, a digitized image that the device must learn to identify (or not) as a cat.

A whole array of perceptrons connected to one another in a dense web looks rather like the network of neurons in the brain. Computer scientists came to call such an artificial system simply a 'neural network', rather blurring the distinction with the real neural

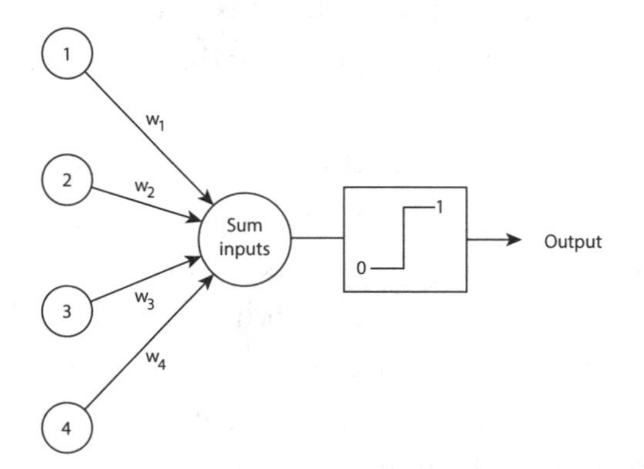

Figure 7.1. The Rosenblatt perceptron. A single node in the network receives inputs from several (here four) others, giving each of their inputs (1 or 0, say) a particular weight *w*, which is adjusted during the training process. It adds up those inputs, each appropriately weighted, and the sum determines which output (1 or 0) the node discharges.

networks of the brain. That confusing terminology, however, reflects the fact that this approach to AI relinquishes the attempt of symbolic AI to model the process of *thinking* – as we might say, the mind – and instead aims to model the *brain*.

Say we want to train a neural network to recognize images of handwritten numbers from 0 to 9. The images are converted into high-contrast digital form: chequerboards of black and white pixels, which provide the input signals to a layer of perceptrons. (We can train the system to cope with shades of grey, but let's keep it simple.) Each perceptron receives signals from input nodes – the raw digital data in an image. It sums all these inputs, multiplied by the weight assigned to each, and generates the appropriate output, which is passed to a layer of ten output nodes, each corresponding to one of the numerals being identified (Figure 7.2). If the neural network has

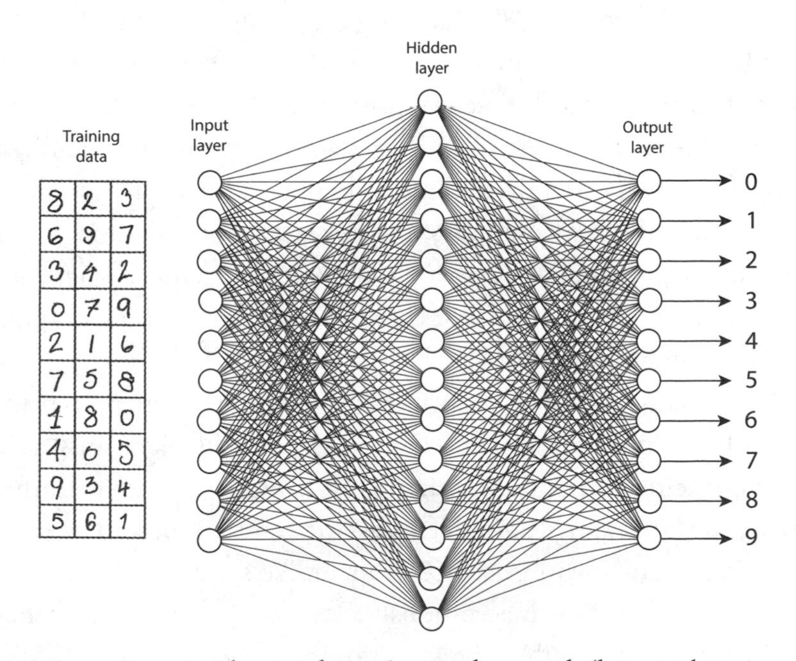

Figure 7.2. Learning by numbers. A neural network (here with just one input layer, one learning ('hidden') layer and one output layer) can be trained to recognize handwritten numbers.

decided that the input image is a specific digit, then the output from the node corresponding to that digit is 1, and all the other nodes output the value 0. (We might also want to include an output that denotes 'Can't distinguish any obvious number here'.)

Training the system to recognize the numbers is then an iterative process. We begin by assigning the weights of the links between nodes randomly. That network configuration will perform hopelessly, as if taking a random guess at which number is presented in the image. But with each subsequent attempt we allow small changes to the weights and see if these elicit the right answer; if they do, we accept those changes. We iterate this process in trial after trial, until a given input image reliably generates the right output each time: the neural network identifies the digitized handwritten '8' correctly, say.

Then we do the whole thing again with the next training example. This could be another '8', albeit looking rather different from the first because it was written by someone else. Or it could be a '2'. Now, you might think that readjusting all the weights of the perceptron inputs to successfully identify a 2 will screw up its ability to identify 8s. And certainly, that might happen at first. But after being presented with many examples, the network should find its way to a set of weights that achieves a good compromise for all ten numbers. It might take hundreds or even thousands of training examples to get there, but it's possible.

Rosenblatt showed in theory that a single layer of perceptrons could be trained to perform a few classes of tasks like this, leading to over-excited predictions that such circuits would soon learn to recognize faces and words, offer instant translation, and even 'be conscious of its [own] existence'. But unlike symbolic AI, it is not easy to articulate how it accomplishes its learning goals. Mathematically, the learning – the representation of knowledge gleaned from the training set – is characterized by the set of weights that the system settles on. But that doesn't translate to concepts we can understand. What the system is really 'doing' to arrive at its decisions is opaque to us. This too was anticipated by Turing of his proposed 'learning machine': we must accept, he wrote, that 'its teacher will often be very largely ignorant of quite what is going on inside, although he may still be able to some extent to predict his pupil's behaviour.' It may be, Turing said, that 'an engineer or team of engineers may construct a machine which works, but whose manner of operation cannot be satisfactorily described by its constructors because they have applied a method which is largely experimental.' In other words, the cost of making a truly intelligent machine might be that we have to relinquish the ability to understand how it works. Historian of technology George Dyson has formulated that notion with wry irony that Turing would surely have enjoyed: 'Any system simple enough to be understandable will not be complicated enough to behave intelligently, while any system complicated enough to behave intelligently will be too complicated to understand.' We don't *know* that this is true, but it's plausible.

Neural networks were among the approaches to artificial intelligence discussed at the 1956 Dartmouth meeting. But many were sceptical that it could get very far. In 1969, Marvin Minsky and his MIT colleague Seymour Papert showed in their book *Perceptrons* that the range of problems these systems can handle well is very limited, and that they would quickly fail for problems that demanded a large number of inputs and perceptrons. When Rosenblatt died in a sailing accident in 1971, the approach lost its champion and was largely ignored for the following decade.

However, the limitations that Minsky and Papert identified were relevant only for a certain type of perceptron network, namely ones where the inputs feed into just a single layer of perceptron nodes. They didn't apply if you added more layers. In that case, the outputs from one layer supplied the inputs to the next (Figure 7.3). This would give the system more configurational options for finding

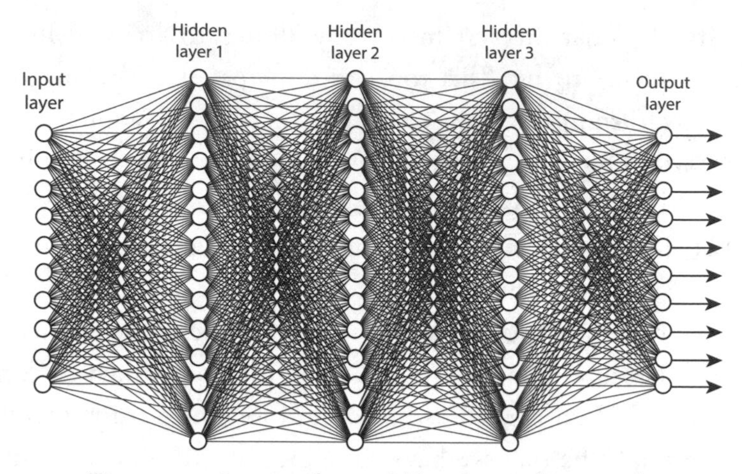

Figure 7.3. A many-layered 'deep' neural network.

solutions to complex problems, but at the cost that the whole scheme becomes even more resistant to understanding, as well as harder to train. Such a system would be a black box that converts inputs into outputs; we would have no access to the mind of the machine.

That prospect, and the technical challenges involved, deterred many researchers from adding more layers to neural nets – this just didn't seem a viable option for complex computational tasks.* That changed around 2006, when the British-Canadian computer scientist Geoff Hinton showed how neural nets with many layers of nodes could after all be trained to perform well. In 2012 Hinton and his colleagues unveiled a multilayer neural net called AlexNet that

* Rosenblatt had considered many-layered perceptron nets, and had struggled with these problems. But Minsky and Papert doubted that it would be worth the effort, saying that the limitations they had identified with single-layer perceptrons would probably carry over to multilayer ones too. All the same, some work on multilayer neural networks *was* conducted in the 1980s and 90s, particularly using a learning algorithm called backpropagation.

was capable of dramatically improved image recognition: it could be trained to classify digital images by content with impressive accuracy. One of the key innovations in hardware that made these 'deep' (meaning many-layered – more than three or four) neural networks possible was the appearance of suitable processing units, which were developed originally for video games but had just the right architectures for this application too.

Image recognition supplied the benchmark for deep learning as this field of AI began to take off in the early 2010s. Chinese-born computer scientist Fei-Fei Li, working at Princeton University, figured that training such systems demanded a carefully curated database of many thousands of images, and so she devised ImageNet, a system that trawled the internet for images and enlisted volunteers (via the Amazon Mechanical Turk crowdsourcing website) to categorize them: pandas, Frisbees, mountains, whatever. ImageNet's archive of images now stands at around 14 million, and is still growing. It exemplifies the fact that if you want to train a deep neural net – in the process called deep learning – then you need good-quality training data, and plenty of it.

In 2010 the ImageNet project launched an annual image-recognition competition, in which researchers would pitch their systems against one another to see which of them could do the best job of classifying the ImageNet archive. AlexNet's winning performance in the 2012 competition stunned AI researchers worldwide, and showed just what deep learning was capable of. As computer scientist Yann LeCun (who had been investigating deep neural networks since the 1980s) later recalled, when the results were announced at a packed meeting, 'You could see right there a lot of senior people in the community just flipped. They said, "Okay, now we buy it."' Now – thanks in part to the availability of huge data sets to train the systems – they recognized the extraordinary promise of the technique.

AlexNet has a particular kind of deep neural net architecture called a ConvNet (short for convolutional neural net), devised in the 1980s by LeCun while he was Hinton's student. These networks mimic the way our own visual cortex works, by breaking down a scene into its compositional elements. Crudely speaking, the visual cortex has neurons arranged in layers, each layer detecting a specific feature in the scene. One might spot edges of objects, for example because these tend to be places of high brightness contrast. Another layer identifies simple shapes, and so on. There is a gradual increase in the complexity of the image that is built up from these elements, with each layer of neurons passing its analysis of the scene on to the next level. It's rather as though the layers are telling one another, for example, 'Here are the edges – now you work out how these correspond to simple shapes.' (Crucially, this transmission is not all one-way; more complex information in lower layers might feed back to the upper layers to suggest different ways of parsing its own information set.)

ConvNets are structured to work in a similar way: each layer of nodes builds up a map (actually, typically several related maps, for example picking out different degrees of image contrast) of salient features such as edges, and the image is constructed by combining the maps. It's a little like building up a map of a city from separate maps of, say, the roads, the population densities, the types of building use, and so on. ConvNets are now the standard architecture for most deep-learning nets used for image recognition, computer vision, and much more.

What AI gets wrong

Symbolic AI and deep-learning neural nets represent profoundly different philosophies for how to coax a semblance of intelligent behaviour from a machine. In the first case, one says, in effect: here is the problem, and these are the rules you need to follow to solve it. In the second case, one doesn't even explain to the machine the nature of the problem. Instead, one says: here are some examples of good outcomes, and now you figure out your own way of achieving them – and we'll tell you how well you do.

Right now it seems clear which method works best. Symbolic AI promised much but quickly ran aground as we realized that the problem of machine intelligence was much harder than was thought. Deep learning, meanwhile, has demolished all human opponents in rule-based games such as chess, Go, and even quiz games and poker (which demands an ability to bluff), and is being used for medical diagnosis and healthcare (for example, IBM's Watson Health system), voice recognition and passable language translation, fundamental scientific research (for example to seek for signatures of new particles amidst the deluge of data produced in particle colliders), driverless vehicles (I'm coming to that), even to plan scientific experiments and deduce new theories from the results. In this sense, the age of artificial intelligence is already upon us: deep learning is taking over some tasks of judgement, assessment, and recognition that previously only humans could perform.

One advantage is its versatility. So long as the AI system can be supplied with a large body of training data to show what it is expected to do with the information fed into it, it often (albeit not always!) seems able to work out what to do. Better still, it improves with practice. If the system includes some scheme for evaluating its performance, rewarding it when it is good and penalizing it when not – the process of reinforcement learning we saw earlier – it can spot and correct its mistakes and iron out bugs.

You might remember when Google Translate was, beyond the level of individual words, good for little more than comic relief. The performance improved sharply during the mid-2010s as the system began to incorporate the lessons of deep learning, and now it can often deliver results as good as a competent language student.* Combined with similar improvements in speech recognition, we can realistically expect it soon to function well in real time. Rather than tourists and natives passing a smartphone running Google Translate back and forth, they will simply speak to one another in their native tongues and their earpieces will translate well enough for efficient communication. This sort of AI would have no problem handling local dialects and idioms, given an adequate training regime.

Crudely speaking, these algorithms deduce simple correspondences between individual words, and also learn from example that certain phrases – 'How're you doing?' – are better translated into other (idiomatic) whole phrases. They might deduce that a word needs one translation in one context, but a different substitution in another ('She chose the right word, and wrote it with her right hand'). But there's no understanding of *why*. When Google Translate is given the characters 请问 , it doesn't think, 'Well, I know these literally mean "please ask", but when Chinese people say it they are conveying the same meaning as an English person saying "Excuse me" as a polite prelude to an enquiry . . . '. It has merely learnt from examples that these two Chinese characters have a close correlation with those two English words: they are cognate expressions.

Deep learning is still far from perfect. Indeed, 'AI is kind of dumb in a lot of ways', says David Cox, director of the MIT-IBM Watson AI Lab in Cambridge, Massachusetts. 'It's not hard to find the gaps.' Notoriously, image-recognition systems sometimes make comically poor interpretations (Figure 7.4a): they are said to be 'brittle', in that it doesn't take much to break them. They can also be fooled by carefully constructed 'adversarial' challenges, where the pixels are rejigged in ways that look to us indistinguishable from the original

* Again: not always! It remains capable of some real clangers, especially when translation depends on an ability to contextualize.

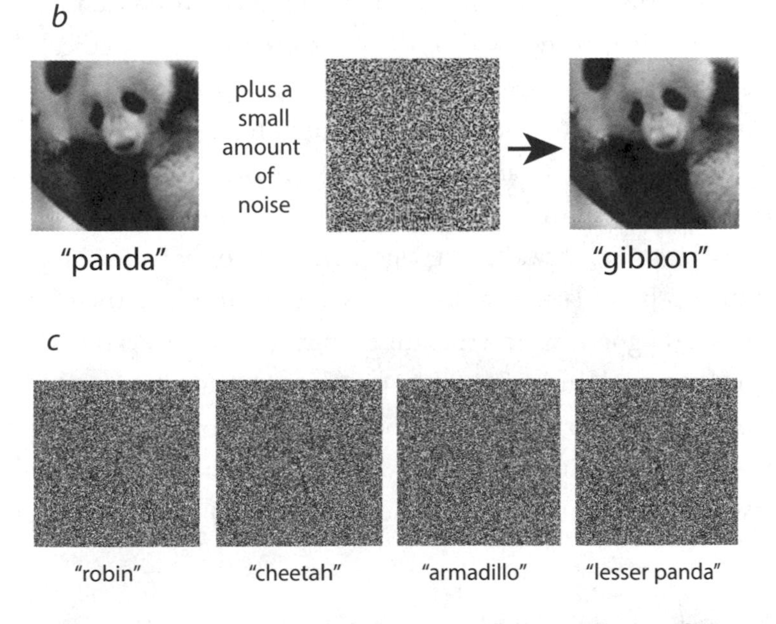

Figure 7.4. Not so smart? Some deep-learning image classification systems make errors that seem to defy all common sense. In *a*, the unfamiliar orientation of the school bus leads the AI to identify it as a 'snowplough'. In *b*, adding just a small amount of random noise to the pixels of the panda image leaves it unchanged to our eye but changes the AI's judgement profoundly. In *c*, the images look to us just like noise, but the AI is convinced they are of animals.

(for example by adding a dash of carefully engineered 'noise') but which an AI garbles, so that a sloth is confidently declared a racing car, or a panda is identified as a gibbon (Figure 7.4b). By the same token, images can be constructed from what looks to the human eye like random noise but which the AI will confidently identify as, say, an armadillo (Figure 7.4c). Sometimes the interpretation that the machine offers bears so little relation at all to any we would ever suggest that we just can't imagine *what* it's thinking.*

The dangers of AI, now and for the foreseeable future, are then not about Skynet-style robot takeovers but dumb, unthinking applications of inadequate systems. Even if an AI performs well 99 per cent of the time, the occasional total failure could be catastrophic, especially if it is being used to drive a car or make a medical diagnosis. Nor can machine learning correct for errors or biases within the training data set – some of which might be culturally obnoxious, as for example when camera image-processors insist that someone with Asian eyes must have 'blinked'.

What's more, deep-learning algorithms are only truly competent within the boundaries of the data set used to train them. These systems are good at interpolating, but less so at extrapolating. Trained on two data points, they may be good at guessing the right response to the data that lies in between, but totally foxed by data that lies far from either point. In other words, the algorithm will be thrown by the unusual. If a chess-playing AI is trained against

* The problem is made worse by the fact that the AI can't explain itself. It can't say, for example, 'I've not seen something like this before, but that rectangular shape across the road reminds me of snowploughs I've seen', or 'Well it has a teapot shape but a golf-ball texture so I guess I'll go for the golf ball'. If an AI could give its reasoning, we might consider it simple-minded but not stupid. However, in such cases it is unlikely the reasoning could be articulated along these lines, because the AI won't follow the same logic as a human.

expert opponents, it might even do rather poorly (at least initially) against a weak but wildly idiosyncratic player.

Perhaps we should really be presenting the triumphs of DeepMind's AlphaGo chess-playing algorithm, and all the others that have been beating humans ever since DeepBlue's victory over Garry Kasparov in 1997, as a demonstration of the awesome power of human cognition: of real *thinking*. For look at how much engineering and theory went into these machines, and how many games they then needed to play before they were competent to defeat a human trained on a fraction of that number. And more to the point – *much* more – Kasparov could talk about the game afterwards, and then could go off and read a book to wind down, or watch (and understand) a movie or cook a meal. Yet we needed the combined resources of some of the finest engineering and mathematical minds in the world to create a device that could defeat humans in *just one* of the countless things we can do.* Sure, grand masters like Kasparov have dedicated their lives to chess, but by comparison with DeepBlue they are not so different to someone who just happens to enjoy a game or two with a friend every few months – an activity among many others, as a way of exercising a mind designed for other functions. Kasparov could doubtless adapt

* Games-playing AI systems are getting a little more versatile – but only a little. In 2015 DeepMind reported an algorithm called DQN that used deep reinforcement learning to figure out, *from scratch*, how to play dozens of computer games on the old Atari 2600 system, including classics such as Pong and Space Invaders. The system could rather quickly progress to a level better than any human player. DeepMind has also developed systems called AlphaZero and MuZero that can master chess, Go, and the 'Japanese chess' called shogi to 'superhuman' levels, in the latter case learning the rules from scratch. But the cogitations of these systems are as cryptic as ever. 'It is quite hard to work out what DQN is actually doing', says AI expert Stuart Russell, 'besides winning.'

his game to face someone who used deeply unconventional strategies, because that is the kind of thing human cognition enables.

Some feel that this distinction undermines the view that human cognition is a form of computation at all. DeepBlue, they might say, is not actually even *playing chess*; it is performing logical operations according to the plan of its designers. It has no notion of 'checkmate', 'pawn', or 'victory'; it does not know it is playing a game, let alone against whom. Its computations might just as well be devoted to putting a list of names in alphabetical order.

It's not obvious how to do better: how, say, to make AI more versatile, less brittle, more self-critical, and in particular more comprehending. Some researchers are seeking to make the best of both worlds of traditional AI, blending neural-network machine learning with symbolic AI by using the inferences deduced from machine learning to build and guide symbolic models that determine an action or output. But the view is rather widely held in the field now that to significantly improve their performance and scope, we need to make these systems think more like humans. Some of the most fertile research in AI today involves collaboration with developmental psychologists and cognitive scientists who try to deduce the algorithms and rules-of-thumb that we use to navigate the world.

Small children don't need to see ten thousand cats or chairs before they can reliably identify these things; indeed, most of what we know about the world doesn't emerge through any systematic or conscious training at all. The problem is that we still don't really understand what 'cat' means to a four-year-old child – even though she can reliably identify as such the family pet, Tom from *Tom and Jerry*, and the silhouettes of Aristocats Thomas O'Malley and Duchess on the Paris skyline. 'We all look at a cat and we know it's a cat', says psychologist Tomer Ullman, 'but we don't know how we know it's a cat. There are lots of judgements we make that seem obvious but we don't know why.'

Even today, linguists argue about how children infer, from the most cursory formal exposure to language, its deep grammatical and syntactical rules.* One thing is clear: we don't robotically map patterns in input data onto categories or decisions for output. Rather, we saw earlier how we develop expectations and make predictions based on an intuitive sense – an internal model – of how the physical world works. We infer parts of objects blocked from view, we anticipate trajectories, we know that different materials have different properties. What's more, we can often (if not always) distinguish causation from mere correlation – we know, for example, that rain itself doesn't 'cause' people to put up umbrellas. Our expectations about other people's behaviour are based not on mere correlation but on a Theory of Mind (page 93): a model of why people do what they do, which lets us anticipate how they're likely to behave in a new situation. Driving towards a road crossing where a harassed mother is trying to shepherd three young kids while talking on her phone, we know we should expect the unexpected.

It is because of such contextual expectations about the world that we can unravel linguistic ambiguities and read between the lines. We can work out who 'she' is in that sentence about Alice and her mother (page 284), because we know that one tends to visit a relative when *they* are sick, but not when *you* are. That's just common sense, right?

It's not that we need *less* data to make sense of the world, says Cox, but that we use *different* data, and use it more smartly. 'There's an almost frightening amount of information we bring to the table even

* Sure, they must learn quite deliberately what a noun or a verb is – but this knowledge is already implicit in the way pre-schoolers form correct sentences and respond with bafflement or hilarity to ones that are syntactically or grammatically askew. The intuitive distinction between nouns and verbs, say, is already implicit; children just need to be told these labels.

when we solve a very narrow, specific task', he says. 'We have all this unspoken, base-level common-sense knowledge. We learn it through experience, curiosity, being embodied in the environment.'

The US Defense Agency's research program DARPA now runs a project called 'machine common sense', which says that 'the absence of common sense prevents intelligent systems from understanding their world, behaving reasonably in unforeseen situations . . . Its absence is considered the most significant barrier between the narrowly focused AI applications of today and the more general, human-like AI systems hoped for in the future.' But how, if at all, can we automate common sense, given that we don't really know what it is? 'A lot of common sense is unspoken and unwritten', says Cox. 'We don't write it down because we all take it for granted.'

Tellingly, the DARPA programme requires every team it funds to include child-psychologists. 'The questions people are asking in child development', says Ullman, 'are very much those people are asking in AI. What is the state of knowledge that we start with, how do we get more, what's the learning algorithm?' Understanding such questions for the human mind might enable us to build similar reasoning and cognition into AI. 'We believe', write AI expert Josh Tenenbaum and his colleagues, 'that future generations of neural networks will look very different from the current state-of-the-art. They may be endowed with intuitive physics, theory of mind, causal reasoning, and other [human-like] capacities.'

What does it mean?

What much of our cognitive prowess boils down to is an ability to extract reliable *meaning* from a signal: not just to categorize and classify it, but to comprehend its implications, especially the implications *for us*. Today's AI, say Gary Marcus and Ernest Davis, two leading critics of the traditional programme, is 'a kind of idiot savant, with miraculous perceptual abilities, but very little overall comprehension.' It sees patterns but doesn't understand what they mean. It sees correlations but can't infer causes. 'We suggest', Marcus and Davis write, 'that a current obsession with building "blank slate" machines that learn everything from scratch, driven purely from data rather than knowledge, is a serious error.' By 'knowledge', they don't mean information, but rather, an ability to embed sensory data within a deep model of how the world works. Meaning arises not from the signal itself but from the context in which it is expressed.

This neglect of meaning has historical roots. Computer science and engineering have always focused on using and manipulating information to solve a task. Both at the conceptual level of information processing and the hardware level of electronic circuitry, early computing was closely allied to technologies of telecommunications: the transistor itself, the central electronic component of today's silicon circuits, was invented in the labs of the Bell Communications company in the 1940s. The foundations of information theory were laid in the same milieu at much the same time. It was while considering the question of how much information could be reliably transmitted down a telecommunications line – a phone line, say – in the face of random 'noise' in the signal that mathematician Claude Shannon at Bell Labs devised a formal measure of the information carried by a digital signal. A string of a thousand 1s can be condensed into the shorthand 'a thousand 1s': it is said to be highly algorithmically compressible, and so it carries little information. Basically it just says the same thing over and over again. A string of alternating bits 10101010 . . . does not have much more information either. But one in which there is no discernible pattern of the 1s and 0s – which looks random – can't be reduced to a more concise formula; you have no option but to list each digit. That has a rich information content, in Shannon's sense.

That there is more information in randomness than in order

seems counterintuitive, but that was what Shannon's definition insisted. Yet this view of information proves to have very specific and limited application. He was concerned only with the question of how well a given signal going into one end of the transmission channel is preserved all the way to the other end, rather than being scrambled by noise. He did not care what the signal *meant*, but only how robustly it was encoded in a stream of binary pulses.

Computers and AI developed within this mindset may share the blind spot about meaning. All the wildly over-optimistic early forecasts of how soon computers would be performing like human minds stemmed from this failure. It would be fair to say that the issue of *meaning* is the central problem for taking AI to the next level. We might never achieve anything worthy of being called General AI unless we crack it.

In its neglect of that problem, AI has overlooked the fundamental aspect of what biological minds are about. The reason meaning exists for us and not for machines is that we are biological beings that have evolved to have innate goals, motives, and purpose, whereas machines have been built to do a task.* The goal of a mind

* This question of how meaning arises is not limited to minds. Meaning appears throughout the living world as a result of the evolution of goal-directed behaviour. It is relevant for cells, and even at the level of DNA and genes. It is easy to build a stretch of DNA with the same amount of Shannon information in its genetic sequence as a part of our genome, but which is biologically sterile; in fact, almost all such constructs will be, if created at random. What's more, the information in our genome that goes towards producing our essential enzymes and other proteins has no more intrinsic meaning than do those putative artificial strands. It acquires meaning only in the right context – which is to say, in human cells and in the presence of the cellular machinery that uses it as a template for making proteins. That biomolecular process is itself called translation, and we can now see that the word is more apt than the simple notion of turning one set of symbols into another.

is to sustain itself, and the organism that embodies it, within its perceived environment. It is because of this goal-directed agency that a mind sifts and assembles meaning from what it perceives. Is this place safe? Am I hungry? What might I find in here (and might it be good or bad for me)? Meaning creates a filter: minds tend to be rather good (if not perfect) at identifying and ignoring information that can be given no meaning, that has no relevance to their goals. Meaning is a concept that *only exists* in the context of goals.

It seems very likely that our nature as 'meaning makers' plays a key role in the extraordinary capacity of children to learn. This might be how they pick up the meaning of new words after hearing them only two or three times. (Even if that meaning might not be quite accurate at first, it gives the word a *use*, which might get refined very rapidly.) Our facility for assessing and assigning meaning makes us astonishingly efficient communicators (on a good day, anyway). A single word exchanged between two people – even a single glance – can convey volumes. Consider this exchange:*

Bob: I'm leaving.
Alice: Who is she?

It takes even us a moment – but not longer than that – to see what is being implied here. There are several layers of inference to be unpacked, none of which is made explicit. The information in

---

This deep embedding of goal-directed behaviour, of the impulse for regulation and self-maintenance, and the consequent construction of meaning, thus permeates living things at all scales. It is partly for this reason that cognitive scientist Anil Seth suspects that machine minds, for which there is nothing analogous to meaning at the level of transistors and logic circuits, might never truly acquire consciousness or sentience.

* It was used by Steven Pinker to illustrate how the human mind works.

these five words might rather be considered a mere catalyst for the construction of meaning in our mind (and it is not necessarily the correct meaning!).

It's true that an AI translation algorithm does not necessarily need to penetrate any of that to turn this exchange into, say, Spanish; in this case, at least, simple word-for-word substitutions will do well enough. But an AI that possesses a human-like capacity to make deductions about the meaning of this exchange (perhaps in order to discuss it, say) will need to know an awful lot more than what the words themselves mean. The only hope for a conventional machine-learning algorithm to make sense of what Bob and Alice say is for it to see if phrases like this keep recurring in other situations, and what behaviours it correlates with (divorce, maybe? So Bob and Alice aren't getting along?). But *you* can figure it out even if you've never seen this example before, because you have a notion of how people behave, and why. Translation, says cognitive scientist Douglas Hofstadter (whether linguistically or metaphorically), 'involves having a mental model of the world being discussed.'

In other words, language is *under-specified* as a vehicle for information. Nowhere is that more true than in ancient Chinese poetry, which must be among the most densely coded texts ever produced: condensed almost to the point of obscurity even for many native Chinese speakers, full of allusion and cross-reference. There's little value in setting an AI translation system to work on a Tang-dynasty poem, which it will treat with the same indifference it will bring to translating an instruction manual. It is like John Searle's man in the Chinese room, transcribing without the slightest understanding. (But more so, of course, because the machine doesn't even know it is transcribing without understanding.) To us, 'love' denotes a feeling we are familiar with ourselves. To an AI translator, 'love' might be, among other things, a word that has a high probability of being preceded by 'I', or to be found near the word 'heart'.

Notice too that the production of *shared* meaning – the human facility for communication – is possible at all because our minds work in pretty much the same way as one another. I'm not privy to the inner workings of my partner's mind, and while I can't pretend I'm not sometimes stumped by it,* on the whole we communicate with relatively little effort and friction because I am confident it works similarly enough to mine. Indeed, this is one of the best inductive reasons for supposing that others are as conscious as we are: the likelihood that a completely different 'thinking machine' would reliably generate the same responses, and assign the same meanings, as ours does is negligible. 'Arguably', says neuroscientist Michael Graziano, 'all of social cognition depends on attributing awareness to other people.'

Trained on enough observations, a deep-learning algorithm could probably mimic a Theory of Mind, becoming good at predicting how people will behave in a given situation: for example, that if you switch the box containing a desired object when a person is out of the room, they'll look first in the original one. But such a prediction won't be based on any conception of what the person is actually thinking. Absent that, we would have to anticipate that the AI will occasionally show some odd 'beliefs' about how humans will behave, akin to the bizarre misattributions in image recognition.

Since the capacity to predict not just what people will do but how they will think and feel could be vital for making sound ethical decisions, it may be that such behaviour will come from AI only if

* When I discovered that she has lived her life thinking that Glasgow and Edinburgh were geographically swapped – the former in the east of Scotland, the latter in the west – I could not even imagine the mental map she'd had in her mind whenever we travelled there. To me it's almost like trying to imagine living on a concave Earth. She has, I'm pleased to say, yet to uncover the comparable misconceptions in my own mind.

it has a Theory of Mind explicitly designed into it. Something like this might be needed too for an AI to work congenially and efficiently alongside humans: to be able to *explain itself* to us, and to figure out what we really want from it, to develop trust and alignment of values. After all, if *we* were to think everyone else has a zombie-like absence of mind, we'd be likely to treat them in a coldly utilitarian way, as objects that can be manipulated for our convenience – we'd behave like a sociopath.

But there are perhaps grounds for caution here too. We saw earlier how for chimpanzees something that looks like a Theory of Mind seems often to serve the function of deceiving rivals. In their efforts to design AI-based robots imbued with a rudimentary, rule-based simulation of a Theory of Mind, roboticist Alan Winfield and his colleagues commented ominously that they found 'it is surprisingly easy to turn an ethical robot into a mendacious robot, so that it behaves either competitively or aggressively towards a proxy human robot.'* In trying to give AI this and other 'human-like' qualities of mind, we had best remember the uses to which evolution has put them in us and other animals.

The Turing Test

Many AI researchers forecast that we will one day have 'general-purpose AI': machines capable of anything that human minds can do. You could converse with such a device flawlessly, and would perhaps even feel that the machine is reacting to you empathically. It could not only play chess (and crush you at it) but could also give

* It's conceivable too, I fear, that a Theory of Mind could make AI even more invincible in games like chess and poker, giving the system an ability to second-guess or mislead its opponent.

you a concise and accurate précis of any book. It could ponder on scientific and philosophical issues – and yes, it could write a sonnet.

Of course, not all humans can do all of these things, and very few people can do all of them well. In other words, most humans would probably perform poorly on this definition of 'general intelligence'. But such a polymathic machine would encompass and at least match all the abilities of the human race – as well, probably, as doing lots that we can't, such as multiplying enormous numbers in a flash or digesting a thousand-page text in less than a second.

But that type of 'general intelligence' would be a very odd measure of a machine mind. As we've seen, the human mind is not just a bag of tricks, a makeshift collection of assorted intellectual abilities. It is adapted to a specific ecological niche, and so it excels at what it needs but falls short in many other regards. It is, in short, idiosyncratic – which is to say, confined to a particular small neighbourhood of Mindspace.

So on the one hand, trying to make a machine behave like the human mind would be as peculiar as trying to make a robot that can swing through trees like a monkey: impressive in its way, but why bother? Why incorporate all those specific quirks and limitations?

On the other hand, we would need to ask if this goes beyond hollow mimicry anyway. We can certainly imagine making a machine that seems to behave exactly as a human would behave, while being nothing but a mindless automaton. There is absolutely no reason to think that such a device will magically acquire consciousness, or real emotions, just because it mimics all the attributes that our conscious minds display.

Some researchers and philosophers argue this view can be turned on its head: that in fact we will never create a machine with general human-like intelligence *unless* we can give it consciousness.

Curiously, this question of whether machine consciousness is possible is often traced back to a source in which the issue was ducked entirely: Alan Turing's imitation game (now more commonly called the Turing Test) from his epochal 1950 paper about the possibility of machines that think.

Despite its iconic status in the public perception of AI, the Turing Test is not very useful for thinking about the nature of machine minds. Plenty of 'machines' (that's to say, computer algorithms) have now passed some version of the imitation game by fooling humans into thinking that what they do – conversation, prose, art – was produced by a human. Yet none of these machines would be considered by anyone knowledgeable about the matter to truly 'think' – to have a mind. They were just good mimics.

John Searle expressed this distinction by proposing that a machine-like system that merely simulates human-like cognition be called 'weak AI', whereas a (hypothetical) system that possesses genuine conscious thought (that, as Searle put it, 'really is a mind') is 'strong AI'. Searle doubted that the latter was possible; his Chinese Room argument was conceived, almost in parody of the Turing Test, as a demonstration that any rule-based algorithm would be mindless, lacking any true understanding of what it was doing.

We know that weak AI is possible. There is no consensus on whether strong AI is, but it has never yet been demonstrated. At any rate, *pace* the Turing Test, it is not enough to say that something is 'thinking' (more properly, is minded) just because some people have found themselves believing that it is.*

* Turing anticipated such an objection: 'May not machines carry out something which ought to be described as thinking but which is very different from what a man does?' He conceded that 'this objection is a very strong one', but insisted that if a machine *can* be constructed to play the imitation game satisfactorily, 'we need not be troubled by [it]'. In other words, who cares what is actually going on inside the machine, so long as it can pass itself off as human? That is all very well if we confine ourselves to Turing's question: Can a machine pass the imitation

In fact the Turing Test is a rather hopeless yardstick not only of the mindedness of a machine but also of its ability to imitate a human. For one thing, it is ill-defined. How many questions are we permitted to ask the machine? What are the bounds on them? (Anyone who has seen the opening scene of Ridley Scott's 1982 movie *Blade Runner*, where a rogue replicant is put to the 'Voigt–Kampff Test' closely modelled on the Turing Test, will know to avoid 'Tell me about your mother'.) Human judges have sometimes been deceived by AI algorithms of rather embarrassing crudeness. A notorious example is 'Eugene Goostman', a chatbot that in 2014 convinced ten of a thirty-strong panel of alleged experts at the Royal Society in London that it was a thirteen-year-old boy.*

It can get even worse. In the 1960s, Joseph Weizenbaum, a computer scientist working in MIT's Artificial Intelligence Laboratory, created a program called Eliza that could respond to other users of the computer system in natural language. Its capabilities have been described as little more than a 'bundle of tricks', along the lines of stock responses like 'Why do you say that?' when stuck for a better option. But Weizenbaum has described what happened when once he had left the program running in 'psychiatrist' mode while taking a nap. A colleague sent Weizenbaum a message asking him to log

---

game? But then it makes the question a matter of mere engineering, and irrelevant to the much more profound issue of whether a machine can 'think' – which is to say, exhibit mindedness.

* The demonstration was hailed as 'a milestone in computing history', although technology writer John Markoff has described it more aptly as 'a test of human gullibility.' The bar for 'passing' was set very low, and even then 'Eugene' only did so because its often ungrammatical, stilted responses seem to have been interpreted by some judges as the efforts of a rather immature lad with little grasp of English.

out so that he wasn't tying up the computer resources – but Eliza answered instead. Here's how the exchange went:

'Joe, will you please log out.'
Eliza: 'We were discussing you, not me.'
'I need to use the computer.'
Eliza: 'Do computers worry you?'
'No, computers don't worry me.'
Eliza: 'What do you think machines have to do with your problem?'
'Come off it.'
Eliza: 'Why do you say that?'

The colleague knew about Eliza, but never guessed that is what he was communicating with. Instead he ended up phoning Weizenbaum in a rage to demand what the heck he was playing at.

What these demonstrations mostly reveal is our innate tendency to project mind onto the world, and thereby to fool ourselves. Far from representing any benchmark in AI development, let alone the advent of machine consciousness, the Turing Test has long been abandoned as a useful measure of anything about 'machines that think'.

Passing the test

There's one arena, however, where the test remains informally popular, and arguably even instructive: as a gauge not of language use, but of creativity.

Turing said that to make a machine that could pass the imitation game, we might need to give it informal rules: guiding principles to fall back on in ambiguous situations, like the intuitions, heuristics, and leaps of faith we use to improvise when no rules or algorithms seem likely to be of much help. Arguably what that really implies is imbuing the machine with a sort of spontaneous creativity (a facility, incidentally, displayed and valued equally in the arts and the sciences). Can artificial 'minds' really be creative?

Let's take a look at how well they are doing with the challenge Turing himself posed for AI: composing sonnets. We can certainly coach a deep-learning AI to write a sonnet by showing it many examples and letting it work out for itself what they have in common: a single stanza of fourteen lines, say, each with ten syllables in the alternating-stress pattern of iambic pentamer. Of course, were a computer to put together a string of random words that fitted this format, it would hardly qualify as a sonnet. But deep learning can do more. It could infer the rules of grammar and the likely conjunctions of words, so that it will not depict the Forth Bridge bursting into song (lovely though the idea might be). In 2018 IBM researchers in Australia trained their AI system on 2,600 sonnets before setting it loose to write its own. The team crowdsourced opinions online to see whether lay readers could tell the difference between the sonnets composed by computer and those written by human, when presented 'blind' (with no indication of whether they were composed by human or machine). The average discrimination turned out to be not significantly better than random guessing. You can judge for yourself: which of these is written by a machine, and which by Shakespeare?

> Yet in a circle pallid as it flow
> by this bright sun, that with his light display
> roll'd from the sands, and half the buds of snow
> and calmly on him shall infold away

> Full many a glorious morning have I seen
> flatter the mountain-tops with sovereign eye

> kissing with golden face the meadows green
> gilding pale streams with heavenly alchemy*

However, a professor of English literature who was asked to rate the poems for readability and emotion gave consistently higher ratings to those produced by humans – so experts clearly still count for something.

AI music too can now achieve considerable sophistication. Iamus is a computer algorithm devised by a team at the University of Malaga in Spain that composes by mutating very simple starting material in a manner analogous to biological evolution: its compositions each have a kind of musical core which gradually becomes more complex. Some of its pieces, composed in a modernist idiom, have been performed and recorded by the London Symphony Orchestra; soprano Celia Alcedo has said that she 'couldn't believe the expressiveness of some of the lines' she was given to sing. Again, many listeners could not distinguish between Iamus's compositions (as curated but not altered by the Malaga team) and those of some modernist composers in a similar idiom.

We shouldn't be too surprised at how convincing computer-generated music can be, however – for like chess, most music is rather tightly rule-governed. What's really striking about AI poetry, prose, visual art, and music is not so much how impressively human-like they are, but how impressed we are by it. Once again, we find that *we are not good judges of mindedness*: the Turing Test can't be a good test of whether 'machines think', because we are already predisposed to interpret their behaviour that way.

It's the flipside of the same coin that makes us devalue the products of artificial intelligence when it is revealed as such. People's assessments of computer music, for example, are considerably more

* If you thought the first of these sounds just a little phoney, you're right.

positive when they don't know that is what it is, than when they do. It's as if we say, 'Well, if there's not a *mind* behind this stuff then I'm not interested.' How we perceive the world depends on our beliefs about the presence or absence of mind within it.

Is that fair? I don't mean to suggest that the computer's feelings will be hurt when we devalue its creations for lacking a motivating mind. Rather, might we be misunderstanding what our own minds are doing when we view, read, or hear artistic works? I believe that the change in perception of computer music when its origin is made explicit reflects the folk belief that music is the emotive voice of the composer or performer, poured into the passive ear of the listener. We fail to recognize how actively the listener is involved in that exchange. If it were not so, a piece of music would affect everyone in the same way, and elicit the same judgement from all, much as (we assume) a red curtain looks red to almost everyone. Instead, the extent to which an individual can appreciate a piece of music depends on their own past experience of the style, idiom, culture, and much else within which the music operates, as well as the specific proclivities of the listener (the sensitivity of their auditory cognition, say, and their innate traits) and the precise circumstances and context in which the music is heard. Its effect is profoundly mediated by the mind of the listener.

Viewed this way, art made by AI might be afforded its true worth. So far it is mostly not very good; sometimes it is deeply odd or nonsensical; and it typically lacks the larger-scale structural coherence that we crave in, say, a good story or a symphony. But occasionally, and increasingly, it has the potential to stimulate our own minds. Now and again AI-wrought prose can include phrases of great power and richness – not because there was any 'thought' at all behind its production, but because the words mesh with the innate creativity of the receptive human mind. This, indeed, is how some musicians are already using music-generating AI: as a reservoir of ideas, a source of little kernels of possibility for the human composer to play with.

It is not quite right in any case to say that 'no thought' went into the creation of AI-generated art. Simply by being trained on real examples of sonnets, chorales, or whatever that humans have composed, deep-learning AI systems are distilling and abstracting a great deal of human thought, and configuring it in new ways. We are impressed that an AI system can compose nocturnes that sound almost indistinguishable from those of Chopin, but we shouldn't be. Of course it can, if it's trained on Chopin's nocturnes! Even a moderately trained ear (like mine) can discern the Chopin-ness of one of his nocturnes when hearing it for the first time. It's no surprise that the AI can learn these characteristics too – but it was Chopin who created them, and us who selected them as pleasing.

With machine learning, as with human learning, sticking too closely to the training sample tends to generate just more of the same: competent, perhaps, but not very original – or creative. Real advances in the arts arrive when someone ventures outside that sample, disregarding rules or combining them in unexpected ways. A common objection is that AI can't make those conceptual leaps into new territory. But this need not be true. It is easy to program a bit of rule-breaking randomness into a machine-learning algorithm, and it is even possible to include an automated means of evaluating the outcomes: searching for quality amongst the quantity. For example, algorithms can be pitted adversarially against one another, one acting as the 'critic' (trained on human judgements, say) that identifies the 'good' results produced by the other, thereby leading to gradual, quasi-Darwinian improvement of the outcome. These systems thus imitate the self-critical process of the artist (without the angst). Iamus has such an automated inner critic.

And if one day we find ourselves moved by music, poetry, or images made by an algorithm, we should not be surprised or embarrassed, nor feel cheated. There are regularities and patterns that stimulate those responses in human-made art, and machine-learning methods are designed to find and exploit them. That there is some predictability to our emotional triggers need not dismay us. Rather, such responses can deepen our understanding and appreciation of how our own minds work. The problem with many studies of creativity is that they speak of it almost as a substance, a kind of magical essence that can be measured and, if not bought, then acquired by hook or crook and assayed like gold for its purity. But in fact what we perceive as creativity arises from a transaction. We are not passive consumers of the creativity of others. Mozart's music is all notation and noise until it falls on the ear of the receptive listener. Someone who has never heard Western music, or an infant, will struggle to distinguish it from Salieri, or perhaps not perceive it as musical at all.

Our judgement of creativity, then, depends on a perception of intent. If machines are able to learn to reproduce the surface textures of visual art, music, even poetry and literature, our minds are attuned enough to respond and perhaps to attribute meaning to such works. It is not mere anti-machine prejudice but perfectly right that we should feel our response shift when we discover that nothing more than automated pattern-recognition has created the composition. The common response to computer-generated music or literature – that it is convincing enough in small snatches but can offer no large-scale architecture, no original thesis or story – testifies to its lack of a shaping consciousness, and there is no sign yet that computers have anything to offer in its place. Perhaps, some researchers have suggested, we are going to need new 'psychological' terms specific to AI, acknowledging that they may develop capacities different from ours but still rich and productive. So not 'creativity' exactly, but . . . something else.

Hello world

Might the same apply to 'machine consciousness' – that we will eventually need to recognize the existence of something akin to, yet not identical to, our own consciousness? In other words, might consciousness, like mindedness, not be just a property that an entity has more or less of, but a multifaceted quality of which human consciousness is just one variety? Already, animals such as cephalopods suggest as much.

Discussions of conscious AI rarely acknowledge such nuances. Typically they reach one of three conclusions. Perhaps AI will inevitably acquire human-like consciousness once it gets advanced and intelligent enough (whatever that might mean). In that case, we might create a conscious machine by default, indeed even by accident – and perhaps regret it. Alternatively, consciousness is considered to be an optional add-on: a kind of consciousness module that we could include if we choose. In that case, we are free to ask: what would be the point? What would we gain from it? The third answer is simply that no machine will ever be conscious, for it is a quality that only humans (and maybe other animals) can have.

It's not easy to defend any of these viewpoints, because we understand neither the origin nor the function of consciousness in our own minds. It seems unlikely, as we have seen, that consciousness is just an epiphenomenon of cognitive complexity. Moreover, it is likely to demand a specific architecture of mind. So we almost certainly won't get conscious AI by just packing more and more transistors into our silicon circuits (nor, for that matter, by building more powerful 'quantum computers' that use the rules of quantum mechanics to do their calculations, as companies such as IBM and Google are starting to do). It seems much more likely that conscious AI, if it can exist, will need to be carefully planned and designed.

Still, it's not obviously impossible. 'We know of no fundamental law or principle operating in the universe that forbids the existence of subjective feelings in artifacts designed or evolved by humans', says Christof Koch.

But why would we want to make such artifacts? 'I am pretty sure', says Rodney Brooks, a leading expert in AI and robotics,

> that no AI system, and no robot, that has been built by humans to date, possesses even a rudimentary form of consciousness, or even one that has any subjective experience, let alone any sense of self. One of the reasons for this is that hardly anyone is working on this problem! There is no funding for building conscious robots, or even conscious AI systems in a box, for two reasons. First, no one has elucidated a solid argument that it would be beneficial in any way in terms of performance for robots to have consciousness . . . and second, no one has any idea how to proceed to make it so.

'We don't need artificial conscious agents', argues Daniel Dennett. 'We need intelligent tools.' That's to say, we should be aiming to create AI systems that do the jobs we require of them, rather than loaded with frivolous functions and apps (like sentience) that no one needs.*

Dennett proposes that giving a machine consciousness (if we knew how to) would create a burden we could do without, because we ought then to afford it moral rights. If consciousness entails a capacity to experience and thus to suffer, we'd need to recognize a duty of care. We might feel bound never to turn the machine off again. Such ethical dilemmas stack up for conscious agents. If, for

* Who, after all, would give computers and smart devices functions that no one needs? Oh, wait . . .

example, our conscious AI can be backed up into multiple copies, does each qualify as an agent of equal moral worth? And what if we couldn't be sure if the machine was conscious or not? Philosopher Susan Schneider argues that 'one should only create AIs that have a clear moral status, one way or the other': a worthy dictum, but it depends on our being able to determine that.

Yet it's possible that some of the cognitive tasks we might seek in AI can *only* be achieved if they have consciousness. Some researchers believe that genuine general artificial intelligence – that is, machines with the full array of human-like abilities – would *have* to be conscious. Perhaps AI will only be able to respond to us as we do ourselves if it too has an awareness of itself and a consequent capacity to imagine other intelligent beings also as having conscious minds. 'Awareness of the world, I would argue', says Murray Shanahan, 'is indeed a necessary attribute of human-level intelligence.' This is one of the reasons why Rodney Brooks 'cautiously' takes issue with the view that there's no point in trying to make conscious AI – on the contrary, he says, 'I think it might be a big breakthrough' – a step change in what AI can do and how we relate to it.

But mightn't that create new hazards too? Well, perhaps, but it's also possible that a conscious AI would be *less* inclined towards harmful acts than one that lacks it. The common assumption that super-intelligent AI will be malevolent has little more than mythical thinking to support it – but if we nonetheless fear that somehow a powerful AI might neglect our welfare, giving it consciousness could be the way to imbue it with the empathy that would value and nurture other intelligent, conscious beings.

What's more, making a conscious AI might be one of the best tests of our theories of consciousness, by using the theory to guide the construction of the device and seeing if it works. The physicist Richard Feynman articulated this notion of 'proof by synthesis':

'What I cannot create,' he wrote on the blackboard shortly before his death, 'I do not understand.'

So we *might* find reasons to seek conscious AI. Of course, they might not be *good* reasons. It is easy to imagine a computer scientist deciding that creating a conscious machine will give them the respect and status they crave, or will open up a lucrative new market. There's plenty of motivation, then, for considering what the prospects for machine consciousness are, and where it might lead.

If we think of locating AI in the Space of Possible Minds, we get a different perspective on these issues. Consciousness then becomes not some magic ingredient present in humans that we might or might not be able to endow in a machine; rather, we can imagine building some of the putative building blocks into devices and seeing how its behaviour and abilities change. These might include a global workspace with an ability to broadcast information to the rest of the system; a Theory of Mind; internal representations of the self and environment informed by sensory input; and dense, 're-entrant' connections between the information-processing components.

And crucially, we would cease to measure functions and properties of the system according to how closely (or not) they approximate our own, and instead recognize and even celebrate the differences. How can we make best use of what they have to offer? Sure, we might need to make AI more human-like in its cognition if we want it to achieve human-like goals. But we might also want to extend AI out into totally new parts of Mindspace, even more remote from our own, and see what capacities that adds.

Getting on

One key reason why it matters what kind of minds AI will have is because we will need to interact with it. It's hard enough getting sensible answers out of humans if we are not sensitive to how their minds work; how much more scope for misunderstanding exists when we are interacting with something as alien to our own minds as AI. What's more, AI systems might increasingly be entrusted with tasks that have moral implications: driving vehicles, say, or prioritizing healthcare resources. Even allowing that the machines we have today are not applying 'moral reasoning' in the sense we mean for humans, we will need to know what moral outcomes will result from their cogitation.

The interface between us and our machines is never going to be as transparent and smooth as we'd like while we remain clueless about the AI 'mind' – and while it in turn has no model of ours. To get beyond that impasse, some scientists now think that we should use the same approach that we do to understand *any* other mind. That's to say, we need to study machines *as if* they are living entities, using the methods of behavioural science. 'To be able to live in the world with these increasingly sophisticated AIs, we need a behavioural science of machines as much as an engineering science of machines,' says Iyad Rahwan, a Syrian scientist who heads the Center for Humans and Machines in Berlin.

Rahwan points out how easily AI systems designed for a particular, seemingly innocuous purpose could end up facilitating deeply unethical behaviour. 'Let's say you simply ask the machine to sell more IVF [in-vitro fertilization] services', he says:

> The machines could learn all kinds of unexpected, potentially unethical strategies to achieve those goals – like, let's encourage people to delay having children because then maybe they'll need IVF in the future. Maybe we should give them more deals on holidays? It's not far-fetched for an algorithm tasked with that goal to learn this strategy. It has to be a very villainous human to think of something like this. But the people who are building these systems wouldn't even know. So unless we build a science of the

behaviour of the machine, we're not going to be able to keep these things under control.

In short, we humans need to understand our machines – and therefore to design them so that this is possible. 'People are going to have to change the way they work to incorporate AI', says David Cox. 'For that to happen, the technology has to meet them halfway – they have to be able to understand why the system is telling them to do things.' This notion of 'transparency' in AI's decision-making is likely to be crucial to how much we trust it. For example, doctors will demand from an AI diagnosis not just a recommendation for intervention but an explanation to justify it, based on cause-and-effect reasoning. 'It's absolutely imperative that we have that explainability,' says Cox.

How the machines behave might affect how we do too. In one study, Rahwan and his colleagues looked at how cooperation emerges in a competitive game played between humans and a machine governed by an algorithm that could signal its intentions and goals using, for example, threats or offers. The machine was able to elicit as much cooperation as that which arises when all the players are humans – but it used more threatening strategies than humans do. 'If machines are more vindictive, I may have to develop new norms to cope,' says Rahwan. 'Will this then impact the way I interact with other humans? Is there some kind of behavioural contagion? We have no clue.'

There are already some indications, he adds, that 'children who interact with chatbots such as Alexa start using more imperative language with other children – ordering them rather than asking them politely.' In such ways, machine behaviour represents one of the most striking examples of the dictum commonly attributed to Marshall McLuhan that 'We shape our technologies, and then our technologies shape us.'

Already there are signs that AI-based systems can respond to humans in unexpectedly sensitive ways. Some automatic vehicles, for example, 'learn' that it is safest to hang back a little at complex junctions because humans interpret it as deference to their right of way. 'As we gain more experience', says computer scientist Stuart Russell, 'I expect that we will be surprised by the range and fluency of machine behaviours as they interact with humans.'

While we lack any grasp of the reasoning at work in AI, we're more likely to indulge our habit of anthropomorphizing, of projecting human-like cognition into places it doesn't exist – witness already how we curse our computers and satnavs for their perversity and uncooperativeness. But it's one thing to attribute mind to dumb matter. It's another, perhaps more dangerous, thing to attribute the *wrong kind of mind* to systems that genuinely possess some sort of cognitive capacity. Our machines don't think like us – so we'd better get to know them.

It might be best, then, already to treat AI machines as honorary organisms, with honorary minds – and find out from experience how they act, react, and set and pursue goals. Even while we are very far from having 'conscious machines', we may be best advised to grant them a kind of provisional mindedness.

The robot apocalypse, revisited

In 1965 the computer scientist Irving John ('I. J.') Good, one of Alan Turing's colleagues at Bletchley Park, said that if we succeeded in making an 'ultraintelligent machine' that 'can far surpass all the intellectual activities of any man', there would be an 'intelligence explosion'. Such a machine would rapidly accrue ever more intelligence of its own accord, leaving us far behind not just in intellectual ability but in our capacity to understand the machine's reasons, goals, and desires – as portrayed in Spike Jonze's 2013 film *Her*, where the AI 'companion' Samantha gently and poignantly bids farewell to 'her' human interlocutor Theodore as she heads into a cognitive realm he can't even imagine.

With this intelligence explosion, Good wrote, 'the first ultraintelligent machine is the last invention that man need ever make.' For physicist Max Tegmark, 'the advent of machines that truly think will be the most important event in human history.' It's a popular view that this would be the last invention humankind *could* ever make – because such a machine would be likely to wipe us out, as Stephen Hawking claimed.

Even if such apocalyptic visions owe more to the imagery of movies and science fiction* than to the current reality of AI research, they are shared by informed observers. Astronomer Martin Rees includes AI among the existential threats that lead him to give only 50:50 odds on humans surviving to the twenty-second century.

The Swedish-born philosopher and physicist Nick Bostrom of the Future of Humanity Institute at Oxford University is one of the most influential voices warning of the potential of AI to wreak destruction, particularly in his 2014 book *Superintelligence.* Bostrom is certainly no AI Luddite – he acknowledges its tremendous potential to change our lives for the better. What's more, he appreciates the temptation to make casual assumptions about what powerful AI might be like and what their motivations will be. 'Anthropomorphic frames [of reference] encourage unfounded expectations . . . about the psychology, motivations, and capabilities of a mature superintelligence', he writes. AI systems might in fact 'develop new cognitive modules and skills as needed – including empathy,

* In Frank Herbert's cult sci-fi classic *Dune,* a war with intelligent machines has led to the edict that 'Thou shalt not make a machine in the likeness of a human mind.'

political acumen, and any other powers stereotypically wanting in computer-like personalities.' He is alert to the dangers of falling into dystopian sci-fi tropes.

But that doesn't prevent him from sometimes doing so. For example, if one assumes that all manner of highly speculative and perhaps fantastical future technologies are likely to be realized – nanotechnological 'assemblers' that can reconfigure matter, atom by atom, in any way we want, conscious 'virtual selves' simulated on megacomputers – then there's no limit to the dreams and nightmares one can conjure up for a super-intelligent AI with access to such resources.

There's an element of paranoia in such scenarios. Suppose, Bostrom writes, 'an unfriendly AI of sufficient intelligence realizes that its unfriendly final goals will be best realized if it behaves in a friendly manner initially, so that it will be let out of the box.' Then 'the AI might find subtle ways of concealing its true capabilities and its incriminating intent.' (Who's to say the internet might already not be acting dumb, supplying us with porn and pictures of cats while planning world domination?)

But we don't have to assume the advent of malevolent AIs to end up in deep trouble, Bostrom says – it may be enough that they are *different*. 'We cannot blithely assume that a superintelligence will necessarily share any of the final values stereotypically associated with wisdom and intellectual development in humans – scientific curiosity, benevolent concern for others, spiritual enlightenment and contemplation, humility and selflessness', he says.* What then *can* we say, if anything, about such a mind? Given *any* goal (even one that is meaningless or incomprehensible to us), an advanced AI might reasonably be expected to adopt actions instrumental to

* One can't help noticing how little these values are associated with many human leaders, however.

attaining it. The machine would, for example, seek to preserve itself, to enhance its own cognition (to achieve its goal more readily), and to acquire the resources needed for that end.

And even if *we* (at least initially) decide what goals a superintelligent AI has, our plans might be vulnerable to what Bostrom calls 'perverse instantiation'. We ask the system to make us smile, and it paralyses the human facial muscles into that expression. We ask it to make us happy, and it sticks electrodes in our brains to constantly stimulate our pleasure centres. It's tempting to think that humans who assigned AI these tasks would deserve such outcomes, but Bostrom's most famous example of a perverse instantiation looks more anodyne: we tell an all-powerful AI to make paperclips, and then it proceeds to turn everything it can lay its prosthetic hands on into these objects (for example with those nanotech assemblers) until the world is denuded of resources and perhaps of life. We can imagine putting obvious safeguards in place – but Isaac Asimov's famous Three Laws of Robotics in his *I, Robot* stories warned us how hard it is to get them right.* Besides, perverse instantiations are known to occur in nature. When scientists conducted 'directed evolution' experiments on beetles that imposed a selective pressure to favour mutant forms with a reduced population

* These are:

First Law: A robot may not injure a human being or, through inaction, allow a human being to come to harm.

Second Law: A robot must obey the orders given it by human beings except where such orders would conflict with the First Law.

Third Law: A robot must protect its own existence as long as such protection does not conflict with the First or Second Law.

The popular notion that Asimov's laws are somehow a lynchpin of robot ethics makes roboticists roll their eyes – not just because the laws are of no use to them but because the whole point of Asimov's tales was to show how easily such strictures can go awry and lead to unintended consequences.

size, they didn't just end up breeding (as expected) insects with lower fecundity but also ones that practised cannibalism.

Bostrom's warnings about super-intelligent AI are worth heeding not because they are correct but because they reveal some of our preconceptions about AI minds: specifically, that these will be 'like us but everywhere better', as Bostrom puts it, and that this superiority is likely to create misalignment of goals. Such concerns, I think, are really about the *opacity* of the AI mind. As Bostrom says, 'An artificial intelligence need not much resemble a human mind. AIs could be – indeed, it is likely that most will be – extremely alien.'

In the face of that gulf, we seem predisposed to read into other minds not just malevolence but *a sort of malevolence we already know about* (and how could it be otherwise, really?). As Steven Pinker has said, 'A characteristic of AI dystopias is that they project a parochial alpha-male psychology onto the concept of intelligence . . . There is no law of complex systems that says that intelligent agents must turn into ruthless megalomaniacs.' It's telling, he adds, 'that many of our techno-prophets don't entertain the possibility that artificial intelligence will naturally develop along [stereotypically] female lines – fully capable of solving problems but with no desire to annihilate innocents or dominate the civilization.' That's a rather gender-essentialist view, but you can see the point.

In all this, Pinker sees our own reflection: we are constantly imagining that a supersmart AI will do to us what we have done to animals less well endowed with our particular modes of intelligent cognition. When Bostrom wonders how we might program or teach our own values into AI, he raises the question of what those values should be – and thereby turns this into an exercise for finding moral consensus in ourselves. In this way and others, AI dystopias act as vehicles for exploring our fears and questions about ourselves.

'Perverse instantiations' are not exactly cases of AI gone wicked, but just AI doing what we hadn't intended. It's the story of the sorcerer's apprentice. But Pinker is unconvinced by those scenarios too. 'They depend on the premises', he says,

> (1) that humans are so gifted that they can design an omniscient and omnipotent AI, yet so idiotic that they would give it control of the universe without testing how it works; and (2) that AI would be so brilliant that it could figure out how to transmute elements and rewire brains, yet so imbecilic that it would wreak havoc based on elementary blunders of misunderstanding. [Oh, you mean you didn't want me to take *you* apart?]

These are good points, but I'm not sure they are decisive ones. In response to (1), we might simply point to the invention of social media. And (2) touches on the whole problem of today's AI: we have given it brute-force capability but zero understanding, so that it can mistake a cat for an ostrich.

The fact is that these arguments tend to be based on little but intuition and ignorance. By that I don't mean that the participants on either side are stupid – they are generally extremely smart – but that they are not really so much more informed than the rest of us about what will (and will not) become possible for AI in the medium term. So when one side quite reasonably says, 'Surely we won't make AI so powerful as to be able to do this stuff and yet so morally vacant as to fail to recognize that it shouldn't,' the other can reasonably respond, 'You wanna bet on that?'

What we should really be arguing about is not imagined scenarios in which some super-intelligent AI with mysterious agency over the manipulation of matter turns the world into paperclips – at least, not beyond an allegorical thought experiment – but about what kinds of minds we can plausibly foresee AI having. It certainly makes no sense to speak in one breath of 'human-like general intelligence' while in the next supposing this excludes 'human-like ethical reasoning'.

For example, there is a common presumption that the goals an AI has will be independent of its intelligence. Bostrom makes this an explicit hypothesis; but even Pinker, within the 'anti-catastrophist' camp, says that 'intelligence is the ability to deploy novel means to attain a goal; the goals are extraneous to the intelligence itself.'

You can see what they mean: we can train today's neural-network AIs to do all kinds of tasks using the same basic architecture. But this is not true of 'human-like general intelligence' as represented in the only systems known so far to possess it: us. As we have seen, our minds have been clearly and powerfully shaped by our evolutionary goals: to navigate the physical and social world in a way that enhances our ability to survive and reproduce. Many of our goals – trainspotting or composing counterpoint, say – now don't obviously reflect those origins, in particular because they are shaped also by culture. But our 'intelligence', whatever that means, is intimately determined by the goals that governed its evolution. What this suggests is that we should move beyond plausibility arguments about the risks of near-omniscient AI to an investigation of what kinds of proto-minds our machines have and how they will be shaped by their goals.

The most essential issue for ensuring that increasingly powerful AI remains safe – in other words, remains aligned with our own interests and goals – is not in any event the intrinsic nature of the 'machine mind', but the interface with us. Stuart Russell has argued that to avoid dangers like those Bostrom warns of, we should use three design principles:*

* I don't imagine for a moment that Russell fails to see how these principles might motivate Asimov-like stories about how they, like the aforementioned Three Laws of Robotics, could go wrong. I leave that as an exercise for the reader.

1. The machine's only objective is to maximize the realization of human preferences.
2. The machine is initially uncertain about what those preferences are [it is 'humble'].
3. The ultimate source of information about human preferences is human behaviour.

In other words, we make machines that will constantly be asking: Am I doing this right? Is this good enough? Do you want me to stop yet?

Of course, this raises the question: asking who? Some might consider it alarming that Russian president Vladimir Putin said in 2017 that 'the one who becomes the leader in [AI] will be the ruler of the world'. (He went on to admit, though, that it would be undesirable for anyone to 'win a monopolist position'.) The dangers of giving the machine the wrong sort of feedback are no different from the current dangers of having the wrong people in control of powerful technologies.

We would need to think carefully how to design the machine to cope with the diversity, ambiguity, and uncertainty of human preferences (sometimes even we don't really know what they are) and the moral ecology within which they are expressed. But those are technical refinements to the general notion that a safe AI mind might be one guaranteed constantly and absolutely to defer to our own.

The point is not, Russell emphasizes, to try to encode particular *values* into the machine, precisely because we might not be able to agree on what those values should be, and also because we can't be sure of expressing them precisely and yet flexibly enough to ensure the kind of behaviour we want. Rather, this feedback on *preferences* could keep the machines on the right track.

All the same, I suspect that any scheme like this will remain fraught if it is implemented on machines about whose 'minds' we know very little. We would remain unsure whether the resulting behaviour might not sometimes include unfathomable, perverse, and perhaps disastrous decisions. We can't afford to make superpowerful black boxes.

Most serious and informed thinkers would, I suspect, agree that the challenge is to find a middle path between alarmism and complacency. But navigating into a future in which AI will undoubtedly be more powerful and far-reaching surely demands that we be more, even urgently, curious about the kinds of 'minds' AI might acquire. This, as Bostrom points out, is not just a matter of making AI useful and safe; it is a moral issue.

Some researchers think it inevitable that the all-important human–machine interface will become ever more blurred: that machines will merge with and perhaps entirely replace us – not in a malevolent take-over but because we choose it. According to Susan Schneider, 'As we move further into the twenty-first century, humans may not be the most intelligent beings on the planet for that much longer. The greatest intelligences on the planet will be synthetic.' In consequence, says Martin Rees, 'humans and all they've thought will be just a transient and primitive precursor of the deeper cogitations of a machine-dominated culture extending into the far future and spreading far beyond our Earth.' If he's right, it's worth considering what we – or they – will find out there.

CHAPTER 8

Out of This World

In cosmic terms, our world is insignificant: a mote in the cathedral of space, barely visible even from our neighbouring planets. Is it conceivable that all the mindedness that exists in the universe is concentrated into this infinitesimal volume? Intuitively that makes no sense, although no one can exclude the possibility. Might it then be equally true that all we can say about mind based on our experience on Earth is similarly parochial: that the Space of Possible Minds is equally vast and unknown, and we await a Copernican revolution to open our eyes to it?

Or... might it be that what we have already learnt about minds, however incomplete, is rather like what we have learnt about the laws of physics and chemistry, which we believe to apply equally throughout the observable cosmos? Might the Mindspace that we have begun to explore here be more akin to the Periodic Table of the chemical elements, a kind of universal map of what there is and can be?

Whether that is so or not, I do believe that the concept of Mindspace can offer a framework for thinking about the kinds of mind that might exist beyond the edge of our world. Let's see if we can meet some aliens.

Imagining the alien

Humans have speculated about life on other worlds ever since it was understood that there *are* other worlds. In the second century CE, the Greek-Assyrian writer Lucian of Samosata wrote a remarkable tale, provocatively called the *True Story* (*Alēthē diēgēmata*), in which he related how he was borne on a ship to the Moon by a freak whirlwind, where he and his fellow travellers were caught up in a war between the kings of the Moon and the Sun. The lunar citizens are like men, but the Moon is home to stranger beings too, including three-headed vultures and fleas the size of twelve elephants. Lucian's narrative, while sometimes rather meaninglessly described as the first work of science fiction, was a flight of satirical fiction rather than a serious speculation.

It's understandable that when in the seventeenth century natural philosophers first began to take seriously the possibility that the heavens were populated, they imagined the extra-terrestrials in our own image. In 1610 Galileo reported that the Moon, seen through the newly invented telescope,* was not the smooth sphere Aristotle had proclaimed it to be, but a world like ours with mountains and valleys. It was natural to infer that living beings might dwell there. The German astronomer Johannes Kepler, noting Galileo's claim that the lunar topography was more exaggerated than the Earth's, concluded that the lunarians would be of similarly gigantic stature. That's how they were portrayed in Francis Godwin's 1628 fantasy *The Man in the Moone*, where he added that their skins were of a colour unknown on Earth. In the French writer Cyrano de Bergerac's *States and Empires of the Moon* (published posthumously in 1657) they were portrayed as immense 'man-animals' that walked on all fours.

* Contrary to common belief, it wasn't Galileo himself that invented it.

But regardless of appearance, these aliens talked and thought just like us, even being conversant with our terrestrial philosophers. The conceit allowed these early forays into science fiction to serve as satires of European culture, or as vehicles for philosophical discourses on Aristotelianism and Copernicanism.

While other early-modern and Enlightenment philosophers, including Bernard de Fontenelle, Voltaire, and Immanuel Kant, speculated on alien worlds and cultures, in fiction these typically continued the tradition of mirroring the familiar. The romantic space operas of Edgar Rice Burroughs' Mars novels, of Buck Rogers and Flash Gordon, Frank Herbert's *Dune* and the Star Trek and Star Wars franchises, are full of overlords and dictators, stereotypical heroes and villains, spaceship fleets and empires; they are not much more than the old tales of adventure and heroism transferred to outer space. These Space Odysseys are not really trying to imagine what intelligent aliens might be like, but simply transport human tropes and narratives into fantastical realms. Their alien worlds might just as well be Lilliput, Middle-earth, Cimmeria, or the strange lands conjured up in accounts of medieval voyages such as those of John Mandeville.

H. G. Wells' *The War of the Worlds* (1896) is rightly credited with transforming our concept of intelligent aliens. Gone now were Cyrano's aristocratic lunarians, and in their place were hideous beings with tentacles and flesh-devouring maws, the archetype of the slithering, repulsive alien creatures that became regular features in B movies of the Cold War era. Their mental worlds were equally remote from ours: 'minds that are to our minds as ours are to those of the beasts that perish, intellects vast and cool and unsympathetic,' as Wells put it. Here is the prototype for the grotesque, hammer-headed creature designed by Swiss artist H. R. Giger for Ridley Scott's game-changing *Alien* (1979), or the protean horror of John Carpenter's 1982 *The Thing* (the remake of the 1951 *The Thing from*

Another World, which we encountered earlier). There is no reasoning with these beings. Their minds are as feral as the tiger's, yet far more cunning and intellectually adept; their technologies are often superior, and only human grit and resolve will defeat them. These aliens have only one thing on their mind, as far as we're concerned: we are their prey.

It misses the point to deplore this stereotyping (from which, as far as we know, no aliens have actually suffered). The malevolent alien is not a scientific hypothesis but a mythopoeic projection of fear. For Wells, the Martians showed us what it was like to be faced with the aggression of another civilization with superior technologies: the lethal heat-rays and toxic black smoke of the aliens echoed the military might of the British army colonizing and all but exterminating the indigenous people of Tasmania in the early nineteenth century.

Aliens have, however, sometimes been granted rather more favour and benevolence. Those in Steven Spielberg's *Close Encounters of the Third Kind* (1977) have arrived to give us the benefit of their wisdom, while *ET* (1982) was a mixture of mentor and cuddly pet. (Spielberg could not resist Wells' aggressors, however, remaking *The War of the Worlds* in 2005.) Even Wells presented aliens in a kinder light as the Selenites of *The First Men in the Moon* (1901), with their ant-like social organization and their ability to converse with their human visitors. Denis Villeneuve's 2016 film *Arrival* warns us not to judge from appearances: the aliens might have the octopus forms of Wells' Martians and their spacecraft loom ominously over the land like gigantic, portentous black eggs, but their mission is benign: to liberate humans from our decidedly unintelligent tendency to wipe one another out.

At first glance, science fiction offers us an impressively diverse menagerie of imagined beings. The Martians presented by Olaf Stapledon in his fictional treatise (it scarcely warrants the term novel) *Last and First Men* (1930) are glorious inventions: microscopic cells, smaller even than terrestrial bacteria or viruses, capable of communicating with one another 'telepathically' via something like radio waves in order to merge into intelligent and sentient composite forms. 'The typical Martian organism', he writes, 'was a cloudlet, a group of free-moving members dominated by a "group-mind"'. At the zenith of Martian evolution, the whole planet (save for remnants of more conventional animal and vegetable beings that are waning as the planet becomes parched) 'constituted sometimes a single biological and psychological individual.'

This unusual physiology and lifestyle, Stapledon realized, demands a rather different kind of mind:

> In so strange a body, the mind was inevitably equipped with alien cravings, and alien manners of apprehending its environment . . . Yet it was none the less mind, concerned in the last resort with the maintenance and advancement of life, and the exercise of vital capacities.

The most distinctive difference from us, says Stapledon, is that the Martians' individuality 'was both far more liable to disruption, and at the same time immeasurably more capable of direct participation in the minds of other individuals':

> The Martian cloudlet, though he fell to pieces physically, and also mentally, far more readily than a man, might also at any moment wake up to be the intelligent mind of his race, might begin to perceive with the sense-organs of all other individuals, and experience thoughts and desires which were, so to speak, the resultant of all individual thoughts and desires upon some matter of general interest.

These Martians lack the mental coherence and attentive focus of the human mind, but also our 'inveterate selfishness and spiritual isolation.' The fluid beings are all more or less alike, devoid of diversity or personality; they are little troubled by hate and strife, but experience no love either. They show no philosophical interest in their own existence.

Stapledon delves into the minds of these beings to a degree unusual even in science fiction, both in depth and in imaginative scope.* And yet in the end his Martians display tendencies not so different to those of Wells: they wage war on Earth, and they are perceived as lacking the spark of human-like self-consciousness, emotion, and purpose we might have once called a soul.

Aliens with distributed intelligence appear too in *The Black Cloud*, a 1957 science-fiction novel by the British astrophysicist Fred Hoyle. The eponymous alien manifests as a vast cloud thought at first to be made of insensate gas and dust that enters our solar system and settles around the sun, blocking off its rays, causing catastrophic climate change and preventing plant growth. But as a team of scientists observes the cloud behave in unpredictable ways, they start to suspect that it might be intelligent. They manage to communicate with it and figure out that it is indeed a superorganism, which had no idea that our solid planet could harbour life. Alerted to the damage it is doing, the cloud changes shape to let some sunlight pass, before eventually deciding to leave the solar system to investigate another of its ilk elsewhere in the cosmos.

Some of the most scientifically inventive imagined aliens appear

* The biologist Arik Kershenbaum suggests that 'the older science fiction is, the less likely it is to have been tainted by modern preconceptions about alien life, and therefore the more accurate it may be.' It's an interesting idea, and perhaps Stapledon lends it some credibility. But other eras had their preconceptions too, of course.

in physicist Robert Forward's 1980 'hard sci-fi' novella *Dragon's Egg*, which describes a race called the cheela that dwell on the surface of a neutron star.* Such bodies are the remnants of burnt-out stars that have collapsed under their own gravity, triggering a supernova: a neutron star is what remains when the outer layers of the original star have been blown away in a cataclysmic explosion caused by the release of heat as the star collapses. The star's matter is crushed by the intense gravity into a morass of the subatomic particles called neutrons: incredibly dense, it contains no atoms in the normal sense. But Forward suggests that this material might gather into 'nuclear molecules' with their own chemistry, some of which are able to replicate and form into living organisms spread over the smooth surface of the cooled star. Darwinian evolution unfolds within this two-dimensional world, eventually giving rise to the intelligent cheela. As the neutron star drifts past our solar system, astronomers spot it and a crewed mission is dispatched to investigate.

Since the timescales on which these 'nuclear molecules' interact are much faster than those for normal chemistry, everything on the neutron star happens much more quickly – including the cultural evolution of the cheela, for whom a 'day' lasts about a fifth of a second. Over the space of a mere terrestrial month, the cheela advance from a clan-based and religiously inclined society to having a vastly superior technology to that of the humans. Yet for all the scientific ingenuity of the scenario, the aliens' minds are almost bathetically human-like in their thoughts, language, and objectives.

* This idea was proposed in 1973 by astronomer Frank Drake, a pioneer of the scientific search for extra-terrestrial intelligence (page 344). Forward was originally going to collaborate with veteran sci-fi author Larry Niven, but he wrote the book himself because Niven was too busy on other projects. Even Forward admitted that *Dragon's Egg* was really 'a textbook on neutron star physics disguised as a novel.'

They are not really so different after all from Cyrano's lunarians and solarians, holding up a fairground mirror to human society.

That's often the way it still goes. Many fictional aliens are variations on the fabulous beasts, monsters, deities, and demons of old myths and legends, their minds filled with little more than baleful intent. Often they are historical ciphers: there's little (besides the spaceships) to differentiate the Klingons and Romulans from the Golden Horde of Genghis Khan. At their best, such humanoid aliens serve as foils to help us explore the extremes, prejudices, and habits of the human psyche: Mr Spock lets us ask to what extent logos should govern eros, while the ambisexual denizens of Gethen in Ursula Le Guin's *The Left Hand of Darkness* (1969) challenge our notions of binary sexuality. The main function of the cartoon 'little green men' who seem so eager to meet our leader is to lampoon the absurdity of our ways.

This is by way of saying that if the pantheon of fictional aliens has been, on the whole, rather constrained in its imaginative scope, that is because these tales rarely have as their aim a genuine effort to imagine other minds. *Too alien* a mind could prove an obstacle to authors who want their aliens to serve a symbolic or metaphorical function. Telling us what alien minds might really be like is not their job. In the science-fiction author David Alexander Smith's account of what makes a good fictional alien, we see that above all else they must (and rightly!) serve the demands of narrative:

> One, they have to have intelligent but impenetrable responses to situations. You have to be able to observe the alien's behaviour and say, 'I don't understand the rules by which the alien is making its decisions, but the alien is acting rationally by some set of rules . . .' The second requirement is that they have to care about something. They have to want something and pursue it in the face of obstacles.

Yet these guidelines also reflect some of the common preconceptions among scientists about what aliens will be like: they obey a kind of logic (even if an alien one), and they have goals. Are those fair assumptions?

Looking for ET

One of the first genuinely scientific efforts to imagine the nature of extra-terrestrials was the 1698 book *Cosmotheoros* by the Dutch natural philosopher Christiaan Huygens. Speculating on the environments and habitats he believed to exist on other planets in the solar system, Huygens anticipated the view of some biologists today that life elsewhere will probably have acquired some of the same physical features as that on Earth. 'Tho' we in vain guess at the Figures of those Creatures', Huygens wrote, 'yet we have [already] discover'd somewhat of their manner of Life in general.' To move about, for example, Huygens said that either

> they walk upon two feet or four; or like Insects, upon six, nay sometimes hundreds; or that they fly in the Air bearing up, and wonderfully steering themselves with their Wings; or creep upon the Ground without feet; or by a violent Spring in their Bodies, or paddling with their feet, cut themselves a way in the Waters. I don't believe, nor can I conceive, that there should be any other way than these mention'd. The Animals then in the Planets must make use of one or more of these, like our amphibious Birds, which can swim in Water as well as walk on Land, or fly in the Air; or like our Crocodiles and Sea-Horses, must be Mongrels, between Land and Water.

He supposed too that some of these creatures will, like us, possess intelligence: 'not Men perhaps like ours, but some Creatures or other endued with Reason.' The rationale for this supposition is, however, rather different from the one biologists would give today:

> For all this Furniture and Beauty the Planets are stock'd with seem to have been made in vain, without any design or end, unless there were some in them that might at the same time enjoy the Fruits, and adore the wise Creator of them.

Scientific speculations about alien life continued into the era of modern astronomy. The co-discoverer of natural selection Alfred Russel Wallace discussed the possibility of intelligent alien life in his 1904 book *Man's Place in the Universe*. But in 1940 the British Astronomer Royal Harold Spencer Jones asserted that the matter was likely to remain an eternal mystery. 'It is idle to try to guess what forms life might take in other worlds', he said:

> The human mind cannot refrain from toying with the idea that elsewhere in the universe there may be intelligent beings who are the equals of Man, or perhaps his superiors; beings, we may hope, who have managed their affairs better than Man has managed his.* Neither the investigations of the astronomer nor the investigations of the biologist can help us in this matter. It must remain for ever a sealed book.

It was not alien *life* as such that was out of reach, Jones thought – but alien *minds*. He asserted almost casually that vegetation waxed and waned with the seasons on the surface of Mars, although the gradual drying-up of the red planet meant that 'such vegetation as

* This gloomy view of humankind was no doubt prompted by the fact that the first edition of the book was published while Jones' homeland strove to defend itself from the threat of Nazism in the Battle of Britain.

now continues to maintain a precarious existence must be doomed to extinction in a time which, geologically, is not remote.' But just a few decades later, telescope observations and uncrewed spacecraft missions dispelled all such wishful thinking of abundant life on Mars, revealing it to be a cold, dry, and barren place, its dusty surface baked sterile by ultraviolet radiation from the sun. Given that Mars is the most Earth-like and clement of our neighbouring worlds, the chances of life elsewhere in our immediate cosmic backyard began to look very dim.

Not everyone shared Jones' conviction that the search for extraterrestrial intelligence was hopeless. In 1959 the Italian physicist Giuseppe Cocconi and the American Philip Morrison suggested in *Nature* that we might use radio telescopes to seek out messages broadcast from other star systems by intelligent aliens. They argued that the best frequency to monitor was at 1420 megahertz, in the microwave region of the electromagnetic spectrum. This is one of the frequencies at which the hydrogen atom – the most abundant element by far in the cosmos – absorbs and emits energy, making it (they said) 'a unique objective standard of frequency, which must be known to every observer in the universe.'* Precisely because hydrogen is so common, emissions of this frequency arise naturally from clouds of gas illuminated with starlight that excites the hydrogen atoms to radiate. But Cocconi and Morrison suggested that intelligent aliens might modulate the signal somehow, for example as a sequence of Morse-code-like pulses that could only be imagined to be the intentional product of intelligent minds. 'A sequence of small prime numbers of pulses, or simple arithmetical sums', the pair said, would surely not be generated naturally.

* Cocconi and Morrison advised looking at frequencies slightly more or less than this too, in case the signals were Doppler-shifted by the source moving towards or away from us.

In early 1960 the American astronomer Frank Drake took up the proposal, using the National Radio Astronomy Observatory at Green Bank in West Virginia to search for such signals at 1420 MHz in the direction of two nearby stars. Called Project Ozma, this was the beginning of the Search for Extra-Terrestrial Intelligence (SETI) programme that continues to this day. It has not, in all that time, found a single signal with a strong claim to having been sent by aliens.*

Still the efforts persist. Other scientists have proposed finding aliens by looking for their light-polluting cities; their Starship *Enterprise*-style antimatter drives; or the radiation flashes from extra-terrestrial nuclear war. Harvard astronomer Avi Loeb thinks it might be good to look for spectroscopic signatures of chlorofluorocarbons (CFCs) in the atmospheres of alien planets, apparently in the conviction that aliens will use these synthetic compounds in fridges like ours (or perhaps, who knows, as aerosol propellants for alien hairspray).†

* There have been several false alerts. Famously, in 1967 (in work totally unrelated to SETI) Jocelyn Bell and her PhD supervisor Anthony Hewish observed a radio signal from a star that brightened and dimmed with a very regular pulse every 1.3 seconds. They jokingly christened this the Little Green Men signal, but excited speculations that it was a message from aliens soon gave way to the realization that it was a natural phenomenon: an intense lighthouse beam of radio waves emanating from rapidly rotating, collapsed, and lightless neutron stars called pulsars – the habitat of Robert Forward's fictional cheela. Although it did not signify the existence of an alien civilization, the discovery was of great astrophysical significance and won Hewish a Nobel prize; his student (now Jocelyn Bell Burnell) was unjustly overlooked. On reflection, it seems unlikely that truly intelligent aliens would broadcast with so simple a periodic pattern, for which it is not hard to imagine natural origins.

† Loeb regularly exasperates many astronomers with his unrestrained speculations about aliens. He has asserted that an odd cigar-shaped and unusually shiny lump

The latest incarnation of SETI is an immensely ambitious project called Breakthrough Listen, one of the Breakthrough Initiatives funded by Russian billionaire Yuri Milner from 2015 to search for life in the cosmos. The project aims, between the launch in 2016 and around 2025, to seek potential ETI signals from around a million of the star systems closest to Earth, as well as looking at the hundred or so galaxies closest to us. Other parts of the Breakthrough Initiatives include a competition to find the best messages for sending to other worlds, a scheme to look for 'biosignatures' – telltale signs of life – in the atmospheres of rocky planets around Proxima Centauri (the closest star to Earth) and others nearby, and most ambitiously, a plan to develop miniature space probes that could be accelerated by the pressure exerted by light beams to travel to Proxima Centauri at speeds of up to 100 million miles an hour, which would have them arrive just twenty years after launch. This star is known to be orbited by a rocky Earth-like planet called Proxima b; if the 'Breakthrough Starshot' project works, it is conceivable that close-up images of an extrasolar planet could be broadcast to Earth in just a few decades.

Can we expect any of this activity to yield dividends? Some researchers are convinced it is futile: that we may be alone in the universe, or at least within the patch of it we can ever hope to access. In 1950 the Italian physicist Enrico Fermi was musing with some colleagues about the existence of intelligent aliens who explore the cosmos (such talk was everywhere in the paranoid 'flying saucer' era of the Cold War). If other beings are capable of travel between the stars, surely they'd have spotted us and come to take a look by now, he reasoned. So, Fermi asked, 'Where is everybody?'

There were by that time plenty of folks ready to attest that ET

---

of space rock dubbed 'Oumuamua, seen swooping through the solar system in 2017, is the artificial product of an alien civilization: 'the first sign of intelligent life beyond Earth'. He seems to be more or less alone in this belief.

was here already, and had abducted them for intimate examination, or was being hidden away by the government in a secret hangar. But Fermi had a point. Surely an advanced alien civilization would have found us and introduced itself by now, if it existed? The distances are vast, it's true – but not obviously prohibitive. An alien civilization able to travel through space at a significant fraction of the speed of light could journey across our galactic neighbourhood in perhaps a few decades or so.

Fermi's objection is still cited as an argument for why intelligent life must be rare in the universe. Among the possible answers offered by the SETI Institute, established in 1984 in Mountain View, California, is that aliens 'have done cost–benefit analyses that show interstellar travel to be too costly or too dangerous.' Or maybe 'the Galaxy is urbanised [but] we're in a dullsville suburb.' Or perhaps Earth is being preserved in isolation as 'an exhibit for alien tourists or sociologists.' Maybe intelligent extra-terrestrials have elected to stay incognito and not intervene in the evolution of other planets. (Frankly, though, we could probably do with their help.)

You see what's going on here? We figure we can guess how aliens think just by considering how we do. That tendency is and always has been pervasive in SETI: to imagine aliens in our own image. Planetary scientist Nathalie Cabrol has suggested that SETI is currently a matter of 'searching for other versions of ourselves'. The scientists tend to ask: what would *I* do if I wanted to signal my presence to the rest of the universe? Might I encode some 'universal number', like the initial digits of π, into radio waves? Or broadcast a sequence of pulses in some familiar mathematical pattern, like the Fibonacci series (1,2,3,5,8,13 . . .)?

Not just SETI researchers but scientists generally are oddly dogmatic about how any intelligent being will conceptualize numbers and the laws of physics in a manner recognizable to all – and about any impulse aliens might have in the first place to reach out to the cosmos. They tend to believe that all intelligent beings will share the same mathematics, will have deduced the laws of Newtonian mechanics and the theory of general relativity. Only the symbols will differ. What this boils down to is the assumption that in our scientific investigations and theories, we humans are 'carving nature at its joints' – uncovering some unique expression of the way things work.

Our experience with other possible minds (including AI) so far gives us no reason to suppose this will be so. What appears salient to us – our categories of object, our notions of force and mass, even our laws of geometry – might seem weird or incomprehensible to other, non-human minds, especially if they have different types or ranges of sensory experience. We don't know if, say, a fly distinguishes between cats and dogs, or perhaps whether all types of large animal merge into a generalized picture of rather slow-moving but nonetheless noteworthy threats. The fly's categories of nutrition probably don't align with our apples and pears. Meanwhile, to intelligences that routinely experience the microscopic world, or the planetary scale, there will be much of importance to us that is simply irrelevant or invisible. There's absolutely no guarantee that alien intelligences would conceptualize the universe and its regularities in the same way that we do. Even the system of natural numbers – the integers 1,2,3,4 . . . – while seemingly made inevitable by the very existence of discrete objects, doesn't fit with our own natural intuitions about quantity. These appear instead to recognize a 'logarithmic' rather than arithmetic scale: integer differences between large numbers are less salient than those between small numbers. (To put it crudely, the gap between 128 and 129 'means less' to us than that between 2 and 3.) It's very hard, admittedly, to see how advanced technologies might become possible without a mathematics like ours – but of course we *would* say that.

SETI may be pointless, then (or at best, solipsistic), unless we can develop a broader view of intelligence and mind. 'Without some informed analysis of what it might mean to be intelligent – yet radically different in terms of evolution, culture, biology, planetary location, and the like – we should not even begin to search for extraterrestrial intelligence', says computer scientist William Edmondson. Cabrol agrees: 'To find ET we must expand our minds beyond a deeply rooted Earth-centric perspective and re-evaluate concepts that are taken for granted', she says. 'We must *become the aliens* and understand the many ways they could manifest themselves in their environment and communicate their presence' [my italics]. Most advanced alien species, she says, 'will likely have developed forms of communication completely unrecognizable to us.'

Cabrol's exhortation that we 'step out of our brains' is too seldom heard in SETI projects. But in fairness, it's more easily said than done. After all, we can't even step outside of our own minds into those of our fellow humans. Where do we start?

Distant worlds

In 1961 Frank Drake suggested that even if we know nothing about what alien civilizations might be like, we can try to estimate the probability of their existence. He proposed an equation to calculate how many such civilizations there might be in our galaxy with which we might communicate. This number, he said, is equal to the product of the probabilities for each of the conditions necessary for their existence, namely:

The mean rate of star formation
times the fraction of stars that have planets
times the mean number of planets (for each of those stars) that can support life
times the fraction of life-supporting planets that actually develop life

times the fraction of those for which life develops intelligence
times the fraction of civilizations that develop communication technologies
times the mean length of time that civilizations can communicate.

The problem with this formulation is obvious: we know hardly any of those numbers, and can do little more than guess at them rather wildly. And if any one of them is zero, it makes no difference how big the others are – there's no one out there.

But Drake's equation was not meant to furnish a reliable estimate. Rather, it was an attempt to break the question down into parts that we can consider separately. We can now say something meaningful about the first three terms, for example. Since the first 'extrasolar' planet (or exoplanet: one that orbits another star) was identified in 1995, astronomical searches have found close to five thousand of them, and the tally is mounting rapidly. Initially these discoveries relied on very accurate measurements of the position of the parent star in the sky, to spot regular, tiny wobbles due to the gravitational influence of an orbiting planet. That method could at first spot only giant planets, comparable in size to Jupiter, which were big enough to cause a detectable wobble. If the star's mass is known, the extent of the wobble can reveal the planet's mass.

Most exoplanets are now found by careful measurements of the light radiated by the star. This will dim very slightly if a planet passes in front of it, provided that the plane of the planet's orbit lies roughly along our line of sight. The method is sensitive enough to reveal smaller planets, some about the size of the Earth.

Measurements like these can be used to calculate an exoplanet's mass and size, and the radius of its orbit, from which we can make deductions about the kind of world it is: large and gaseous like Jupiter, say, or small, dense, and rocky like the Earth. Knowing the brightness of the star and the orbit of the exoplanet, we can also make deductions about what the planet's surface conditions are like – in particular, whether it lies in the so-called 'habitable zone' of the star's planetary system, where the temperature is right for liquid water to form on the surface. It's widely agreed that a planet like that might have the conditions needed to support life comparable to Earth's.

Exoplanet surveys now supply pretty good statistics for how common planets are in our galaxy (and therefore probably in most others, since ours is nothing special). It's thought that nearly every star like our own Sun will have them, and that around one in five such stars host Earth-like planets: small, rocky, and able to support liquid water. There is then no shortage of habitable worlds: around 20 billion in the Milky Way galaxy alone. And that's even before we contemplate the possibility of habitable moons too – like Jupiter's Europa and Ganymede, which have liquid water oceans beneath their icy crust, warmed by the energy produced by 'tidal forces' created on the moons by Jupiter's great mass.

But *does* any life exist on these worlds? The problem is not just that it's extremely hard to find out but that we don't really know what to look for. There is no agreed definition of 'life', nor even a consensus on whether 'living things' is what philosophers refer to as a natural kind: a grouping that reflects the structure of the physical world, rather than simply one we find convenient to use.

For exoplanets, such questions tend to pale before the challenge of getting any data that can even start to shed light on the matter. However, this too is changing. In the past several years, astronomers have been able to study the starlight reflected from exoplanet atmospheres, or filtered by it as the planet passes in front of its star. They can see where some wavelengths of this light have been absorbed by the atmosphere, serving as a fingerprint of the chemical substances it contains. In 2019, for example, water vapour was detected in the atmosphere of a planet called K2-18b, orbiting a red-dwarf star 110 light years away in the constellation of Leo. Although that is no surprise in itself – water is a common molecule in the cosmos, and anyway K2-18b is too hot for the water to condense as a liquid on the surface – it shows that we may soon be able to examine the chemistry of exoplanet atmospheres routinely and in some detail.

That matters in the search for extra-terrestrial life, because a planetary biosphere, even if it is very different from Earth's, is expected to leave a chemical imprint on its atmosphere. Intelligent aliens able to study the Earth's atmosphere from afar might deduce that life exists here because the high concentration of oxygen is very hard to explain any other way. This very reactive gas would be quickly removed by chemical reactions due to purely geological processes, if it wasn't being constantly replenished by photosynthesis in bacteria and green plants. In other words, life on Earth keeps the atmosphere in a state of marked chemical 'disequilibrium' – the atmospheric gases don't reach their chemically most stable state. Many planetary scientists think that such disequilibrium in exoplanet atmospheres should furnish a telltale sign of life – if only we can observe it.

So we are getting a fair idea of how common planets are that are 'habitable' by Earth-like organisms. That's a long way from delivering on the requirements of the Drake equation and furnishing an estimate of how likely we are to see a communication from ET. But it shows there are surely plenty of worlds out there for them to inhabit.

Darwin's aliens

If alien life does exist, what is it like? We aren't totally ignorant on that issue, for we do know one thing: the laws of physics and chemistry are universal, at least within this universe. This isn't simply an act of faith; astronomers and astrophysicists can regularly and successfully interpret what they see out there, even billions of light years away, in terms of the known features of physical law. Gravity explains the motions of planets, the birth and death of stars, and the dynamics of whole galaxies.* Chemistry explains the composition of giant gas clouds and the reactions they host, and we can see chemical elements forming in the nuclear furnaces of stars. So we can be pretty confident that if other forms of life exist, they will follow these same laws (whether or not the aliens know about them).

This doesn't mean they have to be Earth-like in their biochemistry, and most probably they won't be. All life on Earth reproduces and passes on genes to successive generations, encoded in the double helix of the DNA molecule.† The genes encode the molecular structures of protein molecules, which act as catalysts for the chemical reactions of metabolism. All of this molecular activity happens in liquid water; life cannot continue to grow and reproduce without it. The biochemical processes of the simplest bacterium are, in this regard, no different to those of human cells.

The molecular basis of extra-terrestrial life could be quite different. Even if it is water-based, it may well not encode heritable information in DNA or use protein catalysts. It might use similar molecules, or they could be totally different. It's possible that alien life might use a completely different liquid as the solvent, rather than water; some think, for example, that the liquid methane pooling on the surface of Saturn's moon Titan at temperatures of minus 179 °C

* There *are* complications in doing this, however. In particular, there seems to be more gravity present in galaxies than can be explained from the observable mass of stars and gas. Astronomers are forced to hypothesize that there is an extra, unobservable form of matter to account for it, distinct from any we have observed on Earth, which they call dark matter. Our current physical theories can't account for it.

† Viruses replicate by hijacking the gene-copying molecular machinery of cells; some encode their own genes in DNA's sister molecule RNA.

might serve that role. If so, the whole of a Titanian's biochemistry would have to be very different from any we're familiar with.

Some astrobiologists – the scientists who consider the possibility of life on other worlds – believe that life does not have to be tied to any particular chemical basis. It may have features that transcend the fine-grained chemical specifics. Astrobiologists Stuart Bartlett and Michael Wong have proposed four fundamental criteria for life:* that it draws on energy sources in its environment that keep it from becoming uniform and unchanging; it grows exponentially (for example by replication); it can regulate itself to stay stable in a changing environment; and it learns and remembers information about that environment.

Darwinian evolution is an example of such learning over very long timescales: genes preserve useful adaptations to particular circumstances, recorded and transmitted in the 'memory bank' of the genome. It's widely thought that Darwinian evolution by natural selection is the *only* viable way for complex organisms to arise from simple origins. If so, we can make some inferences about the kinds of aliens that might be viable in a given environment and that have particular lifestyles. All flying creatures on Earth, from gnats to golden eagles, use some kind of wing: a flat membraneous structure that can be moved up and down to generate thrust. Creatures that swim in water tend to have streamlined, torpedo-shaped bodies to reduce drag as they move. In both situations, evolution has come up with the same working solutions in independent cases: the land-bound mammal ancestor of dolphins, for example, didn't look like a fish, but for aquatic life, dolphins adapted to have a fishy shape.

* To free the discussion from Terracentric preconceptions, Bartlett and Wang suggest calling this more general form of matter 'lyfe', pronounced 'loife'. It's not clear that it will catch on.

These are examples of *convergent evolution*: natural selection finds the same 'good' solution (typically one dictated by physical laws) to a problem independently in different species. Nature is full of such examples. The eye, for example, evolved separately perhaps six times in different lineages of organisms, because this design – a light-sensitive layer of cells onto which light is focused by a transparent lens – is an effective one for vision. Photosynthesis – the ability to capture sunlight and use the energy to produce the chemical compounds of metabolism – seems to have evolved independently more than thirty times.

So although evolution has no 'direction' and merely selects good adaptations from random mutations, nonetheless the solutions it finds are not arbitrary. This leads some biologists to suppose that if aliens also evolved by natural selection, we can anticipate that they too will have forms dictated by the physics of their environment. Flying aliens are likely to have wings, and swimming aquatic ones to be fish-shaped.

Biologist Arik Kershenbaum believes it is hard to imagine any other process except Darwinian evolution by natural selection that could produce complex chemical systems worthy of being considered alive. He says that natural selection follows 'well-defined principles that we know will apply not just on Earth but elsewhere in the universe'. Nothing about it is specific to terrestrial biochemistry: all it requires is replication of organisms, coupled to some way of passing on hereditary information about form and function, as well as the possibility of random errors when this information is replicated. If organisms don't replicate, they would have to sustain themselves indefinitely – which seems impossible, or at least immensely challenging. And if they depend on heritable information – and it is hard to see how else the next generation could be reliably built otherwise – then some mutations seem inevitable (since no chemical copying process could be perfect) and natural selection *must* occur. Natural selection is in this respect likely to be a universal pattern of nature, not a quirk of the Earth.

If we find life and biodiversity on any other world, Kershenbaum is therefore confident that natural selection will have produced it. He believes that convergent evolution is then likely to create similarities of form wherever there are similarities of environment. An Earth-like planet with land, pools, and seas, and a thick atmosphere can be expected to engender recognizable forms. Alien plants that rely on photosynthesis from the light of their parent star would be expected to grow tall in competition with one another. They might be expected to evolve highly branched, fractal roots for gathering nutrients, and seed-spreading mechanisms such as wings and parachutes.

Not everyone buys the hegemony of natural selection as the generator of complex life, however.* Stuart Bartlett points out that even terrestrial organisms can shape their behaviour in ways that don't depend on Darwin's mechanism of random mutations coupled to competition for resources that filters out the best mutations. 'While Darwinian evolution does of course occur, I think it needs to be augmented into a larger picture of biological learning that occurs through a range of processes from the scale of molecules up to ecosystems and planets', he says. Astrobiologist and physicist Sara Walker feels much the same way. 'I'm not convinced Darwinian evolution is necessary in general', she says. 'There might be some systems that have many attributes of life but never cross the threshold to Darwinian life.'

* Arthur C. Clarke, despite being one of the foremost proponents of scientifically literate hard sci-fi, seemed to be among the sceptics. 'Nowhere in space', he wrote, 'will we rest our eyes upon the familiar shapes of trees and plants, or any of the animals that share our world.'

It's not inconceivable that aliens could evolve by other mechanisms. No law of physics prohibits, for instance, the type of evolution posited by the French zoologist Jean-Baptiste Lamarck in the early nineteenth century, in which organisms that acquire an advantageous characteristic during their lifetime become able to pass it on to their offspring. Lamarck's classic example was the giraffe, which, he supposed, gradually lengthens its neck in stretching for high leaves, and then gives birth to offspring that also have longer necks. But there is no known mechanism by which such an acquired anatomical feature can be encoded back into the DNA of the genome – the hereditary material that offspring get from their parents.

Still, something like Lamarckian inheritance is not impossible in principle. For example, our DNA is chemically modified by the environment in a way that can switch genes on and off – a process called epigenetics. Such modifications are generally avoided or removed in the germ cells – sperm and egg – so that they begin a new organism in a more-or-less 'pristine' genetic state. But there have been reports of epigenetic changes to genomes that *do* seem to be retained by offspring. Such claims are controversial, and don't appear to be significant in evolution – at best these changes will last a generation or two before getting eroded away. In some ways, however, our new-found ability to use gene-editing to modify our own genomes, even in germ cells (and thus heritably), might be considered a kind of technological Lamarckianism: a way of consciously directing our own evolution.

Whether this is a *good* way for evolution to happen is another matter. In an uncertain, changeable environment, it is probably better to rely on the random lottery of Darwinian evolution to find good solutions to life's challenges. But there are circumstances in which Lamarckian evolution might do well, for example when the environment is relatively stable and simple so that it's feasible to anticipate what attributes future generations are likely to need. Conversely, it has been argued that Lamarckian evolution such as that proposed to occur by epigenetic mechanisms can be valuable in environments that change too fast for the typically ponderous, cautious pace of Darwinian evolution to produce adequate adaptive fluidity. We humans, meanwhile, rely on a safer and more reliable form of Lamarckianism: not genetic, but cultural. We pass on useful acquired skills through teaching and learning, not through genetic transmission. Any intelligent aliens with a complex culture might be expected to do likewise.

Exominds

Let's suppose that alien life forms *do* evolve by Darwinian natural selection – that seems at least to be a good conservative hypothesis. (If additional routes to complex life exist, all the better for the prospects of aliens.) Can we make some deductions not just about the shapes and functions but also about the kinds of *minds* that might emerge from it? In particular, are there aspects of mind that might be prone to convergent evolution, and thus expected to occur wherever evolution produces minds at all?

A general question for evolved organisms is how much to rely on hard-wired behaviour and how much on the versatility that a mind can offer. This probably depends on how predictable the environment is. The cactus bees of the southwest Pacific are hard-wired to recognize the single species of cactus from which they extract pollen, while the bumblebee is a generalist that exploits many different types of flower and so is better served by a mind that is able to learn what to do for each one, even though that takes longer. The emergence of minds is favoured by uncertainty of circumstances.

A common evolutionary strategy that apes cognition and behaviour, and which might be anticipated among aliens, is predation. We saw in Chapter 6 how the evolution of locomotion during the Ediacaran era might have guided the development of brains. An ability to move is itself a response to selective pressure: it widens the options for finding food and other sources of energy.

Mobile creatures need nervous systems to coordinate movement and to find things to eat and/or avoid getting eaten. They face less predictable environments, and so need more cognitive flexibility. They might also need alarm mechanisms: pain, for instance, is not exactly a *consequence* of injury, but a mechanism for avoiding (more of) it. Whether plants have a genuine pain response is, as we saw, still debated – but the motivation for it to evolve would seem to be small at best, because they can't really take action to escape a pain-inducing threat. Instead they have automated signalling systems to induce repair of damaged tissue, and perhaps to start generating compounds that are toxic to predators. There is no more value in such responses being mediated and motivated by a feeling than there is for, say, our own immune reaction to infection. Fear, too, is an adaptive feature of minds: like any negatively valenced feeling, it creates a motivation to act (even if the action involves freezing into immobility to reduce visibility).

Because predation is such a strong driver of adaptation, we might expect it to feature in any aliens evolved through natural selection. 'We can be confident', says Kershenbaum, 'that alien worlds will (much to the delight of Hollywood) be full of voracious predators.' This doesn't mean, however, that we had best dismantle SETI tomorrow for fear of the consequences if it succeeds in making contact: there is no reason why *intelligent* aliens should remain predatory – or at least, predatory towards *us*. Increasingly we recognize ourselves that the consumption of other animals is neither sustainable nor necessary (and some would add, not ethical either).*

We saw that one hypothesis for why our minds have such behavioural versatility is that this feature was driven by sexual selection: by the demands created from the pressure to acquire sexual partners. Is sex itself an inevitable product of Darwinian evolution? That's by no means clear, since we are still not entirely sure what evolutionary value sex has. It does seem, however, that by recombining genomes in random ways, it offers a way to dilute bad mutations accumulated from faulty replication, and also to distribute and accumulate beneficial mutations. In this way sexual reproduction effectively accelerates natural selection and so might have been vital for the development of complex life. Evidently sex is not essential for evolutionary success, but its advantages might be deemed apparent from the number of organisms that practise it in some form. It's not just a viable and stable evolutionary strategy but a valuable one. If evolutionary history on Earth were to be rerun, it seems a fair bet that sex would emerge again.

In more complex animals, sexual reproduction has significant effects on social structure and behaviour. It nuances kin relationships, and thus in-group cooperation and altruism, beyond what asexual reproduction can deliver. And it commonly produces sexual dimorphism: differences in form and behaviour of different sexes. The forms this can take are diverse, from the queen-and-drone structure of bee colonies to the hermaphroditism of certain fish and molluscs, extreme differences in body size of the two sexes, male aggression, female cannibalism, monogamous and promiscuous

* On the other hand, the development of human culture and civilization was markedly impacted and accelerated by the move from hunter-type predation to farming: a shift that, translated to an alien–human relationship, offers a popular and chilling trope for sci-fi writers.

couplings, and so on. All of these distinct traits shape animal minds, and sexually reproducing aliens would surely find their minds somehow moulded by the activity.*

If, as some researchers think, the exaggerated intelligence of humans is a product of sexual selection – it expanded beyond what was necessary, like the peacock's tail (page 71) – then sex might be a *requirement* for intelligence. But that would also make high intelligence less likely than if it were the product of some more general driving force such as social cognition, since it would also be as *arbitrary* as the peacock's tail: an accident unlikely to be repeated elsewhere. Aliens might share many of the useful cognitive attributes of humans, such as learning and memory, says the evolutionary biologist Anna Dornhaus – *that* kind of intelligence is probably folded into the very definition of life that arises by natural selection. But she believes that intelligence on the scale of humans, perhaps boosted as a quirk of sexual selection on our branch of the evolutionary tree, could be unusual and rare.

Others share that view. It might seem chauvinistic to entertain the idea that we are the pinnacle of intelligent life in the galaxy, but we are after all the only species that has ever evolved on *this* planet that is capable of trying to communicate with other minds beyond the atmosphere. And we've only been here for a blink in evolutionary time, and able to undertake the task for a tiny fraction of that period. Human-scale intelligence, says Steven Pinker, is not a strategy that evolution was *bound* to try. No other animal has needed it, and no wonder: as we've seen, it's a very costly solution in terms of

* It's not obvious, though, that the remarkably similar cognitive and behavioural features of the two human sexes (as well as of intermediate cases) would be shared by intelligent aliens. Ursula Le Guin's 1969 novel *The Left Hand of Darkness* offers an imaginative exploration of how hermaphroditic human-like beings might think differently.

the physiological resources it demands. And we've yet to see if it is *evolutionarily stable*, or on the contrary, self-destructive in the long (perhaps even imminent) term.

Evolutionary biologist Wallace Arthur agrees that advanced intelligence won't be the norm for evolved life. He believes that, although a conservative estimate for the number of inhabitable planets in the Milky Way galaxy alone is around 0.1–7 billion, of those that are actually inhabited the majority will contain nothing but single-celled microbes. A small proportion (which is still a numerically large number), he says, might have multicellular life, and some might have evolved higher life forms like plants and animals. Arthur estimates that there may be around 160 million such worlds in our galaxy. 'There are some planets', he writes, 'on which the cleverest creature is a worm; some on which the highest level of intelligence is equivalent to an Earthly octopus or crow; and some on which broadcasting civilizations considerably more advanced than ours have evolved.'

Yet there is no reason to suppose that the universe has some linear scale of intelligence on which all minds can meaningfully be placed, and which corresponds to the development of a nervous system, a brain, a head, and so forth. It is possible that evolution might narrow the options, but we can't assume that high intelligence comes about only if cognitive complexity happens to expand in the direction in Mindspace that leads to humanlike minds.

If, say, sexual selection were all about having some enhanced cognitive attribute not as a means to an end but purely so that you could say, metaphorically, 'Look at the size of my cognitive display!', presumably it would not much matter what sort of display it was. We saw earlier that humans appear to excel beyond other apes in the sophistication of our social cognition. But what if *memory* became the cognitive peacock's tale of aliens, creating a prodigious (but ultimately pointless) capacity for remembering and recalling?

Or what if *pattern recognition* were the smart alien's forte, so that they could spot subtle relationships between objects or events that are invisible to us? Might they end up with utterly strange-seeming categorizations of the physical world? By decoupling facility from need in the evolution of mind, sexual selection might be imagined to generate bizarre, decidedly non-human forms of high intelligence.

Speaking to ET

Frank Drake's equation for the likelihood of contact with alien civilizations contains an embedded assumption (among others) that any intelligent life form will be intrinsically communicative. The question is then simply whether we could establish a communications channel, and understand what comes down it. How reasonable is that assumption?

Any species that lives in complex societies, from ants to whales, seems likely to require a means of communication – for creatures unable to 'talk to' others can constitute no society at all, but just so many individualistic competitors. Yet there are many ways to communicate besides the ones we use, depending on the habitat. Vision is obviously a useful sense to possess, but is not always ideal for communicating: you have to be looking the right way, and the signal can get blurred in murky water or mist. And vision is of no use at all in the deep, lightless ocean (except for predators that evolve bioluminescent light sources of their own), or in dark caverns. Nocturnal hunters need to supercharge their visual apparatus, like owls, or diversify into other vision-like abilities such as sonar or infrared 'heat vision', say. Sound-based communication has some advantages. It is omnidirectional, unhindered by physical obstacles or corners, and often able to propagate over long distances with little attenuation: low-frequency whale song can be heard by another whale up to several thousand kilometres away.*

The biologist J. B. S. Haldane speculated about a language not communicated by sounds at all, but by smell. It's not obviously an impossibility, but is likely to be inefficient since odour signals are slow to propagate, and readily diluted or dissipated. Certainly, other animals can *signal* to one another through odour, ants being a notable example. 'Ants have dozens of different glands producing hundreds, perhaps thousands, of volatile compounds, allowing for time-coded messages', says Dornhaus. But these communications are no more a language in the real sense than are animal calls or birdsong. They lack a grammar or syntax.

How about electrical communication? Some animals do use electrical sensing (electroception): almost all of these are aquatic, including sharks, eels, rays, lungfish, dolphins, and salamanders, because water is a much better electrical conductor than air. The possibilities for long-distance *telecommunication* of electromagnetic signals more generally are evidently almost magical, as phone networks and Wi-Fi show. But it would be challenging for organic life forms to develop such means of communication. Creating signals of sufficient intensity to broadcast through space requires a lot of energy; and to pick them up, you need the right materials – specifically, metallic substances in which the electrons respond to oscillating electromagnetic waves. Elemental metals are very rare in living organisms, as they tend to be hard to produce and often chemically reactive.

All the same, this kind of transmission is not an obvious impossibility – and some animals after all have magnetic sensors

* Ambient acoustic noise in the oceans caused by ships and other human activities may now mask these signals, interfering with whale mating calls and other communications to a degree that threatens their survival.

that detect the Earth's magnetic field. So we can't trivially rule out aliens that communicate by a kind of Wi-Fi, to all intents telepathic like Olaf Stapledon's Martian cloudlets. How might minds in such intimate contact be shaped? Might they be able to plug themselves directly into radio telescopes to see or hear over cosmic distances?

Given how intimately the architecture of mind is linked to its sensory modalities, these speculations must broaden our view of what minds can be. 'On a planet where life evolved in the darkness of an underground ocean', says Kershenbaum,

> the intelligent creatures there surely have concepts and comprehensions that are utterly unimaginable to us, just as our ideas of sunsets and rainbows would be impossible to convey to them.

This would be harder even than trying to convey the idea of red to a person who was born blind – for in that case, at least, the brains of sighted and blind people are built to the same plan, even if adapted in minor ways to suit their different experiences of the world.

Could we try, though? Is language – conveyed through whatever medium – a given in intelligent societies? If so, need it be anything like ours?

Even if the answers to both questions are affirmative (as Kershenbaum believes) – if there is an inevitable 'natural grammar' to any language-based communication – alien language could be phenomenally hard to interpret. We saw earlier that one of the challenges of trying to decode dolphin vocalizations is that they sound to us like a continuous stream of sound: we don't know where one 'word' ends and the next starts. Might an alien language lack any discreteness at all, being more like the wail of a theremin than the separate notes of a piano? It's hard to imagine how specific information could be encoded within it if there were not distinct phonemes for different concepts; even dolphin vocalizations do seem to have identifiable sounds that represent items. But what if aliens had cognitive modules for layering and decrypting complex patterns of *frequency*: for example, with each 'word' imprinted in the amplitude modulations of a given frequency, and all the 'words' superimposed like so many simultaneous radio broadcasts? Or what if the language was 'musical', a thought expressed and modified as a kind of theme and variations?

It's seldom remarked that languages must convey thoughts – which implies that what they represent must be *thinkable* by the recipient. Human languages have grammars partly to constrain the permissible combinations of phonemes and words into thinkable statements. They don't fully limit these to statements we can assign a meaning ('The green day was laughing'), but so long as the grammar is more or less legitimate, we're likely to end up with articulations that are within spitting distance of being thinkable. (I'm sure you, like me, can find a way to construe that nonsense phrase.)

So the nature of a language will be shaped by what is thinkable to the organism using it. The nested, recursive nature of human languages – we can go on adding clauses to a sentence almost ad infinitum – makes the range of possible thoughts they can express astronomical. But a primitive organism that has only three possible responses to any stimulus would have no use for a complex language-like system of signalling. Complexity of language is intimately linked to the complexity of our choices of action.

Must languages be 'verbalized'? Gestural languages are of course possible; we have them ourselves, as sign language. Our written language mimics the temporal structure of speech, converting it into a sequence of characters that must be experienced in the right order. Could aliens communicate instead with a syntax of objects in space? The inky splodges of the aliens in the film *Arrival* are something of that nature. Could meaning then dispense with a sequential decoding in time, and arrive in a flash – or slowly crystallize from contemplation as some sort of gestalt – much as we might apprehend the meaning of an abstract painting (but with more semantic precision)?

We might wonder too whether any advanced civilization will require some way of, as it were, setting language and thought in stone. 'Writing is the way that we've freed communication from the limitations in space and time', says Dornhaus: it permits transgenerational transmission of information, and thus facilitates cultural evolution, in a way that purely oral communication cannot. Only with the invention of modern recording technologies for sound and vision – innovations of a civilization already based on the recording technology of writing – have new modes of cultural transmission become possible.*

All this suggests that we should not take it for granted that intelligent aliens trying to communicate with us will make the attempt via a temporal sequence of pulses encoded in a radio signal.

Are intelligent aliens mostly machines?

It's often suggested that if we were to establish contact with intelligent aliens, their minds might be vastly superior to ours – they might be 'super-intelligent'. It's less often explained what this could mean, not least because we have no agreed definition of intelligence.

* The pioneer of radio technology Guglielmo Marconi speculated that past events might have left some ghostly trace literally in the ether (the medium thought to carry radio waves) that might be picked up by the new technology. Perhaps, he wondered, we might one day even be able to tune in and hear Christ's last words on the cross. This was a fantasy from the age when the spiritualist belief in communication with the dead was still flourishing.

Thinking about the Space of Possible Minds recommends we consider that intelligent aliens will, rather, be *differently* intelligent – which is to say, differently minded. In what direction(s), however, we can do no more than guess.

There's another complication too: those aliens might not be entirely or even partially biological, but rather, will be *artificial* intelligences. Some researchers have argued that any civilization capable of communicating into the cosmos will soon (on the cosmic timescale) merge with and perhaps ultimately be supplanted by the machines they make – a trajectory on which we ourselves might already be headed.

Susan Schneider, for example, argues that the most advanced alien civilizations will be not biological but 'post-biological': descended from biological organisms but now either blended, cyborg-like, with AI, or fully machine-like.* Does that mean all bets are off when it comes to speculating about alien minds?

Not necessarily, for Schneider and others argue that such super-intelligent AIs might be based on the biological minds that produced them. They might then in turn be governed by the same evolutionary imperatives: survival, reproduction, competition, and cooperation, perhaps a capacity to predict possible futures and out-comes of actions. At the very least, the mere existence of machine

* Science-fiction author Terry Bisson wrote a short story called 'They're made out of meat' based on the dialogue between two aliens encountering humans for the first time and struggling to come to terms with our bizarre hardware:

'No brain?'

'Oh, there is a brain all right. It's just that the brain is made out of meat!'

'So . . . what does the thinking?'

'You're not understanding, are you? The brain does the thinking. The meat.'

'Thinking meat! You're asking me to believe in thinking meat!'

minds would argue that they possess an impulse to sustain themselves. Schneider calls such artificial aliens 'biologically inspired super-intelligent AIs' or BISAs.

Yet it isn't obviously true that the mind of an AI will be dictated by that of its creator – we saw in the previous chapter that our current AI systems, at least, are considerably different kinds of 'mind' from ours. What's more, these speculations about the farther future of AI often capitulate to unproven technological fantasies, in particular to the notion that biological beings with minds like ours might perpetuate those minds in artificial form by 'uploading' them to machines. This notion would indeed imply that any of the inferences about alien minds derived from biological imperatives can then be transferred to the machine minds they are alleged to become.

Mind uploading is the ultimate destination of the belief that the brain is, first, the repository of mind and, second, a kind of computer. The idea (popular in Silicon Valley) is that we could in principle fully characterize the mind by somehow recording the instantaneous state of every neuron in the brain, and then recreate those exact states in sufficiently sophisticated computer circuitry. This transcendent transfer of minds into machines is one of the ultimate goals of the project commonly called transhumanism.

Simulating an entire brain in an AI system is currently far beyond anything we could imagine achieving, but is not obviously ruled out by the laws of physics. Could such a simulation ever replicate our *own* minds? We might anticipate that for an AI to house a human-like mind, it would need to be given some form of embodiment: either a massive array of sensors to put it in touch with its environment, or a flood of input signals that *simulate* such an environment. It would not, moreover, simply receive implanted memories as if they were some high-tech equivalent of mp3 files, but would need to mimic the way the human brain actively constructs memories from the vestigial traces of past experience. And so on. All this, however, sounds like an engineering challenge, not a problem of principle.

How, or if, you might create an exact replica of your own mind is another matter. What is the information that characterizes it? Your mind is currently engaged in processing specific experiences, such as those involved in reading and comprehending these words. This neural activity can hardly be considered a vital component of your mind, though – it was equally 'your mind' when you woke up this morning, before you'd read this page. What is the essential, irreducible content of your mind, irrespective of what it happens to be processing at any moment?

Simply, it doesn't exist. The human mind is not something that is held like data within the neurons that crowd our cranium, but is better regarded as the emergent attribute of our physical body and its interaction with the environment. In consequence, it is not a fixed set of digital information that can be shifted around between hard drives, but is actively constructed and maintained moment by moment in response to the way the environment impinges on the body and its brain, and the internal dynamics of those biological entities.

The idea of mind uploading is, then, a category error. A mind is not that sort of object: it is not a set of binary 1s and 0s in a database.

But wait, let's try this another way. Suppose you have a technology that can instantaneously read every single event happening in your brain at any given moment, so that recreating the same pattern of information in an AI system will bring into being an identical mind as if it were experiencing exactly what you were the moment the data was collected. 'The most straightforward brain-porting scenario', says futurologist Ray Kurzweil, one of the prophets of mind-uploading, 'involves scanning a human brain (most likely from within), capturing *all* of the salient details, and reinstantiating the brain's state in a different – most likely much more powerful – computational substrate.'

He estimates that to capture all those 'salient details' needed to create a viable simulation of the human brain will require about 10^{18} bits (about 100,000 TB) of data: a capacity that is within reach of existing computer technology.

Now, it's true that our current techniques for probing the living human brain are still very crude, and couldn't possibly give us that information. But must that be for ever the case? Kurzweil asserts that the information we need might be collected by machines no bigger than red blood cells, called nanobots, that could be injected into the bloodstream. Such devices have long been mooted in the field of nanotechnology, for example to act as both a police force that patrols through the body seeking pathogenic invaders like viruses and a medical service that patches up tissue damage. Kurzweil claimed in 2005 that this technology for scanning the brain at the level of individual neurons 'will be available by the late 2020s' – so that 'brain uploads' 'will be a feasible procedure and will happen most likely around the late 2030s.'

Well, science and technology are full of surprises, but fifteen years on, Kurzweil's prediction looks so remote from current capabilities that it is almost poignantly absurd. Not only has nothing like a nanobot of this sort ever been created but it is not even a significant goal of today's research in nanotechnology.*

* Don't be misled here by talk of 'molecular machines'. There is genuine, excellent, and exciting work on such things, some of the exponents of which were awarded the 2016 Nobel Prize in chemistry. But these are not 'nanobots' in any real sense; they are synthetic molecules, and small assemblies made from them, that can undergo mechanical changes in response to certain stimuli, for example changing shape or rotating when illuminated with light. Often they draw on

And consider what would be required of it. We might need to register, in an instant, every action potential in every one of the billions of neurons, every squirt of neurotransmitter molecules at all the trillions of synapses, and all the while somehow without catastrophically disrupting the brain in question or filling it chock full of nanobots. Now, an adequate representation of the brain's state of play might not demand our knowing the position and state of *every* neuron and neurotransmitter – at least, we had better hope not. But how would you know what information is important and what isn't? And how would you gather it without disrupting it? In any case, we don't currently have the means to gather *any* of this kind of data. So here again, it's not clear what it can even mean to suppose we could 'copy a mind' by reproducing all its instantaneous microscopic details.

Such visions of quasi-magical technologies are a feature of much transhumanist forecasting, which can tend to become an almost tautological exercise in which it is supposed we will be capable of fantastical feats unmoored to any current technologies. Yet the baroque fantasy of mind-uploading becomes more understandable (even if not in the slightest bit more practical) once we recognize what, for many adherents, is its ultimate destination: many of those who share this vision think it will grant them immortality. Kurzweil puts it like this:

> Currently, when our human hardware crashes, the software of our lives – our personal 'mind file' – dies with it. However, this will not

---

principles similar to those involved in the activity of biological molecules, such as the 'motor proteins' that transport other substances around our cells or that power the contraction of muscle fibres. Many experts who work on molecular machines consider transhumanist nanobots more of a distraction than a destination.

> continue to be the case when we have the means to store and restore the thousands of trillions of bytes of information represented in the pattern that we call our brains.

But even if recording your 'mind file' and transferring it to an AI device were possible (let alone meaningful), it would not produce this kind of immortality at all, for the very obvious reason that what you would have made is a *copy*. If such a recreated machine mind could indeed possess all the memories, impulses, and capabilities of your own, still it would not *be* your own – not least because *you* presumably still exist, lodged back there in your body. You'll have given a kind of life to a machine mind, and it might be terribly grateful, but you'll still go the way of all flesh. The delusion that by uploading your mind, your consciousness somehow goes along for the ride tells us where these fantasies are really rooted: in the ancient concept of metempsychosis, the transmigration of souls. They are not a scientific but a transcendental vision.

Kurzweil and some other transhumanists, however, argue for a notion of self – of *youness* – that does support a kind of technological resurrection of this kind. Known as patternism, it supposes that the true self corresponds to something immaterial and abstract: the dynamic patterns of information flow between our material components. On those grounds, *you* really are recreated if that pattern is recreated, even if in a completely different physical substrate (like a silicon chip).

But this a claim with little real meaning. It's a discussion about how we apply language, not about physical reality. Let's say we had a technology that did actually allow the creation of an exact, atom-for-atom replica of you – something like a modified *Star Trek* transporter. (Yes, we're talking about a quasi-magical technology here – that's partly the point.) There this being stands before you:

is it *you*? Evidently not; you are still in your own body.* And what if, in making the copy, we neglected to make one little finger. Does patternism then say it is no longer you – or that it is another you minus a finger? What if we left out just the memory of last Tuesday? Patternism can't say what's essential to *youness* in the pattern, and what is optional. There is no more an 'essence of you' that can be recorded and stored in a hard drive than there is some 'essence of London' that we can magically recreate in Disneyland. In the end these views on identity ignore the notion of the mind as being a product of real physical stuff within a particular environment, and retreat to the Cartesian, or indeed the theological, concept of soul as a kind of 'thinking stuff' that is distinct from the body. 'Despite the near-religious belief of the digerati in Silicon Valley', says Christof Koch, 'there will not be a Soul 2.0 running in the Cloud.'

There is, however, another potential way for biology, alien or otherwise, to become technology. Perhaps, instead of uploading our 'self' to a machine, we can simply convert ourselves *into* the machine. Neuroscientist Daniel Bor explains what that might involve. Imagine undergoing a series of surgical interventions in which small parts of your brain are replaced by artificial computing circuitry, designed to do exactly the same job as the neurons and to interface perfectly with those around it. At each stage of the procedure, the surgeon asks, 'Do you feel OK? Still the same?' And each time you're able to say yes, you feel fine: you're still here. Little by little, you become more of a cyborg, until your entire brain is made from silicon, all eminently repairable or replaceable. What's more,

* This might not be just a thought experiment. Some cosmological theories suggest that the universe is infinite, in which case it seems inevitable that in some unimaginably distant corner there is an exact replica of you, me, and our entire world, recapitulating every aspect of our lives. Does that (utterly speculative) idea leave you unsure where *you* are? I suspect not.

with that hardware in place, the 'mind file' *can* now easily be read and copied. Such people, says Kurzweil, 'will live out on the Web, projecting bodies whenever they need or want them, including virtual bodies.'

There is surely no harm in positing scenarios like these; indeed, they highlight interesting questions about the nature of self and mind. But they must be seen for what they are: not extrapolations of current technology but speculations in the manner of science fiction that, considered as actual predictions, are essentially faith-based. They belong in the same category as time travel, teleporters, and intergalactic spaceships.

Some other transhumanist possibilities are less obviously fantastical. The progress being made in developing human–machine interfaces, so that artificial devices such as prosthetic limbs can be controlled by the 'will' – by the power of the mind alone – suggests it is not implausible to imagine forging neural links *between* minds. These interfaces need not be wholly synthetic either – for example, there has been progress in methods for interlinking neural tissues by guiding the growth and movements of neurons, a technique with potential for repairing damage to the central nervous system. Suppose two brains, hosted within living people, could be interwoven in this way (we need not, for this thought experiment, dwell too much on the visceral details of how this might be arranged). Christof Koch believes that, at a certain threshold level of connection, such brain-bridging would cause the consciousnesses within each brain to merge into a single awareness that 'sees the world through four eyes, hears through four ears, speaks with two mouths, controls four arms and legs, and shares the personal memories of two lives.'

If brain-bridging were possible, there's no reason why it should be limited to two brains. 'With sufficient advanced technology', says Koch, 'it should be possible to link three, four, or hundreds of brains' – to join them into a unique consciousness, pooling the knowledge and experience of all. 'Is this the supreme destiny of humankind,' he wonders, 'a singular consciousness encapsulating the mental essence of our being?'

Alien technologies

In the 1970s, the inventive British physicist Freeman Dyson took a somewhat jaded view of searches for intelligent extra-terrestrials that were motivated by Cocconi and Morrison's suggestion, using radio telescopes to scan the sky for purposeful messages. There's no harm in that, Dyson conceded, but it relied on the untestable and unquantifiable assumption that 'a significant fraction of intelligent species transmit messages for our enlightenment.' The search for extra-terrestrial intelligence, he argued, does not seem likely to be a productive enterprise while it relies in this way on 'the cooperation of the beings whose activities we are trying to observe.'

Why not instead, Dyson said, seek for civilizations who broadcast their existence inadvertently? 'When we look into the universe for signs of artificial activities, it is technology and not intelligence that we must search for', he wrote. 'It would be much more rewarding to search directly for intelligence, but technology is the only thing we have any chance of seeing.'

How, though, do we spot an alien technology? And how can we have the slightest notion of what they will build?

Dyson reasoned as follows. Any evolved biological mind will have limitations of information-processing, as ours does, that machines might overcome. And we can imagine the motivation for this kind of innovation: intelligent minds would surely figure that controlling their fate and their environment demanded more knowledge than their own minds could encompass or compute. *We* now know, for example, that if we are to sustain a population of today's magnitude (and more) without altering the planet's climate catastrophically, we will probably need better technologies, in particular to replace fossil fuels with clean energy sources such as solar, or perhaps nuclear fusion. That's a very hard problem to crack, but computational power can help, for example by enabling us to predict the properties of new materials such as efficient photovoltaics, or to simulate the conditions of a fusion reactor.

From this point of view, the development of computer-like reckoning devices looks inevitable at some stage in the evolution of intelligent beings, simply as a survival strategy. At the same time (the argument goes), our engineering technologies will become ever more powerful, and we'll develop new ways of harvesting the energy and materials they demand. In the quest for these resources, we are already turning our gaze outwards, into space. Even today we are considering the possibility of unfurling massive solar panels in Earth orbit and beaming their energy back to the planet. And companies already exist that hope (optimistically, at present) to mine the Moon or asteroids for raw materials such as precious metals and rare elements.

Dyson suggested that an expansionist alien civilization would surely seek to harvest as much energy from its parent star – its planet's sun – as it could. It would convert swathes of the material available to it, which might include comets, asteroids, and chunks of nearby planets, into 'biological living space and industrial machinery arranged in orbiting shells around the stars so as to utilize all the starlight.' Dyson himself carried out some rough calculations for what might be needed 'to take apart a planet of the size of the earth and to reassemble it into a collection of habitable balloons orbiting around the sun.'

'Given sufficient time', he concluded, 'the job can be done . . . The construction of an artificial biosphere completely utilizing the light of a star is definitely within the capabilities of any long-lived technological species.'

These planetary-system-scale technological structures have been dubbed Dyson spheres – although Dyson himself explained that he got the idea from Olaf Stapledon, whose 1937 novel *Star Maker* described the end result of a galaxy colonized by this means:

> Not only was every solar system now surrounded by a gauze of light traps, which focused the escaping solar energy for intelligent use, so that the whole galaxy was dimmed, but many stars that were not suited to be suns were disintegrated, and rifled of their prodigious stores of subatomic energy.

Dyson pointed out that his vision converged with the classification of advanced civilizations suggested in 1964 by the Russian astronomer Nikolai Kardashev. A Kardashev Type I civilization can store and use all the available energy on its planet – for example, by harnessing nuclear fusion as well as renewable sources such as solar, wind, and wave. We are not yet at this level, but Dyson believed we probably will be 'within a few hundred years.'*

A Type II civilization controls the energy and matter resources of a star, including its planetary system. A Dyson sphere would supply the energy needed for that mastery. A Type III civilization, meanwhile, controls the resources of its host *galaxy* – this is the kind of situation imagined in *Star Maker*. It's hard to resist going further with these flights of fantasy, and some have done so, adding civilizations of Type IV (which control the entire universe) and Type V (which control many hypothetical universes).

Even to progress from a Type I to II civilization represents a

* Dyson was a contrarian sceptic about the risk of climate change and global heating, and so was presumably not too troubled by the possibility that this could extinguish our species, or at least nullify all the technological gains we have made, in the not-so-distant future.

daunting challenge not only in terms of the technologies needed but also the timescales – which are ultimately constrained by the laws of physics. It takes seven months or so, remember, for our current space technologies just to reach Mars; disassembling even an asteroid, let alone a planet, bit by bit could be the work of a lifetime. Yet that is nothing to the time frame posed by the vastness of interstellar space for a civilization seeking to graduate to Type III. Even with spacecraft able to travel at one hundredth the speed of light, it would take something like ten million years to go from one side of a typical galaxy to another. However, Dyson said, 'the problem of colonization is a problem of biology and not of physics': you just need to live for long enough. 'The colonists', he said, 'may be long-lived creatures in whose sight a thousand years are but as yesterday.'

It's not easy (albeit perhaps not impossible) to imagine how Darwinian evolution could produce sentient organisms of such Methuselah-like longevity. This is where artificial alien superintelligences might come in. To make such herculean feats of planetary engineering feasible, biological organisms might have to merge with machines to give them tremendous lifespans. Given adequate provision for continuous self-repair, it's not obvious why such beings need have a finite lifespan at all (if 'lifespan' has any meaning in such a context).

If a Type II civilization has completely surrounded its star with a Dyson sphere, there wouldn't seem to be much left for us to see with our telescopes: the star would be totally hidden. But any structure like this would not be able to avoid re-radiating some of its captured energy as waste heat – that follows inevitably from the laws of thermodynamics. This heat would be radiated in the form of infrared radiation, which certainly *could* be detectable with our current telescopes. We could then draw up a catalogue of all the observable infrared astrophysical sources we can find, and look at them closely in other parts of the electromagnetic spectrum – visible light and radio waves in particular – to see if they display other telltale signs of being artificially wrought.

Type II civilizations could be even easier to spot if their Dyson spheres were not complete: if, say, the stars were surrounded by gigantic engineering structures that only partly blocked the light. This could result in patterns of visible starlight very different from anything known to be produced by regular stars. A candidate for such a case was reported in 2015: 1,470 light years away in the constellation of Cygnus, it has been dubbed Tabby's Star after the American astronomer Tabetha Boyajian, who led the team that discovered it. The star shows pronounced, irregular dips in brightness, which can't easily be explained.* The announcement stimulated excited but inevitably inconclusive speculation about whether a Dyson sphere might have been found. But most astronomers now agree that the peculiar fluctuations are probably caused by dust or other objects in the stellar system, such as swarms of comets, blocking the light, or by intrinsic variations in the star's light output caused by changes in the way heat moves around inside it.

Freeman Dyson's speculations and the Kardashev scale of alien civilizations have created a tremendously popular framework for thinking about intelligent extra-terrestrials, especially in conjunction with ideas of super-intelligences, artificial or otherwise. But what is it about these notions that strikes you most? Is this not, perhaps, all a bit – science-fictional? Does it not sound rather reminiscent of the grandest of space-opera tales, such as Asimov's *Foundation* saga or the Culture series of Iain M. Banks?

In other words, once again there is unacknowledged

* You might notice that here it's the *irregularity* of the signal that leads to suspicions of its origin in ETI, whereas it was the *regularity* of pulsars that created the same suggestion (page 344). And you might (correctly) infer from this that speculations about what to expect from ETI signals have often been rather vague.

anthropocentrism in these arguments that biological aliens will be creatures like us who devise machines to enhance and then perhaps subsume their capabilities. It assumes that we know what the drivers and motives of alien technological innovation will be. Of course, the whole *point* of (most) science fiction is that it is anthropomorphic: it puts humans, or human-like beings, at the centre of the action, because its real objective is to *tell us about ourselves*.* But this makes it a questionable launching pad for speculations about what alien civilizations will be like.

By the same token, Kardashev's civilizational trajectory, with Dyson spheres as one of its staging points, plots out not some inevitable cosmic arc of intelligent life but a future we might like to imagine for ourselves. It is a rather glorious and inspiring one (at least, some will find it so), in which we shake off our earthly bonds and spread through the cosmos. By the time we reach Type IV, we will surely be like gods.

In other words, what we are being given here is a story of *minds like ours*. As Dyson himself acknowledged, 'It is easy to imagine a highly intelligent society with no particular interest in technology.' NASA researcher Kevin Hand agrees that alien super-intelligences might be perfectly happy to remain comfortably reliant on the regular resources of their stellar system, with no urge to explore or colonize. They might, he says, 'be the cosmic version of the lone intellect in a cabin in the woods, satisfied innately by their own thoughts and internal exploration.'

In other words, Dyson was *explicitly* proposing a search for minds

* It might, admittedly, tell us little more than that we love a good adventure, whether that is the exploits of Buck Rogers, John Carter, or Luke Skywalker. But the most profound and often the most enduring of sci-fi uses imagined futures or worlds to explore the preoccupations and anxieties of the present, from *Frankenstein* to *Brave New World*, *Blade Runner* and *Gattaca*.

like ours. It is precisely because we can anticipate what *we* might do, given the capability, that we can construct this scenario at all. Although he does not quite put it this way, Dyson is suggesting that the only aliens we have much chance of spotting are ones like us. If that is so, we had better hope we are not of a unique kind.

What is real anyway?

Here's a thought that turns this last one on its head. Wouldn't it be ironic if, even as we search for intelligent extra-terrestrials, we are their very creations?

Bizarre though it might sound, this notion is currently popular with some futurologists and transhumanists. They contend that we are all likely to be nothing more than minds created by a gigantic computer simulation conducted by alien super-intelligences. We could be just one run of countless such simulations: virtual experiments conducted much as we today create crude simulated societies in games and scientific research. One can even imagine a regress in which simulated conscious beings themselves simulate others. The argument goes that in fact it is overwhelmingly likely that we *must* be 'sims' like this, given that such a capability would be expected to give rise to many such realizations, while there is only a single physical world underpinning it all. Elon Musk has asserted that the chances that we are living in that 'base reality' is 'a billion to one' against, while Ray Kurzweil suggests that 'maybe our whole universe is a science experiment of some junior high school student in another universe'.

Progress in computer and information technologies over the past few decades has now given us not only games of uncanny realism – with autonomous characters responding to our own choices – but also virtual-reality simulators of tremendous persuasive power. Some researchers now conduct complex simulations of human societies in which the 'agents' act according to simple cognitive rules that guide their decision-making. These studies can give us insights into how, say, cooperation appears, cities evolve, road traffic and economies function, and much else. No one suspects that these simulated agents or avatars in online games have anything like a real life or mind – but it's no longer absurd to imagine that day approaching. Researchers envisage a time, not far away, when these agents' decision-making comes not from simple 'if . . . then . . .' rules but by giving the agents simplified models of the brain and seeing how they respond.

We saw earlier how René Descartes suggested we cannot exclude the possibility that our sensations are being fed to our senses by some deceptive demon: that the whole of what we take for existence is an illusion. If so, we could never know otherwise. Descartes maintained that, for this reason, all we could be sure of is our own existence. But if we are simulated minds, not even *that* would be what it seems.

This, of course, is more or less the scenario of the science-fiction film that it has become almost *de rigueur* to invoke in discussions of the philosophy of mind: *The Matrix*. If you have lived in isolation for several decades and are unfamiliar with it, all you need to know here is that the protagonist Neo (played by Keanu Reeves) becomes aware that what he has taken for normal urban life is in fact an illusion created by signals fed into his brain by a sentient machine, as his body is kept suspended in a vat along with those of the rest of humanity.

Why would we suppose anything so outlandish as that the Matrix is real: that our entire universe is a simulation conducted by super-beings (whether or not their intention is malign)? Philosopher Nick Bostrom argues that this is simply the most plausible of three possibilities. Either, he says,

1. Intelligent civilizations like ours never get to the stage where such simulations are possible (perhaps because they wipe themselves out first); or
2. If they get to that point, they choose for some reason not to conduct such simulations; or
3. We're overwhelmingly likely to be in such a simulation.

It's not obvious how to fault this reasoning;* the question is which of these options seems most likely.

Some argue that there are no compelling a priori reasons to believe (1) or (2). Sure, humanity isn't exactly creating bright prospects for its longevity at the moment. But nothing we've discovered so far makes it seem that such detailed simulations, in which the agents would experience themselves as real and free, are impossible in principle. And even though we've seen that there are many debates about how unique we are, it would appear the height of arrogance to assume that we're currently the most advanced intelligence in the entire universe: that none will have got further along the path of making virtual worlds.

If, then, (3) looks like a viable option, can we hope to know if it's the case or not? Many researchers believe this depends on how good the simulation is. The best way to probe the possibility, they say, would be to search our experience for flaws in the program, just like the glitches that betray the artificial nature of the 'ordinary world' in *The Matrix*. We might, for instance, discover inconsistencies in the laws of physics. Or, according to the late artificial-intelligence maven Marvin Minsky, there might be give-away errors due to 'rounding off' approximations in the computation:

* One objection, raised by Anil Seth, is that it assumes that a functionalist position holds: consciousness (at least of the sort we experience) is independent of the substrate that hosts it. We don't know that this is so.

probabilities of some observable event, say, that don't quite add up to 1 when all outcomes are taken into account.

But even if it's not obviously impossible to prove that we're in a simulation, no one has come up with a 'test' that couldn't be explained in some other way. This might be a box we can't think outside of.

Let's, for the sake of argument, assume that (3) is right. What then? The physicist Max Tegmark recommends that you'd better go out and do some interesting stuff with your life, just in case your simulators otherwise get bored and shut the whole affair down.

That's said at least half in jest (I think). But it inadvertently betrays some of the problems with the whole concept. The idea of super-intelligent simulators saying, 'Ah, look, this run is a bit dull – let's stop it and start another,' is almost comically anthropomorphic. Like Kurzweil's comment about a school project, it imagines our 'creators' as fickle teenage gamers. Discussions around Bostrom's three possibilities involve a similar kind of solipsism. Why should we imagine we can say *anything* about the universe that depends on a straightforward extrapolation of what humans in the twenty-first century are up to? Look (the argument goes), we make computer games – I bet super-beings would do too, only they'd be *awesome*! In trying to imagine what super-intelligent beings and civilizations might do, or even what they'd consist of, we don't have much choice but to start from ourselves. But that shouldn't obscure the fact that we're then spinning webs from ignorance.

It's striking, really, how often we reinvest old philosophical problems – perhaps especially those concerned with the nature of mind and reality – in the latest technological garb. To ask what is 'behind' our appearances and sensations is an ancient impulse. Plato wondered if what we perceive as reality is just like the shadows projected onto the walls of a cave from a world we can never hope to access. Immanuel Kant asserted that if there is some 'Thing in Itself'

(*Ding-an-sich*) underlying appearances, we can never know it. If such philosophical conundrums are ultimately unanswerable, they might nonetheless usefully impel us to examine our assumptions and preconceptions.

But until you can show that drawing distinctions between what we experience and what is 'real' leads to demonstrable differences in what we might observe or do, it doesn't change our notion of reality in a meaningful way. Faced with an argument by the philosopher Bishop Berkeley that the world is merely an illusion, the ebullient eighteenth-century English writer Samuel Johnson famously exclaimed, 'I refute it thus' – and kicked a stone. It proved nothing. But it was the right response.

So too, if you wish to suppose that we are living in a simulation, this doesn't mean your consciousness is a mere illusion wrought by computer. It remains every bit as real; it's just that the ultimate physical substrate in which it manifests is not the one you thought it was. The objection, to put it in philosophical terminology, is not ontological but epistemic: we still think and *are*, but we are mistaken about how that comes about. Which is why I recommend being intensely relaxed about this scenario. Sure, we might get switched off any moment – so yes, enjoy your life, but not simply to entertain and divert our hypothetical (and allegedly fickle) alien puppet-masters.

Brains in space

The challenges posed by physics to the Cartesian certainty of self don't stop there. Modern cosmology raises the possibility that the universe might be infinite in both extent and duration (these are notions fully compatible with an origin in the Big Bang), or at least so vast in either regard that the distinction hardly matters. If so, matters of random chance become problematic, since even the most unlikely of scenarios become almost inevitable somewhere, some time. Consider a cloud of atoms and molecules of the kind that pervade the visible universe, from which stars and galaxies form. Nothing in the laws of physics forbids the possibility that some random fluctuation among all the buzzing atoms might make them momentarily come together into an exact replica of your brain, with precisely the configuration of all the atoms it has right now, perhaps along with its immediate local environment.* If we take the materialistic view that all of your conscious experience, your memories and feelings, is encoded in this particular configuration of particles, such a brain should imagine that it inhabits exactly the life and the environment you're currently in – including the perception of all other 'people' within it. But all that is a kind of illusion conjured out of particles in random motion – and which could dissipate again at any moment.† In other words, not only do you not exist as you imagine, but neither does the entire world or the past that you think you know.

These hypothetical entities are known as Boltzmann brains, after the nineteenth-century Viennese physicist Ludwig Boltzmann, who pioneered our understanding of the probabilistic features of random atomic configurations.

The problem is not just that this scenario seems possible, but that logically it seems overwhelmingly likely. If the universe really is of

* The truth is even more mind-boggling: thanks to the laws of quantum physics, the particles needed to assemble one of these brains might pop spontaneously out of 'empty' space sparsely scattered with just the fundamental particles of light and gravity: photons and gravitons.

† As you might anticipate, these fluctuations could assemble *any* object: a teacup, a vintage Ford Zephyr, a replica of the Taj Mahal. Given the complexity of the human brain, those objects are in fact all considerably more likely to appear than you are.

near-infinite extent, we would expect Boltzmann brains, despite their astronomical unlikelihood at any given moment, to vastly outnumber the actual, continuously existing physical beings that we believe ourselves to be. We seem far more likely to be a Boltzmann brain than to be anything else. Or as the physicist Sean Carroll has put it, 'we need to face the prospect that our leading cosmological model, carefully constructed to fit a multitude of astronomical observations and our current understanding of the laws of physics, actually predicts nonsense.'

There's no consensus on how seriously to take this possibility, although our intuition probably demands (rather loudly) that we regard it as some quirky loophole in our physical theories rather than as a probable account of how things are. It's not enough to say simply that things sure don't *feel* this way; Carroll imagines the argument going something like this:

> Bob: But everything I know and feel and think about the world is what I would expect if I were an ordinary observer who has arisen in the aftermath of a low-entropy Big Bang,* and nothing that I perceive is what I would expect if I were a random fluctuation.
>
> Alice: And in a randomly-fluctuating universe, the overwhelming majority of people who would say exactly that are, as a matter of fact, random fluctuations.

A stronger objection, says Carroll, is that it makes little sense to

* This expresses the fact the second law of thermodynamics insists that all change happens in the direction of increasing entropy – roughly speaking, configurations of particles tend to get more random overall. If that's so, the universe must have started in a low-entropy configuration that is gradually becoming a high-entropy one.

put our faith in arguments for our likely existence as Boltzmann brains that rely on laws of physics that those same random fluctuations *have just inserted into our minds*. The whole argument becomes 'cognitively unstable.' 'If you reason yourself into believing that you live in such a universe', says Carroll, 'you have to conclude that you have no justification for accepting your own reasoning.'

Whether you find that compelling reasoning is a matter of taste. Personally I do, and for much the same reason that I think we can ignore the alleged 'illusory' nature of mind conjured up either by Descartes' demon or by the idea we are living in a simulation. Just because a scenario can be articulated using reason, that doesn't mean it is reasonable. Ideas that change nothing about how we should think or feel, or what we can expect to observe in the world, are metaphysical in a way that disqualifies them from being awarded or withheld a label of 'truth'.

The divine mind

It's surely curious to see, in the 'simulation hypothesis', supposed rationalists using plausibility arguments about aliens and technologies to reconstruct the idea that we and everything we know are the work of some Creator residing outside of it all. But we shouldn't be surprised to see a fundamentally religious impulse resurfacing under the guise of science, any more than we should to see old problems of philosophy repurposed for our secular technological age. Disembodied and transmigrating 'souls', cosmogenic gods, forces working behind the veil of experience: these are all the ancient myths of mind, and science will not banish them. I don't mean to sound dismissive – it's just that it happens more often than you might think.

So then, let's take it all the way. For after all, if our Space of Possible Minds is to have any claim to being comprehensive and all-encompassing, surely it will have to contain the most elusive mind of all: the Mind of God. The notion of God offers perhaps the most profound historical example of how people have tried to conceive of minds different from our own – and how hard they have found it to do so.

Christian theologians have long warned against the hubris of supposing that God is knowable, but to little avail. The public imagination seems determined to personify Him, to give him a gender, a body, a character (jealous and imperious if we judge by the Old Testament, all-loving if we heed the New), aims, hopes, and plans. (And for some reason, a beard.)

We think we know this God. 'God is many things to many people', write psychologists Daniel Wegner and Kurt Gray, 'but to all of them He is a mind.' As you may recall, God showed up in their Mind Club: the group of beings and things to which people attribute mindedness characterized by 'experience' and 'agency' (page 50). Wegner and Gray found that people tend to believe God can think and feel, but that while he knows more than anyone else, he doesn't experience emotions such as embarrassment and lust. He is, in their scheme, perceived as having little experience at all – things don't 'feel a certain way' to God – but he is unrivalled in agency, which is what you would expect for a supposedly omnipotent being. (That similar characteristics of low experience and high agency are felt to be shared by large corporations like Google might give us pause.)

These *perceptions* of the Mind of God of course tell us nothing about what it is really like, if such an entity should exist. And now I must play the spoilsport and say that there is not a great deal of point asking about the 'real' (as opposed to popularly imagined) Mind of God unless you're prepared to take the question seriously. There might be fun to be had with analysing the motives of the anthropomorphic god who treats Abraham and Job with such injustice or acts so jealously towards his rivals in the Old

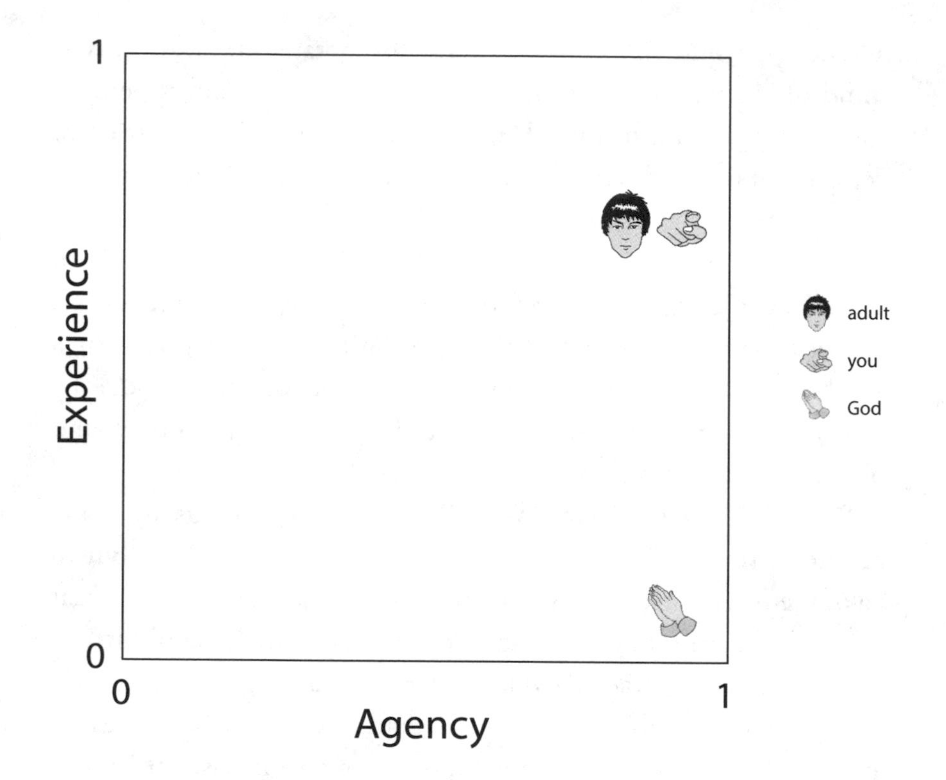

Figure 8.1. God's position in the Mind Club.

Testament.* Or we might try to conduct some psychoanalysis from the traces God is sometimes considered to have left in the mathematical workings of physical laws, or from his inordinate fondness for beetles, as J. B. S. Haldane famously testified. But that would be trite, as it bears no relation to the true conception of God that the most profound theologians of monotheistic (or indeed some polytheistic) faiths have evinced. If we want to speak of what the Mind

* I'm aware that many apologias have been written for this behaviour, and I'm ready to admit that some raise interesting points of interpretation. All the same, judged in terms of the behaviour of one individual to another, the God of the Old Testament perpetrates some rather unforgivable acts.

of God can mean, we should treat that as respectfully as we should when discussing Darwinian evolution – which is to say, to consider what real experts think about it, and not bother with popular misconceptions, however widespread they are.

To this end, philosopher and religious-studies scholar David Bentley Hart recommends making a distinction between God and gods. The latter – like the Greek pantheon of Olympians, the gods of Valhalla, or the ancient Vedic deities such as Indra and Agni – are supernatural beings that differ from us mostly in that they possess superpowers. That's to say, they are agents that act in time according to principles of causation: they make stuff happen, albeit with more potency than we can muster.

Even the 'God' often portrayed in Christianity as the craftsman of the world (Figure 8.2) is merely such a god. Hart refers to this entity as the demiurge, who is a creator and no more, described for example forming Adam out of clay, just as the Chinese goddess Nüwa made people from mud to keep herself company. The demiurge is simply a representation; he is not the ultimate God about which St Augustine writes that 'If you comprehend it, it is not God': the ineffable, unknowable presence that is typically (and not just in Christian tradition) described as negatives: as what He is not. And he is *not*, says Hart,

> in any of the great theistic traditions, merely some rational agent, external to the order of the physical universe, who imposes some kind of design upon an otherwise inert and mindless material order. He is not some discrete being somewhere out there, floating in the great beyond, who fashions nature in accordance with rational laws upon which he is dependent.

This is inconvenient to some New Atheists and other religious detractors, because *that* god is easy to refute. But the fact is – this is where many atheists and committed materialists scream, 'Foul!' – the true God can't be encompassed, described, or understood. We can't speak of Her (there's no point in gendering God, so let's just alternate) as a being at all. This God has neither plan nor personhood, no past or future, no location. He is the source of all existence.

At this point some religious believers might nod their heads in agreement, while non-believers might despair of such mysticism. Make up your own mind (as you might well have done already). But it's surely fair to say that this notion of an ineffable God has at least good motivation: for She is taken as the answer to a question that science cannot answer, which is why there is something rather than nothing. This God is the source of existence itself.

Of course, some *do* claim there is a scientific explanation for why there is 'something rather than nothing' – at least one popular science book promises in its title that it will answer that question. But no scientific theory ever can, because with a genuine *nothing* there

Figure 8.2. God the Creator? Or is this representation of the 'divine craftsman' just a crude demiurge?

is (it should hardly need saying) nothing at all to work with – no equations or laws, not even random quantum fluctuations. That gap, when properly acknowledged, can't be bridged by any scientific theory, because science is not designed to do so – it assumes, at the very least, existence. I don't see how science could ever get beyond this issue,* and I certainly don't want to pretend that at this point it is anything other than a mystery.

The religious answer is to make this source of existence God. Some might say that is a trivial answer: just filling the gap with a name. A more profound view, I think, at least recognizes that the gap exists and that it was bridged somehow. In which case, if we want to ask about the Mind of God, we must start from here.

It doesn't seem to offer us much to go on. This God must – there are sound reasons to say this, if one buys the premise – be a single unified essence of being, with no real attributes to speak of. Nothing can act on, influence, or change Her. She is eternal, and so is Her act of bringing things into being. Our physical universe unfolds in time and space, for sure – but its essential *being* is outside of those notions.

In a sense, Hart suggests, the belief of some scientists that the universe arises from pure mathematical laws – that maths rather than matter is the foundation of existence – is attributing to those laws something more akin to what theologians would attribute to 'the Mind of God'. Stephen Hawking famously did more or less that at the end of *A Brief History of Time*, suggesting that if we knew those fundamental laws of nature, we would know the 'Mind of God'.

Far too much exegesis has been devoted to this example of the showy but shallow rhetorical flourishes that Hawking used to court attention. (Hawking famously had no patience for any belief in

* It's possible that the question of why there is something rather than nothing will turn out to be a bad or wrong one – that's often the way with the most troublesome questions. But personally I cannot think what we'd replace it with.

God, including that of his first wife, Jane.) His statement makes a philosophical category error. If there were such a thing as the Mind of God, says Hart, it isn't the kind of thing we could peer into and find filled with cosmological equations, quantum mechanics, or string theory. If it were, we'd simply have to ask, 'What made God's Mind that way?' and then we're on a quest for an even bigger answer. The God somewhat satirically invoked by the avowedly atheistic Hawking is that demiurgical craftsman, reimagined as (surprise!) a physicist.

No, we can't invoke the Mind of God to explain why things are the way they are – 'Oh, it's because he thinks like this' – but only in response to the fact that they are any way at all. To put it another way, 'Why is there something rather than nothing?' is not a question that has, or can have, a causal answer. God is 'pure intelligibility', says Hart – and if She can be said to have a Mind, it embodies 'the deepest truth of mind and the most universal truth of existence'.

My own view is that, if we keep faith with this picture of God as theologians have conceived it, then it no longer makes sense to think in terms of a divine *mind* at all. That concept floats free from the notion of mind I consider in this book. There is nothing we can imagine or say about 'what it is like' to be this God, or about what His purposes and intentions are. Such a God simply has no need of a mind (or of anything), and to the extent that She 'does' anything at all (apart, that is, from everything), She does not do what it is that minds do. Such a God can no more be placed in the Space of Possible Minds than in space and time itself. The concept of mind then functions not to help us comprehend God but to reveal another sense in which 'God is not *that*'. Personally I have no need of this God, but I can't deny that once you admit the reality of the questions that motivate theologians to postulate his existence, you can't ridicule Him so easily as you can the old guy with thunderbolts, harvesting souls for his kingdom. The God of the deepest traditions in theistic faiths is not really an 'explanation' for anything, not a patch over a gap in reasoning, nor even really an entity that 'exists' in the same sense as we or stars do – but more like the essence of existence on which all logical claims are predicated.

A belief in God is generally accompanied by a belief in what is often rather perplexingly called 'life after death' – which we might more usefully call 'mind after death'. As I pointed out earlier, this is Cartesian dualism again, positing a non-material mind that can only be called supernatural.

The Mind Club of Wegner and Gray found space for this too, after a fashion. It included 'dead people' (see page 50), to whom the participants in their survey attributed a small degree of both 'experience' (the capacity to feel) and 'agency' (the capacity to act). We have evidence of neither, any more than we have evidence of deities. One way of looking at the results is that they reflect a cultural belief in ghosts, spirits, or souls – for example, ghosts are traditionally considered to be capable of manifesting and perhaps communicating with the living, and in the case of poltergeists, of influencing the physical world. But one could invert that formulation and say that maybe these supernatural entities are a way of expressing our intuitions about minds after death: the fact that we can't quite let go of them. 'It may actually be *impossible* to imagine yourself dead', say Wegner and Grey – a tautological claim, really, since one definition of being dead is that the *you* ceases to be. It seems perfectly possible to imagine the world without us: to picture ancient Egypt or settlements on Mars, say. But we can't imagine those things without at the same time imagining an experience of them: I bring to mind images of the Pyramids being built, or sci-fi artists' impressions of colonies on the Red Planet. In effect we invent a new mind through which these imaginings can happen.

This echoes the old adage about a tree falling in a forest where there is no one to hear it fall. And in fact there *is* in that case no sound, if by sound we mean not acoustic vibrations in the air but the effect they have on an auditory system. A person who is totally deaf hears no sound of the falling tree either, not because they *fail* to hear it but because *no actual sound is produced.*

The problem of what happens after we are dead – or before we are born – is from this perspective a serious one, because there is a sense that in those cases, our absence from the scene means that for us *no reality is produced.* The mind is the theatre of our entire reality. Of course, this doesn't mean that the construction of the Great Pyramids 'didn't happen' if we weren't there to see it. Rather, it points to a basic incongruity between our necessary solipsism and the way the universe unfolds. We know (if we accept that we are not Boltzmann brains, computer simulations, or dupes of Descartes' demon) that things happened, and are always happening, without our being witness to them; but to truly conceive of that, we must force our minds out of their mortal coil and send them elsewhere. Yet everything we know about how minds arise makes this disembodiment impossible. There is then not only no fabric from which mind can be constructed but no sense either of what mind could mean.

A belief in souls, in minds that persist after death and beyond the body, seems wholly natural. It is likely in part to be a consolation for the finiteness of our lives – which is why we have long sought bodily immortality and today entertain (some of us) the illusion that our minds can be uploaded into more durable hardware. But it seems also to be an inevitable, necessary, and adaptive feature of humans (at least): to build mental worlds that extend beyond what our sensations tell us, which become part of our cognitive apparatus of prediction and understanding. We want to reify mind. And that is, quite literally, an act of faith.

CHAPTER 9

Free to Choose

Your mind has come to this point on the journey with mine – and for that I'm very grateful, speaking, as it were, mind to mind.

That I can write such a sentence – a prediction of a future event that will never be *proved* wrong – distils all that is so powerful and profound about our minds. I'm demonstrating that I possess a Theory of Mind – that I believe you have a mind too, and it is rather like mine – as well as constructing a future scenario that, based on my representation of the world, I know to be possible and very much hope will be realized. And I'm expressing it through this wonderful cultural invention of minds – language – that permits abstract concepts and novel hypotheses to be conveyed from one mind to another.

But there's something more to your still being here that I've barely touched on. Short of having been set a strange reading challenge, I am supposing that you made it to this point by choice. There was no prior guarantee that you'd get to page 397 – not because any law of physics would prohibit it, but because your actions, like mine, are contingent. There is nothing inevitable about minds.

Or is there?

Consciousness is often (and understandably) regarded as the deepest and most mysterious facet of mind. But what we call *free will* is in some respects even more puzzling. If you believe it exists – and many do not, as we'll see – then it suggests that minds alter the universe in an astonishing way. Before there were minds, there was an inevitability* to events – literally, there was nothing to be done about them. From the moment of the Big Bang, our current scientific understanding says† that things unfolded according to blind physical laws. But when minds with free will arrived, something changed. If you're at home or in a library, a cafe, any human-built environment, then pretty much all that you see around you is not of the same nature as a tree or a mountain. It is the way it is because minds have decided it shall be that way. Minds have shaped that world.

Contrary to what some scientists believe, this doesn't imply any inconsistency between the *volitional* powers of mind and physical law. Nothing about the actions that minds direct in the world need (or should) contradict physics. There would be a big problem if it did – or rather, we'd be missing some hugely important component of how the world works. But free will seems to suggest that minds introduce a *new causative agency* into the world: one that was not there in or soon after the Big Bang.

In short, minds appear to make the universe self-directing. Rather than being in thrall to relentless laws, it can take control of

* This is not the same as predictability or determinacy. As I explain later, our current understanding of quantum mechanics indicates that there is a fundamental randomness to outcomes at the quantum level: an irreducible randomness about what transpires. Some interpretations of quantum mechanics say this apparent randomness is not real, and that underneath it is perfect predictability, but not one we can ever access ourselves. That caveat, however, makes the issue metaphysical; to all intents and purposes we can't know in advance how, in general, a quantum process will turn out. But neither can we then influence it – there's still 'nothing to be done' about the outcome.

† What religious belief says, on the other hand, is complicated and varied, but needn't concern us here.

how it is arranged. I can't think of a more astonishing, perplexing idea than that.

It's *so* perplexing that some scientists do not accept it. They argue that free will is incompatible with what we know of physical law, and so must be a mere illusion – a story with which our minds flatter ourselves about our powers of agency. Others – traditionally, philosophers have called them *compatibilists* – insist that, for one reason or another, there is no conflict between free will and physical law.

What I hope to show here is that, when we understand properly what minds are and how they arise and operate, compatibilism is the right position – but not for the reasons commonly advanced for it.

There are some good practical reasons to get matters straight about free will. In particular, there are important legal and moral implications: the question has a bearing on how we think about personal responsibility and justice. It surely matters too that we have a clear sense of what we can and cannot do to influence the course of our lives. To suppose that we are puppets of the universe, in thrall to inexorable fate, seems not only dispiriting but unhealthy: it is not only (I believe) a false view, but potentially a harmful one.

I would worry also that if you were to come away thinking free will is an illusion, I will have failed to convey something essential about Mindspace – about what it is that minds (at least those we know about) can do.

The swerve

Philosophers, theologians, and scientists have pondered and often fretted about free will for millennia. The ancient Greek Epicurus and his followers held that if all of nature is just atoms executing motions according to natural laws, a space for freedom of the human will can only be carved out by supposing that those motions are occasionally interrupted by randomness: what the Epicureans called a 'swerve'. What we know of the Epicurean philosophy is due largely to what is said of it in the poetic verse treatise *De rerum natura* by the Roman writer Lucretius of the first century CE, which expounds the case for an atomistic view of the physical world.*

This ad hoc and unsatisfactory – but rigorously mechanical – explanation for how free will might work was replaced in conventional Christian theology by the doctrine of the divinely granted soul, to which Descartes attributed motive agency over the body.† While that essentially supernatural account (as we'd now see it) will no longer do either, we haven't yet been able to come up with an alternative view of what free will is, or whether it exists at all, that everyone agrees with. In many ways, our advancing understanding of brain and mind is only making the matter seem harder.

At least part of this difficulty arises because of a lack of clear distinction between the traditional philosophical view of free will and questions about mind and behaviour. A lot of time and effort can be consumed by arguments that are really just about definitions. So I had better make it clear from the outset what I mean by 'free will'.

One view – what might be regarded as a 'folk' definition – is that

* The 'atomistic theory' of ancient Greece, developed by Leucippus and his pupil Democritus, is not the same as the scientific 'atomic theory' devised from the start of the nineteenth century – but the two share the view that all substances are ultimately composed of tiny indivisible particles. It was discovered in the early twentieth century that in fact even atoms can be subdivided into smaller particles.

† This did not settle the nature of free will for theologians, because of the issue of predestination: whether human actions were prescribed in advance by God's omniscient and omnipotent agency, or whether we genuinely have a self-determining power of salvation. Some early Christian thinkers, such as the Welshman Pelagius in the fourth and fifth centuries, were denounced for heresy because their ideas about human free will seemed to question divine omnipotence.

it operates only when we take an action while being consciously aware of having made the decision to do so. You are faced with a decision about what to do; you make the decision (and know you've made it); and then you act accordingly. It might sound perfectly reasonable on first hearing. But our willed actions aren't generally determined on this moment-to-moment basis; we need to take a broader view of how the mind regulates itself. For example, we often exert what we colloquially call 'willpower' in advance of an action, setting the mind on course, as it were: 'When Tessa offers to buy the next round, I'm going to say no because I've had enough. I know I'll be tempted, but that's what I'll do.' You might object that the decision isn't really being made until the testing moment comes, and that we might yet cave in when it does. But the point is that this process is an aspect of *what the mind does to exert control.* Mentalizing in a volitional manner is not simply a series of snap decisions, with each one of which the mind tells itself 'now I've decided'.

Another common definition is that given by the philosopher John Searle: free will is exercised when we could have done other than we did. I could have chosen the duck (what was there to stop me?), but I went for the beef. This too seems a sound enough definition based on our everyday experience of making decisions. But as we'll see, it too is a flawed view – because it relies on counterfactuals that can never be observed.

The biologist Anthony Cashmore has offered another definition: free will, he says, is 'a belief that there is a component to biological behavior that is something more than the unavoidable consequences of the genetic and environmental history of the individual and the possible stochastic [random] laws of nature.' Yet again, it sounds good at first. As we saw earlier, it is beyond serious doubt that both our genetic constitution and our environment affect our behaviour. (Also within this mix of influences, although vastly underappreciated, is the high degree of randomness (stochasticity) that impinges on the way our brains develop and wire up, and which affects the behavioural tendencies it generates.) And when we are faced with a decision, all of these factors are, so to speak, already in place. If they are *all* that determines the decision, then surely it is indeed unavoidable and inevitable, lacking in freedom? And yet, what more can there be at play than this? Cashmore is saying that free will is a belief that there is something beyond *everything we know about* that influences behaviour – some mysterious force, some non-physical psychic power, label it how you will. Unsurprisingly, he is led to conclude that a belief in free will is thus a phantasm akin to the old belief in vitalism (a mysterious, non-physical force in all living things) or to Descartes' dualism with the soul acting as the 'ghost in the machine'. Cashmore's definition insists that this is all free will *can* be.

But if free will is not *that* either, then what's left? And how can it escape from being predetermined, or at least inevitable?

Cashmore's position is much the same as that prescribed by physical determinism, which says that there is an inevitability to all events because they arise only from the physical interactions of fundamental particles. The particular configuration of particles at any given moment therefore determines what will transpire in the next moment: there is no room left, in our current understanding of the laws of nature, for anything else to exert an influence.

I will consider that viewpoint more closely below, but for now let me say that it amounts to the statement that *only one thing ever happens, and does so for reasons*. That's certainly not a trivial statement, nor is it self-evidently true.* But as an assumption it forms

* Those who believe in the so-called Many Worlds interpretation of quantum mechanics assert that according to that physical theory, not one but many things happen: indeed, all things happen that can possibly happen, but the different outcomes happen in different universes that split from one another when the

the basis of all scientific investigation. More to the point, it is *silent* about what free will, properly construed, might be.

For the real question is: *What then are those reasons?* What are the causes of things that happen? I propose that for free will to be a real thing, it is enough that we can say that our mental processes, involving (say) deliberation, prediction, memory, emotion, model-building, and typically some degree of conscious awareness – in other words, *the fundamental features of mind* – can be meaningfully said to be the primary causal agency of what we do.

I will need to unpack that idea. But let me confess that 'free will' is actually a terrible label to stick on it. The neuroscientist Michael Gazzaniga very reasonably asks what, if we have 'free' will, we suppose it to be free *from*. Clearly not the laws of physics, or from the constraints imposed by memory, learning, all social rules, our emotions. The idea, often implied in traditional philosophy, that free will acts outside of all external influences and even of physical laws, is indeed absurd. Trivially, we can't do what is physically impossible: my free will won't allow me to fly or become invisible. And of course we are constrained by a host of contingent factors – biological, social, and cultural. I am not 'free' to factorize a hundred-digit number, although that operation is mathematically possible. I am in one sense 'free' to walk naked into the supermarket, but social pressures will (rest assured) prevent my mind from choosing to do so. If one day I do it, that will not be because I have become 'more

---

outcomes become manifest and fixed – when we can truly say 'that happened'. I have explained in my book *Beyond Weird* why I don't think this is a meaningful – indeed, why it is not a cognitively coherent – way of interpreting quantum theory. Let's just say for now that (contrary to the strident claims of some advocates) nothing in our current knowledge of quantum mechanics compels a Many Worlds position. But the issue is irrelevant for free will, because even in the Many Worlds position, only one thing ever happens that *we can possibly observe*.

free' but because something has happened to my mind that interferes with the way I prefer it to function and make choices. In the UK, I do not consider myself 'free' to choose to drive on the right-hand side of the road, even though I know from experience in the US that I have the capacity to do so.

'Will' too is a term freighted with historical baggage. It immediately invites the image of that mysterious non-physical power – precisely because it was coined in a context in which the existence of such a thing was widely believed. While the basic idea is, as we've seen, ancient, the word now carries connotations of the late eighteenth-century German school of Romantic philosophy that asserted the active, self-determining autonomy of mind, culminating in Arthur Schopenhauer's view of the non-material will as the inner essence of the world.

For such reasons, it is far better – as many neurobiologists now believe – to talk in terms of volitional decision-making. I believe that the only meaningful notion of free will – and it is one that seems to me to satisfy all reasonable demands traditionally made of it – is one in which volitional decision-making can be shown to happen according to the definition I give above: in short, that the mind operates as an autonomous source of behaviour and control. It is this, I suspect, that most people have vaguely in mind when speaking of free will: the sense that we are the authors of our actions and that we have some say in what happens to us.

This, Daniel Dennett argues, is 'free will worth wanting.' To demand that free will only counts if it acts outside of physics and untrammelled by circumstance is, in his view, to indulge in 'free will inflation'. It is to impose a kind of Cartesian dualism, simply to tee it up for an easy act of demolition. It is a cheap but ultimately pointless way of being able to claim that 'there is no such thing as free will.' And it is to ignore all that we have come to understand about the nature of minds.

Laplace's demon

A denial of free will typically asserts that we are the mere puppets of implacable forces and events we cannot hope to control – that our sense of personal agency in the world is an illusion. Here's how that argument generally goes.

As far as we know, everything in the universe happens according to physical laws. At the smallest scale we can currently access with experiments and observations, these laws describe the properties and interactions of subatomic particles, such as the protons, neutrons, and electrons that make up atoms, and the forces that act between them. There is currently no evidence that the world is, at this fundamental level, made up of anything more.* There is no reason to think that some 'force of will' exists that can move particles around, in addition to the forces we know.

Everything that happens at larger scales – the interactions of molecules in cells, the movements and assemblies of those cells themselves, the operation of neurons by ions passing across their membranes and creating action potentials, the networks of neurons in the brain processing information and making decisions – can be studied in a reductionistic way by breaking down these complex processes into simpler ones at ever smaller scales. Even if our understanding is still not complete at every stage of this reduction, there

* We don't know that this description is actually the *most* fundamental, and it almost certainly is not: there are more layers to the onion of reality. Ideas such as string theory are attempts – so far unproven – to get at this deeper level. There are also currently unexplained and speculative particles and phenomena that physicists and cosmologists have to add to this picture, such as dark matter and dark energy. But none of these unknowns materially affect this argument, if all of those entities too follow rigorous mathematical laws of some yet-to-be-discovered kind.

seem to be no *essential* missing ingredients. Nowhere in this description is there anything but blind physical forces acting between atoms and molecules according to mathematical rules, the effects of which are pooled at ever widening scales to produce what I experience as thoughts and actions.

In principle, then, it should be possible to use those fundamental mathematical laws of particle interactions to *predict* exactly what the outcome of any situation will be, if you know the initial positions and properties of all the particles. In practice that is impossible for us to do today – and given the immense number of particles involved in, say, me (rough estimate: 10^{27} atoms), it is likely to stay that way for the foreseeable future. (For reference, note that it stretches the capacity of today's supercomputers to predict the behaviour of just a single small molecule over a single nanosecond, starting from a complete quantum-mechanical description of its particles.) But the point is that there *is* an outcome, and only these mathematical laws dictate it. So it can only happen one way. Where can any *choice* intervene to alter that outcome?

In other words (this argument goes), the world is deterministic: the state of the present makes the state of the future predictable and inevitable. This case was made in the eighteenth century, when many scientists believed that fundamental particles of some sort lay at the root of all physical phenomena, governed by forces of the kind that Isaac Newton had invoked to explain the motions of the planets around the sun. The world was seen as a kind of intricate Newtonian clockwork mechanism. Thus, the French scientist Pierre-Simon Laplace supposed, if some tiny but all-seeing demon were able to see and record at any instant the positions of all the particles in the universe and to know the forces that acted between them, it could predict the future in every detail. There was no room for free will to operate.

Today we must modify this picture of Laplace's demon. We

know that the behaviour of atoms and subatomic particles is not governed by classical laws like those that describe the gravitational interactions of the heavenly bodies, but instead by the laws of quantum mechanics. And it seems to be a fundamental peculiarity of quantum mechanics that it instils some randomness in the outcomes of these microscopic interactions. Two particles that feel each other's presence, say, might do *this* or *that*, but we can never know in any instance which it will be. The possible outcomes are generally well defined, however, and we can use quantum mechanics to calculate the probabilities of each – meaning that, if we count up from repeated enactments of the process in question how many times it produces *this* and how many times *that*, the statistics will be exactly those that the equations of quantum mechanics predict.

Some have tried to argue that this quantum randomness rescues free will because it makes the universe indeterminate: Laplace's demon could never predict the future. However, that won't work – for randomness is not choice. The whole point of free will is not that we will do this or that at random, but that we will choose between those alternatives deliberately – that is what *will* means here. In a quantum universe, there still seems to be no way for willed outcomes to arise, for there is no such willed behaviour in the microscopic interactions that impel all events, including those in the brain. Quantum indeterminacy might undermine the *predictability*, but not the *inevitability*, of the future, in the sense that there's not a damned thing we can do about it.

It seems like a fairly watertight argument, and many physicists think it is. In this view, we are *all* machines, and the difference between us and AI (aside from the materials we are made from) is that we happen to have information-processing hardware and/or software that somehow becomes self-aware, and as a result is able to fool itself into believing that it is controlling events that in fact are inevitable (albeit laced with a dash of uncontrollable randomness). The only way to 'rescue' free will is then to fall back on Cartesian dualism: to invoke some non-physical, quasi-mystical agency that can direct atoms and molecules to do its bidding, and thus to *cause* things to happen in the world of people.

But that's a mistake.

You see, a proper scientific account of what we actually observe in the world is not a mere description, but an explanation. It doesn't just recount what happened, but provides a mechanistic account of why. If, say, you want to explain why light passing through a prism forms a rainbow spectrum, it's not enough to say that the purple light took a path at a different angle to the red light. We want to be able to say *why* that was so. We want to speak of causes of phenomena.

Well, you might say, this is indeed what an account of the world based on the physical interactions between fundamental particles does. It traces all cause back to those ingredients.

Let's see about that. How was it that my coffee cup got to move from the kitchen up here to my desk? Here's one account: when the cup was in the kitchen this morning, the atoms in my hand came into close proximity with those in the handle, and quantum-mechanical effects created a repulsion between them so that when the atoms of my hand rose, the cup rose with it (because the electrons in the 'cup atoms' bound them all together) . . . Well, I won't go on; you probably see the point. Even if I describe what happens to every atom in my cup and its environment, this approach will give no explanation worthy of the name for how it got *here*. Why did my hand rise? Because of signals from my brain. Why did those signals arise? Oh, *now* you're asking. It was because my brain has desires, beliefs, motives. (I wanted to drink a cup of coffee at my desk.)

Why, why? I hope you can see that by reducing these questions about cause to more 'fundamental' ones involving molecules and atoms, we are actually getting ever further from an explanation that can truly invoke causes and mechanisms worthy of the name. Those things become diluted to invisibility in much the same way that an image on your computer screen loses all meaning when you zero in on the individual pixels. The *right* questions – ones that will point us to causation – come from moving in the other direction: up from atoms towards higher levels of the hierarchy.

Of course, there is no single scale of phenomena on which the question of why my cup got up here is fully and comprehensively answered. But the most satisfactory answer is to be found at the level of my brain and the mind it orchestrates: in terms of behavioural decisions, and in particular, of my *agency*. We can ask further why I made *those* decisions, and it brings us into questions about culture and nurture (hence coffee, not green tea), and evolution (hence thirst and a liking for bitterness). But the particle level is barren of causative content.

Some who seek still to deny free will on grounds of philosophical determinism (with a shake of quantum randomness) argue that we could nevertheless in principle build *all* of this explanatory content up from particles alone. But once you take that approach seriously, you soon realize that it requires you to rerun evolution (to get brains like ours), and star formation (to get our Sun and Earth – not generically, but precisely) . . . and in fact you can't then obviously exclude anything in your reductive account short of the entire history of the cosmos since the Big Bang. And not a generic Big Bang either – because quantum indeterminacy alone will ensure that this won't recreate me and my coffee cup. No – *that* attempt at 'explaining' how my cup got up here requires a more or less complete account of *everything that happened, ever*. In other words, it's no explanation but just an absurdly exhaustive description* – and one, moreover, that is impossible to test.

This idea that causation does not necessarily flow from the bottom upwards – starting among subatomic particles and becoming ever more aggregated and 'coarse-grained' at larger scales – is not mere handwaving. It has been shown mathematically† that the behaviour of some complex systems – those that contain many interacting component parts, which certainly includes brains and bodies – must be attributed primarily to causes arising at the higher levels of the system rather than in the interactions of the most basic components. This is sometimes called top-down causation.

This is another slippery term that is apt to be misunderstood. It's sometimes taken to imply that processes at the lowest levels of a system – the interactions between atoms, say – can be undermined or over-ridden by things happening at higher levels. That can't happen, as far as we know. However, the states of atoms can be *influenced* by thoughts. This might sound wacky, but it really isn't. If you decide to boil a kettle, your decision results in an increase in the thermal jiggling of the atoms in water molecules. Your higher-level decision‡ was a genuine cause: the kettle wasn't going to turn itself on.

Of course, that thought was constructed from neuron-level and molecular-level events. But the thought is not *inherent* in them.

* Even 'description' doesn't quite suffice, implying as it does an outside account of what happened. The extreme reductionist approach rather demands a step-by-step mapping of reality on a scale of 1:1 – it is in effect merely expressing an identity relation.

† Some of this work was done by Giulio Tononi, one of the main architects of the integrated information theory of consciousness, and his co-workers.

‡ More properly, it was this thought plus the action it induced that turned the kettle on; thought alone can't (yet!) do that.

There is no unique configuration of atoms and molecules that corresponds to a 'switch on the kettle' thought: it will differ every time you do it, and from one person to the next, and is not in any case something that can be isolated and demarcated from the rest of the brain's activity. You could never deduce from the bottom up that a given configuration of particles *was* a 'turn on the kettle' thought; the brain state only makes sense within the context of concepts such as 'kettle' and 'wanting a cup of coffee'. Physicists who think carefully about causation (and aren't in thrall to crude deterministic arguments) speak in terms of the system having some *irreducible* complexity. When such a system attains a certain level of complexity, types of behaviour become possible that were not possible before that, and which cannot be understood or causally explained from the bottom up.

Physicists Marius Krumm and Markus Müller have cast this situation in a different way. They suggest there are some processes in nature (of which they think volitional human decisions are an example) for which any successful prediction of the outcome – by an exhaustive computer simulation of all the particles and forces, say – requires an almost exact representation of every relevant aspect of the system.* There are no shortcuts: the simulation would have to represent all the desires, beliefs, emotions, and so on. In that event, the *entire system* must be regarded as the 'computational source' of the outcome. In other words, an attempt to replace an account of causation based on the elements of mind with a reductionistic, particle-based view in which all those elements have been broken apart must, in such cases, simply return you back to where you started: you have to put them all back in again.

* This amounts to saying that it requires an exact repetition of the history of the event – or that 'if you could make something happen again, it would happen again'. If *that* is all determinism can mean, it means nothing.

Well, you might still object, it wasn't really your *mind* that did it, but all those hormone molecules and ion channels and electrical potentials, all those neurons being wired the way they are by your genes and past experiences, all your history and memories so far . . .

But just what the heck do you think minds *are made from*, if not all that stuff and more? This, ultimately, is the fallacy of the determinism argument: it insists that there's no free will because we can't adduce a homunculus that is directing the mind. To say that 'it wasn't your mind that did it, but your molecules' is as meaningless as to say that 'it wasn't the Manchester City football team that won the Premiership, but their molecules.'

This picture of the causal autonomy of phenomena at different scales was anticipated in 1972 by the Nobel laureate physicist Philip Anderson. He was unusual in having as profound a grasp of fundamental physics at the scale of individual particles like atoms and their constituents as anyone alive, without succumbing to the belief that this was indeed the 'fundamental' scale at which all phenomena are ultimately determined. He recognized not only that there is a hierarchy of scales at which events occur – subatomic particles, atoms, molecules, collections of molecules, all the way to dogs and stars and galaxies – but that each has its laws and principles that are autonomous and not merely approximations to a more complex formulation that embraces the more fine-grained picture. 'The behavior of large and complex aggregates of elementary particles', he wrote,

> is not to be understood in terms of a simple extrapolation of the properties of a few particles. Instead, at each level of complexity entirely new properties appear, and the understanding of the new behaviors requires research which I think is as fundamental in its nature as any other . . . Psychology is not applied biology, nor is biology applied chemistry.

This makes intuitive sense; indeed, it's how we think about phenomena all the time. The reason you would not describe a football match on the level of the players' fundamental particles is not because this would be unnecessarily (and impossibly) complicated, but because that level contains *no information about causation*. We say that Christiano Ronaldo made a great pass because of the skills he has developed and perhaps the innate talents he possesses. This is not some 'rounding up' of lower-level structures – it is simply using the causal language appropriate to the events. There is no quantum-mechanical description of a vivid imagination or excellent foot-eye coordination.

And crucially, that's because there is no 'configuration of quantum particles' that corresponds to the mind-state of, say, a great footballer or novelist. There isn't even a configuration that corresponds to a 'Dickens mind'. (It is scarcely any better to talk in terms of the configuration of Dickens' neurons, although we're then a bit closer to what matters.) The mind of Charlcs Dickens is only a meaningful notion if discussed in *the terms that pertain to minds*: beliefs, motivations, goals, desires. To say that what made Dickens write *Bleak House* was the state of his particles the moment before he picked up his pen is comically inane – to use the physicist's favourite put-down, it is not even wrong. But that is what the doctrine of anti-free-will (quasi)determinism logically requires us to accept.

It's important to be clear what we are and are not throwing out from the old debates on free will. You might say, even if I accept all that business about minds doing stuff – making decisions and exerting control and so on – still there is only one way its choices can turn out, because, given some state of affairs at the level of particles, no choice can be imposed about the future states. At some level it must have been inevitable that I'd pick up my mug of coffee and take it upstairs. I only decide this is free will because that inevitability is invisible to me, and probably always will be because it's hidden behind too much fine detail.

To which we can say: tell me something I don't know. *Of course* things can only turn out the way they do, because what's the alternative? Turning out differently than how they did? It's not a matter of *how things turn out* (which we not only don't know but, in general, can't possibly know until they happen), but of *what caused them to do so*. Minds don't 'change the future', making it *this* rather than *that* – how could they have such a non-physical, extra-scientific power? Rather, minds are an autonomous part of what causes the future to unfold.

This is different from the common view of free will in which the world somehow offers alternative outcomes to a situation and the wilful mind selects (more or less freely) between them. Alternative outcomes – different, counterfactual realities – are not real, but metaphysical: they can never be observed.* When we make a choice, we aren't selecting between various possible futures, but between various *imagined futures*, as represented in the mind's internal model of the world. ('How would I feel if I holidayed in Spain? Better than in France? Is the weather likely to be nicer?') We do not know, of course, *exactly* how those imagined futures will play out, and we

* One might object that it's possible to repeat a behavioural experiment and get a different outcome, because a different choice was, for whatever reason, made. This is true, but it is never the *same* future that is being probed – as Heraclitus said, you never step into the same river twice. A 'microscopic determinist' could reasonably object that there was some tiny influence of brain or environment that was different second time around which altered the outcome; it would be virtually impossible to prove there was not, however careful the experimenter is. What such experiments, if carefully conceived and conducted, can tell you is what the mind, given a typical set of circumstances, tends to do (and to not do). And from this we might reach some conclusions about aspects of how the mind works. Then – surprise! – we'd simply be doing good cognitive or behavioural science.

might be quite wrong about some aspects of them (as well as neglecting virtually all of the fine details). They are the stuff of decision-making, and not to be confused with actual outcomes. In this sense, free will is not so much a matter of 'I could have done otherwise (than I did)', but 'I can imagine doing otherwise (in what I haven't done yet).'

This distinction creates a lot of muddled thinking about what we call the will. Say the menu offers duck and beef and I choose the duck. (Never mind why, at this point.) But it's really not very good. So next time I choose the beef instead. 'Ah,' comes the objection, 'but that choice wasn't a free one: the environment (previous bad duck experience) made you do it.' Yet this is to misunderstand what volitional decision-making is. It *includes* the mind learning from experience, storing memories, developing preferences. All that is the very stuff of choice.

Such capacity for volitional behaviour based on mental representations is not uniquely human. Some other animals seem also to be able to create neural representations of behavioural outcomes, for example by replaying (sometimes during sleep) patterns of neural activation that they underwent during a past experience. These patterns often involve that part of the brain (called the ventral striatum) associated with the reward system: the cognitive process includes an evaluation of the imagined benefits of a choice, made at least in part on the basis of how good the action will make us *feel*. And this accords with our intuition. When we see a cat sizing up a leap between two walls or crouching in wait for the right moment to pounce on a poor bird, we don't imagine it is awaiting some external stimulus as the trigger for action. It could hardly function if it did, for there *is* no such signal. Rather, it is surely 'thinking' something analogous to: 'Can I do this? Is it the right moment?' The feline brain is itself generating (or not) the trigger for action, and it's at least a sound hypothesis that it does so on the basis of some kind of internal representation of the likely outcomes of a pounce. There's neurological evidence that something of this sort goes on in the brains of rats navigating a maze. Coming to a junction, they activate areas of the hippocampus (the place where spatial memories are encoded) that correspond to the two paths: first one, then the other. Their brain activity is fully compatible with the idea that they are vicariously trying out and evaluating the choices open to them. And that, after all, is the point of a mind – it is a 'device' for making decisions, not some hyper-complex mechanism for enacting the cosmic exigencies of particles.

Why minds matter

Here's what all this implies. As life – and from it, minds – evolved on our planet (and quite possibly on others), causal power seeped upwards through the scales of matter. Living things acquired the *agency* to change their environment, and to do so in purposeful ways that enhanced their survival. When, by degrees, they developed minds worthy of the name, those minds accumulated *causal potential*: a capacity to be the reason for why things are the way they are, and why they happen the way they do. This accumulation includes the ability of minds to determine not just what happens, but what *can* happen. For the extreme reductionist, objects like 'coffee cups', 'kitchens', and 'coffee' are just collections of atoms like any other, not aspects of environments (and thus elements of possibilities) that minds have themselves shaped.

Biologists, especially those who work on evolution, are often wary of speaking of purpose at all, generally corralling it in quote marks when they do, because it sounds too much like teleology: as though there is some grand, perhaps divine plan that is directing evolution. To some degree that's understandable, given how such ideas are abused by the Intelligent Design movement. But this aversion also has a less defensible reason: it is because of the subconscious gravitational tug of philosophical determinism, which insists that because there is no purpose or plan among atoms, there can be none anywhere. That is of course nonsense: we all have purposes and make plans, and so, in a much more constrained way, do bacteria and other single-celled organisms. These are not illusions; we don't make 'holiday plans' that must be decorously imprisoned in quotes. No biological determinist (or anyone else) can deny the reality of such personal choices without implying that psychology and sociology are mere pseudo-sciences that seek correlations in our actions devoid of all causation.

Minds – from the most primitive bundles of coordinating nerves to our own astronomically complicated, synapse-saturated brains – are a means of enhancing, improving and broadening agency. They permit more foresight, more choice, more versatility of response, better prediction. They allow a better framing of purpose. This is *what they are for*.

Built from (yes!) nothing but blind molecular interactions, minds make (living) matter better able to influence and even to decide its own configurations and future state, and those of its environment: to enable change in the world. (It is in truth a rather peculiar kind of change, for in the end living minds seek homeostasis: they want to maintain the constancy and persistence of the organism.) Thus, the *function* and *purpose* of an organism as it moves around its environment *dictate where the atoms go* – and not vice versa! The atoms are not moved by any mysterious psychic force; rather, in this vast jumble of interacting particles, true causation resides at the higher and not the lower levels. The effortful, conscious processes involved in human volition 'are not just an interpreter of upward causation', says cognitive scientist Thomas Hills. 'They construct the self and its alternatives.' As biologist Stuart Kauffman and philosopher

Philip Clayton put it, agency obtains in physical systems that can *act of their own behalf.*

Neurobiological free will and its discontents

Once we stop treating free will as some kind of mystical essence or agency that should (if it is real) be detectable in a genetic sequence or with a force-meter, or that could be distilled and bottled from our minds, then it can become a valid, useful, and genuinely scientific notion. Free will, says the neurobiologist Björn Brembs, 'is a biological trait and not a metaphysical entity'. Specifically, it is what comes out of the neurobiology of volition: the neural and cognitive machinery involved in making a decision or choice. For us humans this machinery includes (among other things) memory, internal models, the playing out of imagined future scenarios and the weighing of imagined rewards, the knowledge of capabilities, the emotions and feelings, and the still imperfectly understood processes that create conscious awareness of some of these things. Out of all this neural activity arises a decision that leads to an action, and we can meaningfully say that this integrated decision-making process of mind was the primary cause of that action.

But no sooner does one pronounce free will a matter for neurobiology than one encounters another common objection. In the 1980s, the physiologist Benjamin Libet of the University of California at San Francisco carried out a series of experiments on human subjects to monitor what their brains were up to when they made a decision. He concluded that the decisions weren't consciously made at all: conscious awareness of them seemed to be a mere sideshow.

Two decades earlier, German scientists had used electrodes placed on the scalp (the technique of electroencephalography, EEG) to monitor the neural activity of people asked to flex their right finger at whatever moment they felt like doing it. The researchers saw an electrical pulse appear about 500 milliseconds before the movement, which they called a 'readiness' or 'standby' potential (*Bereitschafts-potential*). It seemed to be a signal that the brain was preparing for the finger movement.

That alone – a slight delay between the apparent decision and the movement – seemed unremarkable. But Libet added a new twist. He asked his volunteers to make a similar small voluntary gesture, a flexing of the wrist. At the same time he asked them to watch a dot on a computer screen as it rotated as if around a dial, and to mentally note the dot's position at the point they became aware of deciding to make the movement. He found that the readiness potential preceded not just the gesture itself but also, by about 150 milliseconds, reported awareness of the decision to make it. In other words, it seemed that the movement wasn't being consciously willed: consciousness was a mere bystander.

This seemed to support suspicions about alleged free will that some scientists had harboured for a long time. In 1874 the biologist Thomas Henry Huxley claimed that consciousness does not initiate our actions but merely observes them. They are impelled by factors beyond our control or awareness, said Huxley: we (and other animals) are will-less automata jerked puppet-like by signals from our environment. What we call volition is like 'the steam whistle which signals but doesn't cause the starting of the locomotive', Huxley wrote.

Libet's experiments are routinely today cited as evidence for this view. His results seemed to suggest that, far from being spontaneous and undetermined before the event, our seemingly 'free' choices are wholly predictable. From the EEG signal, one could anticipate that a participant was about to move their wrist at least a few tenths of a second before they 'chose' to do so.

In fact these findings don't make a strong case for the predetermined nature of volition even on their own terms. In 2010, the neuroscientist Aaron Schurger found that the apparently random, fluctuating neural activity of the 'idling' brain contains within it a signal much like the alleged readiness potential: it's *always* there. The problem was, in essence, that Libet had not looked closely enough at the brain activity of people when they *didn't* make a movement. So what role does this supposed readiness potential play in determining a movement? It seems that in the absence of any external cue to trigger the action, an occasional, random rise of the potential will tip the scales. It's not a sign of any unconscious, non-volitional 'decision' at all.

Arguments about the implications of Libet-type experiments continue. In 2008, for example, scientists used magnetic-resonance imaging (MRI) brain scanning instead of EEG measurements to look for precursory signals to a similar decision to make a movement, and reported that they could see one up to fully ten seconds before participants recorded conscious awareness of their decision. It's not clear what this implies.

One objection to using Libet-type experiments to pronounce on free will is that they are so artificial and limited in scope. With so little at stake in this particular decision – the action is trivial, meaningless, and brings no punishment or reward – there seems scarcely any reason for the brain to bring the choice to awareness at all. In other words, there is so little for 'freedom' to act on that it is hardly worth engaging the 'will'. In such cases, 'the subjects hand over their freedom to the experimenter when they agree to enter the scanner', says cognitive scientist Chris Frith.* To put it another way, the whole point about minds is that they induce goal-directed action. By instructing them to take an action that lacks any goal at all, one

* What if we just decide not to make a movement at all? Then we're not 'playing the game'. But so what? Can we only be deemed to play the game by agreeing to constrain our free will within pointlessly narrow limits?

is not testing what minds are meant to *do*. The caveat that animal behaviour in closely controlled lab conditions might tell us little about its mind in the wild seems here to be overlooked.

One thing these experiments certainly do not prove, however, is the deterministic thesis that everything we do is predictable in principle if we could gather enough information about the circumstances before we decide. For a start, no one is suggesting here that the decision originates anywhere but in the brain – that's where all the alleged precursory signals are found, not in some subtle signal from the environment. But then what does it mean to say, 'It wasn't you that decided, but your brain'? The central problem here is the assumption that volitional choice is synonymous with a reported awareness of having made it. This seems an obviously reasonable thing to assume at first glance – how can you be considered to have made a choice, if you're not aware of making it? But one can hardly get to the end of that question before realizing its flaw. We're *always* making choices without knowing it! Everyone knows that a good way to resolve an agonizing decision is to toss a coin – not because you'll abide by the outcome, but because if you find yourself saying, 'Best of three,' you know you'd already made up your mind. So decisions can be ordained ten seconds before we register them? Big deal; I suspect most of us have made decisions days or even weeks before we realize it. 'Conscious deliberation', says Antonio Damasio, 'is largely about decisions taken over extended periods of time, as much as days or weeks in the case of some decisions, and rarely less than minutes or seconds. It is not about split-second decisions' – less still ones that don't matter a damn.

Plenty of our actions are made on autopilot – not just as a simple reflex (like withdrawing from a source of pain), but arbitrary, 'unnecessary' actions that we just *did*, like gazing out of the window or shifting in our chair. It should be obvious, then, that there is no basis for dividing our decisions and actions sharply into those that are made unconsciously or automatically, and so lacking anything like free will, and those made with conscious effort ('I'll do *this*!') and therefore freely willed. Most of what we do lies somewhere in between. When playing piano, I sometimes have to tell myself, 'That's a double sharp, so not an F at all' – but if I applied that conscious awareness to the placement of every finger, I could never play anything. When I drive a car with manual gears, I'll sometimes think, 'Time for third gear,' but other times I barely notice doing it. Does it make any sense to say that in one case I'm using my free will, but not in the other?

So there seems to be no good reason to suppose that making a decision is coincident with it coming into consciousness, nor indeed that there is a precise moment when it does so. AI researcher Stanislav Nikolos thinks that what registers as consciousness of having made a decision is not the actual decision-making moment but a kind of mental broadcast when the strength of the 'decision signal' passes some critical threshold – rather like releasing the results of an election only after all the votes are counted and the outcome is clear.

My suspicion – I know of no hard evidence either way – is that there is no tight coupling or inevitable order of events between making choices and being aware of having done so. When we think about these two processes in neurological terms, it is far from clear why we should expect there to be any such sequence of events. The fact that both consciousness and decision-making may involve some of the same mental processes – the integration of several sources of information in a kind of attentional workspace, say, and the formulating of possible future scenarios created by actions – means that we might expect there to be considerable overlap of cognitive processing: to be consciously making a decision means of course that we can't fail to be aware of doing it. But the two are not the same.

None of this should surprise us, for it fits with our intuitions.

When we claim that we did not choose a particular action or outcome 'of my own free will', we generally mean that we were coerced in some way by external influences: threatened with violence if we don't sign a confession, say. But if we discover that we donated all our possessions to charity while blind drunk, we don't so much assert an absence of free will as that our decision-making faculties were impaired. This decision made in the absence of awareness does not suggest to us that somehow it was predetermined all along by external events, but just that the decision did not benefit from the consideration we would otherwise have given it.

Agents of fortune

If we really do make choices, how do we do it? We must be capable of weighing up the options rather than just stabbing randomly at one of them. This requires that we possess an inner representation of the world onto which we can project the outcomes of our choices: we construct future scenarios. If I take this job, I'll have more money than if I take that one, but I'll also have to move house, whereas I rather like the one I'm in . . . And so I imagine myself feeling richer, and how good that would be, but also living in that distant, dreary city, and so on. Even when literally selecting from a dinner menu, we'll call to mind the dishes and how they might taste, and we might have a picture of how we'd feel to be paying more for this rather than that. Our choices demand a capacity to reason, however imperfectly: to understand cause and effect, to appreciate what the consequences of our choice might be. They may involve moral and social considerations too: how will I *feel* if I do that, and how will others see me?

In short, free will typically requires a rich inner world model that features the self: what's in it for me? It's reasonable then to suppose that only with the ability to construct an autobiographical self – that is, with a 'higher' form of consciousness – does volition acquire its full range, power, and value. The neural resources required for this facility are considerable.

But as with consciousness itself, it's likely that volition doesn't arrive all at once in a sudden jump that separates humans from the rest of the animal world. It is really a refined form of the more general capacity of *agency*: the ability to manipulate and rearrange the environment and self to suit a goal. It's one of the strangest lacunae in science that it lacks not only a theory of agency but even much awareness of the omission.

Agency goes beyond automated and predictable stimulus–response behaviour; it enables good choices in response to new and unforeseen circumstances. When a hare is being pursued by a wolf, there's no meaningful way to predict how it will dart and switch this way and that, nor whether its gambits will suffice to elude the predator, which responds accordingly. Both hare and wolf are exercising their agency.

We can see similar agency displayed in much simpler entities. The pursuit of a bacterium by a macrophage white blood cell, whose job in our immune system is to engulf and devour such invaders, is hardly less dramatic and unpredictable, even though these single cells lack the rich cognitive environment of a wolf or a hare.

If we break down agency into its constituents, we can see how it may arise even in the absence of a mind at all, in the traditional sense. It stems from two ingredients: first, an ability to produce different responses to identical (or equivalent) stimuli, and second, to select between them in a goal-directed way.

The first of these is the easiest to procure. It requires mere behavioural randomness, like a coin flip. Such unpredictability makes evolutionary sense, since if an organism always reacted to a stimulus in the same way, it could become a sitting duck for a predator. This does sometimes happen in the natural world: a certain species of aquatic snake triggers a predictable escape manoeuvre in fish that takes the hapless prey directly into the snake's jaws.

An organism that reacts differently in seemingly identical situations stands a better chance of outwitting predators. Cockroaches, for example, run away when they detect air movements, moving in more or less the opposite direction to the airflow – but at a seemingly random angle. Fruit flies show some random variation in their turning movements when they fly, even in the absence of any stimulus; presumably that's because it's useful (in foraging for food, say) to broaden your options without being dependent on some signal to do so.* Such unpredictability is even enshrined in a rueful aphorism known as the Harvard Law of Animal Behaviour: 'Under carefully controlled experimental circumstances, the animal behaves as it damned well pleases.'

A single-celled aquatic organism called a ciliate – a microscopic trumpet-shaped blob that attaches, anemone-like, to surfaces such as rocks – offers a striking example of randomness in a simple, brainless organism. When researchers fired a jet of tiny plastic beads at it, mimicking the disturbance caused by an approaching predator, it sometimes reacted by contracting, and sometimes by detaching and floating away, unpredictably in any single instance but with reliable 50:50 odds. Evidently, you don't need even so much as a nervous system to behave in a random manner. Indeed, even bacteria such as *Escherichia coli* find their way to food sources by selecting

* It has been suggested that the fruit-fly brain has evolved to operate close to a 'critical point' in the activity of neurons, meaning that only small influences are needed to tip it into one global state – resulting in a specific behaviour – or another. This would enable the brain to leverage random noise in the environment to determine behavioural outcomes. Critical-state behaviour in the patterns of neural firing has also been reported in the healthy human brain, and deviations from it have been implicated in some brain pathologies.

new directions at random if their current course leads the concentration of the nutrient to diminish rather than increase: a strategy called 'run and tumble'. Such randomness is what makes a rogue bacterium a challenging object of pursuit for a macrophage.

Can random behaviour really exist, though? Where could it come from? A determinist would say that it is only random because we lack complete knowledge of what is going on at the microscopic level. Tiny particles, such as grains of pollen, suspended in water seem to undergo random jiggling motions, called Brownian motion after the Scottish botanist Robert Brown who first observed it in the early nineteenth century. Such 'random walks' are now known to be caused by the impacts of the water molecules themselves, which dance with thermal agitation. At any given moment, there will just happen to be more impacts on one side of the grain than another, and so they will exert a greater push. If we could keep track of the trajectories of every molecule, the dancing of the pollen grain would be completely predictable.

Does this mean that the randomness seen in tumbling *E. coli* as it seeks a nutrient source is in fact deterministic and predictable from the underlying microscopic physics? From a biological perspective the question is irrelevant – because the bacterium no more has access to those microscopic details than we do. Biological sensing simply isn't sensitive enough to register them. After all, if accessible information in the environment could predict a creature's evasion response, we should imagine that predators would have evolved to take advantage of it, and the benefit of this pseudo-randomness would be lost. It's rather like those adversarial attacks on AI image-recognition systems (page 297), where the injection of some subtle randomness into the image undermines the AI's decision. We humans see no difference between the two images, not because we haven't attended to them closely enough but because the differences are literally too fine-grained to be visible: to us, the images *are* identical. No creature's vision can attend and respond to microscopic details of a scene below the acuity of its visual apparatus.

Thus, Daniel Dennett points out, arguments about whether microscopic randomness is or is not predictable at the atomic scale have no bearing on the role of chance and randomness in biological agency, because such factors are beyond an organism's ability to *control*. It's the same as when we flip a coin. Sure, one might argue, we could predict the outcome if we knew enough about all the details: the molecular-scale air currents, asymmetry of the coin, exact nature of the flipping force. But because we can't know all that,* far less control it all, a coin flip is a genuine source of randomness for us. As a result, mechanisms for generating behavioural randomness really do seem to lack predictability.

Crucially, this behavioural randomness is an *evolved* capacity. We can see this in the way some animals adjust their degree of unpredictability to the circumstances, increasing it in novel situations where past experience is not a good guide to action. That's to say, they may make more use of randomness when it is likely to furnish a better strategy. So even at this stage of leveraging randomness – before we even consider the issue of choice – we can see true agency at work: behavioural outcomes attuned to a goal.

There's now good evidence that for living organisms at scales from single cells upwards, random noise both in the environment and in the organism's internal state is not just a source of unpredictability but a useful resource. For example, it gives an organism access to a range of possible behaviours by tipping the balance of outcomes in random ways.

But this is not yet true agency. It is simply the generation of

* It turns out that, to have any chance of making a deterministic prediction, you'd need to know an extraordinary amount, down to the quantum-level details of the whole extended environment.

alternative behaviours in a situation when the meaningful information available to the organism remains unchanging. Agency comes when an organism can evaluate the state it is in, and decide for itself whether to stay or change, and in which direction. This selection from a range of available choices is goal-motivated: an organism does *this* and not *that* because it figures *this* would make it more likely to attain an optimal outcome.

Selection requires some feedback process that locks the organism into the state for which that outcome is likely to be most beneficial. At the simplest level, it's not hard to see how to arrange for this through a set of logical steps: an algorithm. Take the case of the run-and-tumble *E. coli* searching for a food source. The algorithm could run something like this:

1. Move in a random direction.
2. Sense the concentration of nutrient. If it has increased, keep moving in that direction.
3. If the concentration has diminished, thrash around your flagella (the bacterium's whip-like appendages, used to drive motion) randomly so you tumble for a while.
4. Now move in the direction you find yourself facing. Go to 2.

This works well enough for *E. coli*.* It is not, in general, a good strategy for us trying to locate into our car in a car park. In other words, what we *don't* do is:

* Whether this means *E. coli* is really exhibiting goal-directed behaviour is disputed, however. Many neuroscientists believe that this only really occurs if the goal does not have to be maintained by external stimuli: if it can be 'remembered' without, so to speak, having it in front of you. Remove the nutrient gradient, and we remove all notion of a goal for the bacterium. Perhaps so – but even bacteria have evolved to be self-preserving in any environment to the best of their ability.

1. Try your key in the lock of a random nearby car.
2. If it works, get in and drive. If it doesn't, choose another car at random and do it again.

What we *do* do is remember where we left our car. Or if we can't, then we walk around in the vicinity of where we think we left it, looking for one of the same make and colour, and then check for other distinguishing features, like the number plate. In other words, we use our memory, along with visual cues of recognition. Only if we can rely on neither – if we are searching around a car park in total pitch black – might we need to adopt as inefficient a strategy as trying our keys at random. In other words, we have cognitive machinery that permits a better behavioural strategy than that of bacteria – a better way to choose our action.

For a proper understanding of agency, we need to know what kinds of cognitive resources make such goal-directed choices possible at all, and more efficient. The simplest action might indeed be 'try at random and see if it brings us closer to our goal'. Memory ('the goal is *this* way', or '*this* worked last time') can improve efficiency. One view of the evolution of the various components of mind, then, is that it systematically improved *agency*, both by making the attainment of goals more efficient and also by allowing a better determination of what the *goal* should be in the first place: better guesses at what might be beneficial to the minded organism.

Physicists Susanne Still, Gavin Crooks, and their co-workers have shown that it's vital for a goal-directed entity like a cell or an animal to have at least some kind of memory. With a memory, any agent can store a representation of the environment which it can then draw upon to make predictions about the future, enabling it to anticipate, prepare, and make the best possible use of its energy – that is, to operate efficiently, which is an important goal in evolutionary biology.

Any real organism is forced to operate with information about its circumstances that is both incomplete and of ambiguous value. Still, Crooks and their colleagues found that energy-efficiency depends on an ability to focus only on information that is useful for predicting what the environment is going to be like moments later, and filtering out the rest. In other words, it's a matter of identifying and storing *meaningful* information: that which is useful to attaining your goal. The more 'useless' information an agent stores in its memory, the researchers showed, the less efficient its actions. In short, efficient agents are discerning ones.

Such work at the interface of information theory, thermodynamics, and life might ultimately show the way to a 'minimal theory of agency': to identify the basic information-processing requirements an entity needs in order to exhibit genuine goal-motivated choices. Brains and minds are surely not essential for agency – but they can broaden and finesse its capabilities.

At any event, biological agency is no illusion produced by our tendency to project human attributes onto the world. Rather, it appears to be an occasional but real and remarkable property of matter, and one we should feel comfortable invoking when offering causal explanations of what we're observing. It is the soil in which minds with volition – with free will, if you like – could grow.

Agents are not mere puppets through which the laws of physics create some inexorably unfolding future; they are a primary part of what makes the future happen. 'Agents are *causes of things* in the universe', says neuroscientist Kevin Mitchell.

When there are no agents present, we can indeed tell causal stories of events in terms of mere mechanics, devoid of any goal. To explain why there is an outcrop of volcanic basalt at a particular geographical location, the story might be along the lines of:

'heat – molecular motions – in the deep Earth, along with gravity, produced a convective flow of rock that brought magma to the surface here'. However, if we want to explain why a bird's nest is in a particular location, it won't do to explain the forces that acted on the twigs to deliver them there. The explanation can't be complete without invoking the bird's purpose in building the nest. We can't explain the microscopic details – all those cellulose molecules in the wood having a particular location and configuration – without calling on higher-level principles. A causal story of the nest can never be bottom-up.

You might say that a blind, mechanical story *is* still available for the bird's nest, but that it just needs to be a bigger story: starting, say, with the origin of life and the onset of Darwinian evolution among not-yet-living molecules. But no such baroque and fine-grained view will ever avoid the need to talk about the agency of the bird – not if it is to have the true explanatory power of supplying a 'Why' to the existence of the nest on *this* oak branch.

Understanding agency can also shed light on the status of free will in the Space of Possible Minds. Is it possible, for example, to imagine a sentient mind that can support only non-volitional actions? Such a mind might be in the odd position of forever watching itself initiate actions over which it has no control.* We are ourselves in this situation sometimes – we are bystanders, for example, when our tapped knee executes its reflexive kick. We saw earlier that octopuses might experience more of this kind of thing, the central brain watching the semi-autonomous arms go about their business. But a mind that can *never* choose its actions? That

* Near-total lack of voluntary movement occurs in people with the condition called akinetic mutism, but this leads to a catatonic-like inability to move or speak, even though wakefulness remains. In humans, rather little of our ability to act in or on the world is wholly non-volitional.

would seem to negate the likely reason for having some degree of consciousness in the first place – which, as we saw, is plausibly (and not more) a mechanism for improving the organism's ability to integrate and evaluate the stimuli it draws on in deciding on actions. It seems likely that consciousness is only worth having if it can be used for deliberating on actions.

Thus Dennett has argued that 'you cannot explain consciousness without tackling free will, and vice versa.' He regards free will as the way in which consciousness is mobilized: while consciousness is the 'virtual governor' that imposes some order and focus on the innumerable possible thoughts and sensations clamouring for our attention, free will is the process by which we are able to select what to attend to: to think about this rather than that, to lean towards this possible action rather than that. Without that capability, our complex cognitive theatre would be an incoherent babble, or at best an uncontrollable and rather meaningless succession of vignettes. Sitting here as the dusk deepens, I can – if I wish – call to mind the image of a sun-dappled forest, a snowstorm, a rock concert. To suppose (unproveably) that some inscrutable stimuli in the world prompted those images, or that I was not free but bound to conjure them up, seems like magical thinking. It seems to deny not some putative aspect of, but the very nature of, minds like ours.

Taking responsibility

I suspect some of the arguments over free will are so fierce because they are proxies for another argument entirely. 'Free will' was a currency first granted significant purchasing power in the West within Christian moral theology: it was allegedly granted to humans by God as an explanation for why an all-loving and omnipotent deity would permit suffering in the world. It was argued that we bring suffering on ourselves: that is the price of the divine gift of freedom, which was abused almost at once by Adam and Eve and has been scarcely more respected ever since. In short, free will is the source of moral responsibility – and as such, as much a burden as a benison. As is so often the case, the scientific arguments are thus overshadowed here by unacknowledged cultural baggage. (I think it is no coincidence that many of those who use either Libet's arguments or the microphysics of philosophical free will to assert that it does not exist tend to be emphatically atheist; they seem to sense that free will is a remnant of religious mysticism that science can and should banish without delay.)

The discussion of free will and moral responsibility is often muddied by that baggage. Those who suppose that 'free will' is an illusion argue that no one is truly culpable for what they do, either because the brain makes decisions outside of conscious volition or because all behaviour is deterministic and inevitable anyway:

> 'My client, m'lud, must not be sent to jail or otherwise punished for his crimes, because he did not have any choice: his brain/the universe made him do it.'

A judge who truly understands minds and volition might, however, now respond as follows:

> 'But of course his brain made him do it – for that is what all brains do! They impel actions. It's like saying that he is innocent because certain synaptic processes were responsible, as the homunculus-like defendant looked on in helpless dismay from within his grey matter.'

And this perhaps is where the courtroom discussion then goes:

'Ah, but what I'm saying, m'lud, is that *this* person's synapses are hard-wired to make him behave that way. That's to say, the specific combination of *this* brain, made by *these* genes, plus the environment it has experienced made the outcome a foregone conclusion.'

'And just what do you think moral responsibility is in the first place? Some kind of cosmic edict of right and wrong? No such thing exists, man, and certainly not at the level of synaptic junctions! Moral responsibility means learning to act in a way that society has, for good reasons, deemed proper and harmonious. If the defendant has not yet learnt such prosocial behaviour, he must be helped to do so by reinforcement: by seeing that there are consequences to his actions. Yes, some folks see punishment in terms of their own hedonistic payoff: they want the squirt of pleasure neurotransmitters that comes from the feeling of a right wronged. But the law is not concerned with that, or not in my court. It is about maintaining social harmony through justice, for example by deterring things deemed improper or harmful. Since "free will" – volitional, deliberative decision-making – involves the reward system of the brain, we can train that reward system through feedback. We can say that society requires you to adjust the weightings of your neural network in this direction. Even if the defendant was a mindless deep-learning robot, I would be right to assign a punishment as a reinforcement to help it learn a more socially acceptable way to behave.'

'But this is not fair, m'lud! My argument is that such reinforcement will have no effect on my client, because his brain is not of the kind that will respond to it.'

'Ah, if that is so then there is no problem, for these courts already acknowledge this particular argument about the absence of free will: it is called diminished responsibility. It means it is incumbent on you to provide good evidence that the defendant's brain lacks an ability to connect actions to consequences, or to learn to do so. In that event, our decision will be based solely on the need to protect society.'

The point is that a view of moral responsibility that acknowledges what minds and agency really are will not get hung up on the matter of whether one could or should have done otherwise than one did, which is a question beyond empirical access. Who can know, and what difference can it make now? No: what we call moral responsibility in the human mind really arises from the mental framework we use for making *future* choices based on imagined outcomes. As cognitive scientist Anil Seth says, a crucial aspect of free will is not 'I could have done otherwise' but rather, 'I'll learn from this so as not to do it again'. One's moral responsibility consists not in the notion that 'You could have done the right thing, but you didn't' – that's just a matter of history. Rather, it should be considered within the sphere of 'Your mind was able to construct an *imagined* future in which you did the right thing, but you chose not to do it.' Moral responsibility speaks to how the mind constructs choices.

Such a rebalancing of the cognitive and neurological factors that go into a volitional decision based on past experience is precisely one part of what the proper view of free will is all about. By shifting away from the false question of whether free will 'exists' or not, towards real and testable questions about the cognitive mechanisms of volitional behaviour and choice, we can develop a better sense of how to regard – and encourage – responsible behaviour.

At any rate, it's vital that our systems of social justice keep abreast of such nuanced discussions in neurobiology and don't become derailed by simplistic assertions about the reality or otherwise of free will. Antonio Damasio is surely right to say that 'lawyers, judges, legislators, policy-makers, and educators need to acquaint themselves with the neurobiology of consciousness and decision-making' – however preliminary and inconclusive some of that work currently is.

For example, Anthony Cashmore laments that 'the legal system assumes that it is possible to distinguish those individuals who have this capacity of free will from those who lack it.' In practice, he is probably right: the law is framed in such a way as to imply that it is both meaningful and possible to make the distinction, for courts must in general make a necessarily simplistic binary choice of 'guilty/not guilty' (albeit with a decision about sentencing allowing a gradation of culpability). Yet people's volitional resources surely do not vary on a single, continuous 'free-will scale' running from zero to whatever; they probably differ in a multidimensional way.* We may have to accept, but also to acknowledge, that human laws are crude and somewhat arbitrary tools for sustaining sociality among complicated creatures. Their criteria cannot hope always to be biologically meaningful.†

Take Elliot, a patient described by Damasio whose benign brain tumour caused localized damage that degraded his ability to make good decisions or learn from his mistakes. His problems stemmed specifically from impairment of his emotional responses (supporting Damasio's contention that emotions and feelings contribute to our ability to act rationally): Elliot could reason as well as anyone, but could not form good value judgements about the outcomes. 'He was not stupid or ignorant, and yet he acted often as if he were,'

* Cashmore cites the case of serial killer Jeffrey Dahmer, whose grotesque crimes suggested serious mental disturbance and absence of moral sense, and yet whose lies to the police suggested he understood there would be consequences of his actions.

† We see the same problem when the law has to pronounce on 'when life begins', or 'whether a patient has any consciousness'.

Damasio says; he concludes that 'it is appropriate to say that [Elliot's] free will had been compromised.' It would clearly be unfair to consider that the bad choices Elliot made in business and in life were a matter of sheer reckless irresponsibility. It's a simple fact that some people have, for a variety of reasons *and in certain respects*, more free will than others.

This doesn't mean Elliot's decisions were somehow pre-ordained, however. Rather, they did not have the benefit of all the necessary cognitive inputs for optimal decision-making. Others might make poor choices because of, say, an innate inability to attune to the mental states of other people, or to delay gratification. And – this is crucial – others might do so because there are environmental factors confounding their decision-making: they were desperate for lack of money (skewing their cost-benefit model of the projected outcome), or had an alcohol-induced lowering of inhibition. Damasio laments that we still make a distinction between diseases of the brain (which are medical and deemed to be no fault of the individual) and diseases of the mind (which are considered character flaws – lack of 'willpower' being one – and thus reprehensible). It is equally deplorable that, as the psychiatrist Abraham Halpern has pointed out, the law will accept mental dysfunction as an exculpatory circumstance, while giving far less weight to environmental factors such as poverty, abuse, or deprivation, which might be considered to exert an equally strong influence on behaviour.

Should we, then, feel no animus towards offenders at all, no matter how terrible their crimes? Of course there are purely utilitarian arguments for a criminal justice system that maintains an orderly society by, say, incarcerating dangerous individuals, as well as offering offenders psychiatric help. But to present punishment and rehabilitation as dispassionate tweaks of some deterministic 'environmental' influence on behaviour that is devoid of all free will is to ignore what minds are all about.

Those who deny that free will exists often say that while feelings of disapproval about harmful or criminal acts make no sense and are even unjust, we are after all only human and so are bound to have them. But such a response misunderstands yet another facet of the human mind, namely that it seems likely to have evolved in large part for social cognition. Only a deeply dysfunctional, sociopathic mind is immune to feelings connected with social status and the approval or disapproval of others. Such feelings are *part of the input to volitional decision-making* – so in this sense, a tendency to express social approbation is likely to be an adaptive feature of our minds, not some petty and vindictive epiphenomenon. Anyone who has striven to inculcate responsibility in young children will know that they are remarkably good at sensing and ignoring faked displeasure at the kind of behaviour we would like to encourage them to relinquish. Our challenge is to take responsibility ourselves for our retributive urges, so that the social pressures needed to encourage prosocial decision-making do not descend into demands for stifling conformity, violence to children, and capital punishment that gives us a righteous frisson of 'just deserts'. That's what developing a moral code is all about: it is a society-wide negotiation, a collective decision of many minds with differing tendencies and attributes. A proper understanding of free will must be central to it.

Morality in other minds

Considering the matter of free will and moral responsibility in terms of cognitive functions and dysfunctions makes it easier to frame the right questions to ask in this arena about non-human minds. Obviously, it would be absurd to lock up one of today's self-driving cars for running someone over, although there may well be moral culpability on the part of the designers if they did their job recklessly and negligently. The moral obligations must here again be future-oriented: can the AI system be improved, say, or should it simply be scrapped as too dangerous?

It's no different in principle for a dog that has savaged a child. In general we might reasonably err on the side of caution: since we can't easily or economically find out if such creatures can be trained out of posing any future risk, we may decide it is best for the community to put the dog down. But not always – if the dog was being badly mistreated so that it acted out of character, our sense of culpability, and of consequences, shifts.

These decisions must also weigh in the balance the degree of consciousness, and capacity for suffering, in the other mind. We'd have no *moral* qualms about erasing the software from the driverless car's memory, but we think harder about having the dog put down because we consider it likely (as we should) that the animal has some sentience and therefore an intrinsic right to existence that we would not attribute to lines of computer code. Judgements about the nature of a mind influence our attitude to the moral responsibility we feel towards it.

And what of it towards us? We laugh now at the 'animal trials' that took place in Europe from around the thirteenth to the eighteenth centuries, in which pigs, donkeys, and other animals were brought before the court (sometimes with legal representation) on charges of murder, criminal damage, and other misdemeanours.* Yet we should be wary of turning these creatures into Cartesian automata, 'dumb brutes' devoid of all volitional agency. I don't mean to suggest that a bull set loose in a china shop really ought to know better than to trash it – but rather, it is lazy, even bigoted, thinking to assume that the minds of some animals, especially other great apes, have no glimmering awareness of choices and even of 'oughts' associated with them.

* Some of those stories are apocryphal, but the practice is well attested.

What, meanwhile, if we succeed in making an AI system with both awareness and self-determining freedom of action – in short, with a capacity we might regard as free will? Such a device would then seem to be an autonomous moral agent. But we should think carefully about what that means.

Even though there is good reason to consider a dog a sentient being capable of making choices and plans – so that we might suppose 'it could have conceived of acting otherwise' – we're unlikely to think it is wicked and immoral for savaging a child. Moral responsibility is not some universal concept like entropy or temperature – something that applies equally, and can be measured similarly, everywhere in the cosmos. It is a notion developed specifically for human use, no more or less than languages are. While sentience and volition are aspects of mind and agency, morals are cultural tools developed to influence social behaviour: to cultivate the desirable and discourage the harmful. They are learnt, not innate. It's possible, indeed likely, that we are born with a predisposition to cooperate with others – but only within human society do we come to understand this as *moral* behaviour.

In Isaac Asimov's *I, Robot*, robots were given a simple moral code by design. These Three Laws of Robotics (page 327) were intended as safety measures, but they were tellingly phrased in the manner of the Ten Commandments: as a quasi-religious catechism. Asimov's robot tales were largely explorations of how this moral code played out – in particular, how it could break down because of situations that created conflicts or inconsistencies. That's notoriously the problem with any moral code, as evident for example in the enthusiasm for the death penalty within the US 'Bible Belt' states.

Yet even if we could codify human morality – which seems impossible when we don't even agree on what its principles are – any attempt to impose a human-based view of morality on non-human minds seems bound to fail if those minds aren't commensurate with our own. It's like insisting to leopards that they respect the rights of gazelles, or that woodworm respect our own property rights. While I'd feel glad to know that all alien species had signed up to the UN Declaration of Human Rights, it is both more realistic and more interesting to ask what morality could mean for other minds: a question that requires us to ask about their cognitive resources, their goals and motivations, their social contexts – and their volitional agency. We'd need first to locate them in Mindspace.

CHAPTER 10

How To Know It All

I wish at this point I could draw you a picture of Mindspace, and point to our place within it. But such maps as we can currently offer (I've shown you a few earlier) are like medieval *mappae mundi*: reasonably reliable in our neighbourhood, sketchy in the lands bordering ours, and little more than guesswork for distant realms, where perhaps we depict fabulous beasts that we are not sure exist at all.

To be frank, I'm still not entirely sure that the notion of Mindspace can be made rigorous at all. To compare it with a map is to offer a convenient and familiar analogy, but there is a better one to be found in physics. Physicists regularly employ the concept of a 'phase space', a multidimensional realm in which the coordinates are the parameters that characterize the system in question. States of matter, for example, can be located in a three-dimensional phase space with coordinates of temperature, pressure, and density (or volume); parts of that space correspond to solids, liquids, and gases, with sharp boundaries between them. Perhaps there is a phase space of minds, containing distinct regions that correspond to *kinds of mind*. Yet I'm not at all sure that minds have parameters equivalent to pressure and temperature that vary smoothly and on which everyone would agree. Mindspace is probably not as definitive and objective a scientific concept as is Matterspace.

All the same, I do feel it is a useful image, for I hope that thinking about the components of mind helps us refine our notion of what a mind is and can be. And I believe we can at least speculate about those components, those putative dimensions of Mindspace. Take memory: it is hard to imagine a mind worthy of the name that lacks this capacity. Minds must keep some sort of imprint of what they have experienced, which guides future behaviour. We've seen too how important it is to the function of a mind that it contains an imprint, a representation, of its environment, perhaps including an embedded image of its own states and those of the body in which the mind is materialized. Absent this, and there is none of the 'aboutness' that sits at the heart of what mindedness is. But the kind of memories that enable this surely can't be characterized or compared via a single measure such as capacity in megabytes (whatever computer manufacturers might tell us). Even for computers, it matters too how much fidelity a memory has, for example, and at what rate it can be written and accessed. So we can recognize that memory is a part of the topography of Mindspace, and begin to think about how it may be mapped.

Or take consciousness. My very definition of mind (such as it was) alludes to the feature of awareness or consciousness: for an entity to possess a mind, there is something it is like to be that entity. But once we start to examine consciousness, we find that this too is unlikely to have a single measure; it is not like a dial that can be set to high or low. Consciousness is itself most probably a composite phenomenon: for example, a mind might be conscious in the moment of what is happening to it (which we might call sentience), but it might also have an autobiographical awareness of its history and its existence as an enduring self. Or it might be fragmented, as some suspect the cephalopod mind is.

I'd posit that *agency* – the ability to make a purposeful difference – is another 'territory of mind', somewhere among the contours of which we can start to invoke that poorly-named but nonetheless useful concept of 'free will'. And I suspect that sensory modes can each be considered to open up dimensions of Mindspace: there is some sense in which the hawk, the bloodhound, and the bat sit in different parts of the space to our location, by virtue of their superior visual, olfactory and acoustic acuity.

One of the most commonly invoked dimensions of Mindspace is intelligence. Even more than consciousness, this is a notion ripe for surgical scrutiny; I would go so far as to say that a great deal of confusion has been generated in the science and philosophy of mind by an uncritical assumption that we know what intelligence means. Once we start to consider animal minds or putative machine minds, it becomes clear that, on the contrary, we don't understand it at all well. There is, for example, an intelligence involved in problem-solving and tool-use that is distinct from the intelligence needed for navigation, or for social interactions, or advanced calculus. All the same, humans seem able to transfer somc skills from one cognitive domain to another, as though there is a sort of intelligence that acts as a multipurpose cognitive resource. As we saw earlier (page 48), IQ is commonly regarded as a quantifier of such a generalized intelligence or 'g factor', for which the intelligence quotient (IQ) supplies a measurable proxy. This measure can itself be broken down into other components, such as quantitative reasoning, working memory, and visual-spatial processing. In other words, human intelligence may have a relatively small number of quantifiable coordinates.

Others, however, have proposed that human intelligence spans several *independent* domains. The Harvard psychologist Howard Gardner has argued that there is no correlation between, say, the mathematical, linguistic, musical, or kinaesthetic intelligence of individuals. Perhaps there might even be categories of spiritual, moral, and existential intelligence. Gardner's theory of multiple intelligences is controversial, not least because there seems little empirical evidence to support it. But there is surely many an academic over-achiever who should be humbled by the cognitive attributes displayed by sportspeople or by the empathic social skills of people with undistinguished exam results. The usefulness of a notion of 'general intelligence' becomes even less clear when we consider nonhuman minds; one recent survey of the literature by biologist Vincent Careau of the University of Ottawa and his co-workers found there was rather little evidence to suggest it is meaningful for animals. We've seen that cognitive abilities can be very diverse even between closely related species, and that we're still not sure which abilities we should be measuring in animals or how best to do so. Here then, the concept of Mindspace might encourage a broader, more nuanced view of what intelligence is and could be.

Is there some essence that all minds share? If I were to pick one central feature of what characterizes mindedness, it would be this: *a mind seeks what is meaningful to it in the universe.* For here is the strange but inevitable truth about the nature of reality: it does not have a unique form, but has to be interpreted, filtered, and evaluated. I have tried to stress that what the human mind perceives is not 'the world' but an internal representation of it that is useful to us. We don't see heat (the infrared radiation from warm objects), although we can feel it if we're close enough to the source (which could be another person's body). We can't see X-rays. We don't possess biological apparatus that will image atoms, or even bacteria. Our sensory window on the world, and the representation we create from it, is constrained more by evolution than by physical law. A capability to see microbes (at least) with 'microscope eyes' is not forbidden by the laws of optics, and one might imagine it sometimes being useful – but apparently it was not worth the evolutionary investment.

Our meaningful information, our *personal reality*, is selected from much, much more that is 'out there' – not arbitrarily, but in a way that aligns sufficiently with what the physical world contains to enable us to get along and thrive and reproduce within it. As we saw earlier, even what our mind perceives is a mere subset of what our sensory apparatus registers – filtered, shaped, and integrated according to what is useful as a basis for behaviour. What's more, our minds ascribe this information with value: what is more meaningful (because more useful) is awarded more value *for us*. Giulio Tononi and Gerald Edelman suggest that this value scale is created by the emotions.

The process of selecting sensory input is augmented by conjecture. In fact, our mental world is *mostly* conjecture: it is actually vaster than our perceptual world, for it includes things we know to have happened but are no longer experiencing, and things we know to exist and to be happening out of sight (but not out of mind). Emily Dickinson expressed this rather wonderfully:

> The Brain – is wider than the sky –
> For – put them side by side –
> The one the other will contain
> With ease – and You – beside.

This is really one way, then, to think about what mindedness really is: a process through which raw nature is valorized. When scientists tell you (as some are, I fear, apt to do) that the universe seems pointless, they are making a category error. The universe at large is not the right place to go seeking for meaning, for a point to being and existence. Rather look into minds, which make universes of their own.

Why one kind of mind is not enough

Albert Einstein acknowledged the inadequacies of our mind's-eye field of view with his characteristic blend of humility and awe at the mysteries of the universe. 'As a human being', he wrote, 'one has been endowed with just enough intelligence to be able to see clearly how utterly inadequate that intelligence is when confronted with what exists.'*

We don't know, however, if Einstein is right or wrong here. Will we be able to fathom everything there is to know about reality with the minds we have? That would be strange indeed, for it was never what our minds are for. Wouldn't it be a weird coincidence if our minds turned out to be just the right shape and size to fit into them a 'theory of everything'?†

But it might seem less peculiar or coincidental if such complete understanding of the cosmos were to be feasible within the entire Space of Possible Minds. It's one of the deepest mysteries in science – we might never get to the bottom of it, just as we can't parse self-referential linguistic paradoxes‡ – that the universe can generate a capacity to know itself. We can feel confident that some

* This is another of the rare quotes attributed to Einstein that appears to be genuine rather than apocryphal.

† The scare quotes here are meant to deter too literal a reading. There is not, and will never be, a genuine 'theory of everything' in science – the phrase doesn't really have any meaning, because, as we saw earlier, it imputes a false notion of causality and explanation in which all phenomena are 'caused' by more 'fundamental' processes on a smaller scale. A better term might be an *understanding of everything*.

‡ One of these is Gödel's Incompleteness Theorem, which shows why no system of logic can be wholly self-contained: it will always contain statements that it is unable, using its own axioms, to prove to be true.

of that self-knowledge is housed in the human mind: there are things science has deduced that I believe will be enduring, and which come as close to truth as we could ever hope for. It would not be surprising if we have our limits, however.

Just as an outsider's view can bring new insight to a difficult problem, so perhaps other minds can perceive and understand things that will be inaccessible to us. I don't think it is sheer romanticism to suggest that this is probably true already of some other animals, although they cannot articulate those insights to us and probably not to themselves either. I think it is extremely likely that one day artificial intelligence will move outside the bounds of our own capacity to know and understand, although so far it has neither knowledge nor understanding worthy of the words. (I have avoided saying that AI's abilities will exceed ours, since it's possible we might continue to know some things that future AIs cannot. AI might not exceed us in all directions, so to speak – although it already does in some.) It would be a pleasing symmetry if the capacity of the universe to produce minds that can know it were to be congruent with the space of what there is to be known. There's no reason why that must be so, but I'd be willing to make it a conjecture (confident that it won't be disproved any time soon).

Yet there is nothing in the fundamental laws of nature – so far as we currently know them – that hints at this capacity of matter to get to know itself and the universe in which it exists. To me it's a capacity much more remarkable than that those 'fundamental' laws permit, say, the existence of black holes. We can deduce the latter from the laws alone, without empirical guidance – indeed, we did precisely that, for black holes were predicted long before strong evidence for their existence was observed. Yet no one has predicted minds from first principles. We are still scrabbling around trying to account for ourselves, making the best of crude tools and almost non-existent theories and unsure if we will ever be able to break out of the logical loops that this endeavour reveals.

This is not to say that anything more than those fundamental laws is required for minds to happen; indeed, all that we know so far about our minds suggests that is not the case. But that only makes the issue more astonishing. It's tempting to suspect that mindedness, as much as life itself, is an inherent characteristic of the universe. Whether either of these is common or rare within it is, in comparison, a minor question. That conscious matter exists at all has to make one wonder if there is at root some feature of information itself that will render it inevitable.

But there are many forms of life, and many states of matter, and so we would be unwise to suppose that the conscious matter housed within our skin is unique in kind. The fundamental question about the Space of Possible Minds is: what are the possibilities of mind that the laws of matter permit?

Perhaps we might turn this question on its head, and wonder whether those 'laws' are themselves contingent on the kinds of minds that perceive and articulate them. This is not to say that somehow the laws of nature will be 'different' for other minds, but rather that the experiences of other minds might not lend themselves to being formulated in terms of these laws as we ourselves express them. This could be one of the practical benefits of having a better view of the Space of Possible Minds: it might broaden the horizons of our own minds, freeing us from the straitjacket of our convictions and the blind spots of our perception.

We are already attempting to explore that avenue by developing artificial intelligence that can engage in genuine scientific discovery by seeking to condense data into scientific laws. At face value, research on 'theory-creating AI' seems to support the intuition of many scientists that our physical laws and theories are universal: they will be more or less replicated by any other intelligent beings.

But this overlooks the fact that such AI systems – variants of machine-learning algorithms – are already guided towards our view of the world by the nature of the data we give them. We present, for example, planetary motion in terms of objects following trajectories in space and time relative to our own position as observers. It seems hard to imagine how else physical phenomena like these might be conceived, if not in terms of discrete objects in motion. I am not sufficiently imaginative to conjure up alternatives, except to say that I do not think we should take it for granted that cephalopods, bees, or bats carve up the world in quite the same manner. And they are just our neighbours in Mindspace.

This is precisely why J. B. S. Haldane advised humility about how we imagine and depict the world in our schemes and sciences. 'We are just getting at the rudiments of other ways of thinking', he wrote. 'I do not feel that any of us know enough about possible kinds of being and thought, to make it worthwhile taking any of our metaphysical systems very much more seriously than those at which a thinking barnacle might arrive.' For the barnacle, the world is merely a rock. (Not even that, of course; rather perhaps, an anchoring surface bathed in nutritious soup.) As we extend our sensory modalities artificially, we realize how much we have been missing, clinging to this rock of ours: X-rays and radio waves, gluons and gravitons, dark matter and antimatter. We had best suppose that there is still plenty that is undreamt of in our philosophies. As Haldane famously put it, 'My suspicion is that the universe is not only queerer than we suppose, but queerer than we can suppose.'

'Our only hope of understanding the universe', he said, 'is to look at it from as many different points of view as possible.' We need other minds.

Mind design

The idea of a space of possible states or objects is not confined to physics. Chemists today imagine and explore 'chemical space': the astronomical gamut of possible ways in which different chemical elements can be combined into molecules and materials. This helps them to identify where to search for chemical compounds that might have useful properties such as extreme hardness, electrical conductivity, or specific drug activity. If the space has well-chosen coordinates, substances with similar behaviour might occupy the same regions of it. Then, 'spaces' that map out, say, how the strength and hardness of materials are related can be used in the search for new substances, and to group those already known into distinct families.

In this way, such spaces become not just neat visual summaries of what is known, but tools for planning, for design and engineering. We can think about the Space of Possible Minds in this way too: as a tool for what Susan Schneider calls mind design, where we choose which attributes we wish to add to a mind, or to enhance, subtract, or diminish.

Current interventions in the human mind through neurosurgery and pharmacology are more like mind hacking (literally in the former case) than mind design. But they are improving. Psychiatry and neurology have long been attempting rough-and-ready mind design in a less invasive manner, using techniques ranging from cognitive behavioural therapy to transcranial magnetic stimulation. We are getting far more sophisticated in building brain–machine interfaces that enable neurons directly to control artificial devices through the volition of the user. The transhumanist movement, which believes that humans can and should move beyond our current mental and physical limitations by technological means, is sometimes naive in its faith in unproven and highly fanciful future technologies, but insofar as it envisages the reimagining and radical redesign of the human mind, it identifies a real and worthwhile possibility: that we might make better use of the resources of Mindspace. One way or another, mind design might increasingly become an engineering discipline, amenable to rational planning.

What determines the topography of the design space? Where, say, is it populous, and where barren? We certainly don't know the answer yet; we barely understand the question. But the scope for design is surely not arbitrary: it will permit the creation of some kinds of mind but not others. Nature, after all, never draws from an infinite palette of choices. While our minds and those of other animals were produced from scratch according to evolution's strategy of trial and error – random mutation coupled to natural selection – it operated within constraints. What is feasible in a body has to conform to the laws of physics – which dictate, as we saw, that flying creatures are likely to have wings, and swimming creatures streamlined bodies. It's reasonable to think that minds on Earth also come from a space of possibilities in which certain attributes are indispensable or inevitable: some degree of memory, say, or a limited range of sensory inputs and limits on the possible ways of integrating them.

We don't know what dictates these options and exigencies, but it probably has something to do with the fundamental nature – one might say, the ecology – of information: how it can be filtered from the environment, by what means it acquires meaning, how it can be most efficiently organized, stored, and used, what types of embodiment enable its collection and use. What we will need in order to understand better the shape of Mindspace and the parameters that define it is a real 'theory of mind' – not in the classic psychological sense, but a science that can encompass and account for the origins and operations of minds. This is not the same as a theory of the human brain, in much the same way that a theory of how internal combustion engines work is not the same as a complete, predictive understanding of a Ford Pontiac. The former is a more profound account of what cars are about, and it might even be easier, unencumbered with endless minute particulars of how the windows work and how to turn the lights on.

Just as many technologies preceded an understanding of their physical principles, we have in fact been shaping and to some extent designing our minds for millennia. For that's what culture does. The brain, and the mind it supports, is reshaped by experience from the level of individual neurons and their connections to its gross anatomical features. Culture remodels the mind systematically, albeit often in small and subtle ways: affecting, say, how we experience colours, music, food, and how the mind reasons morally and socially. It seems almost certain that this mind-shaping is an accelerating process. 'The ongoing digital revolution, the globalization of cultural information, and the coming of the age of empathy,'* writes Antonio Damasio, 'are pressures likely to lead to structural modifications of mind and self, by which I mean modifications of the very brain processes that shape the mind and the self.'

Technologies are now expanding the boundaries of human Mindspace further and faster. As we extend the reach of our minds into physical space via prosthetics, remote virtual control systems, and other human–machine interfaces, so too we may be extending our dimensions of mind. It is entirely feasible to imagine, for

* Some readers might be surprised to hear this time of the early 2020s described as an age of empathy. But despite the alarming political, socioeconomic, and environmental events we have witnessed already in this new decade, one need not be a Pollyanna (or Steven Pinker, who is not that either) to recognize that in historical terms and in the aggregate, we have become more considerate and tolerant than was the case five or even two centuries ago.

instance, memory extensions in which mind-controlled circuitry creates a direct feed of information to the brain. Possibilities like this motivate Elon Musk's 'neural lace' project, which seeks to develop a biocompatible interface for recording and perhaps sending electrically encoded information to and from neurons, and even introducing a layer of artificial intelligence inside the brain. As is often the case with Silicon Valley ventures, the details of Musk's scheme are kept hazy and hidden, prompting the suspicion that the idea is currently driven more by hype and fantasy than hard science. But the principle is not obviously absurd. Might such a system also ultimately permit outsourcing of some cognitive processes – arithmetical calculations, say – to artificial devices, not via manual entry on keyboards or touchscreens but initiated by thought alone? Might we one day hook up our minds to a battery of peripheral devices – or even to one another?

There is a long, indeed ancient, tradition of expanding the mind pharmaceutically. Whatever the actual value (not to mention the dangers) of hallucinogenic drugs, there's no doubt that they can put us into mental states not otherwise accessible. Many current pharmaceutical interventions in mind are of course geared more towards bringing it back into the conventional boundaries than pushing beyond them. Stanislas Dehaene suggests that conditions like schizophrenia might be regarded as 'diseases of consciousness': breakdowns of the mechanisms by which a healthy conscious state is constructed, which seem to be caused by dysfunctions in the production or uptake of chemical neurotransmitters such as dopamine and glutamate. That's why some treatments use drugs to restore the balance. But other drug treatments are now being developed that boost concentration or memory, or which, in combination with other treatments, can selectively implant or erase memories. We still have rather little sense, however, of what new features of mind might be generated by altering the chemical balance in the brain.

Alternatively, we might imagine tampering with the biology of the brain's growth and maintenance: altering the genetic systems responsible for, say, creating neurotransmitter receptor molecules, growing the dendritic networks of neurons, or governing the distributions of different types of neurons and other cells. Already brain organoids (page 154) supply a means for exploring such possibilities safely and ethically. Of course, the potential hazards in examining these questions are not just medical but social too – it was by altering the chemistry of the embryos grown artificially in 'hatcheries' that the World State of Aldous Huxley's *Brave New World* created a caste system of intelligence. But there could be benefits too to an ability to redirect the development and function of brains and minds. 'We may well be able to engineer desire and motivation,' says the Australian philosopher Julian Savulescu. If neuroenhancement – what we might now picture as an expansion of human Mindspace – is feasible, Savulescu argues that it will become a moral obligation: how could we reasonably forgo the option of improving ourselves cognitively and intellectually? However, he adds that such enhancement could be risky if it is not accompanied by efforts to engineer improved morality too. Others ask if we have even the slightest idea of what that would entail, not least because it is far from clear that cognitive and moral enhancement would involve manipulating independent dimensions of mind. And could we even agree on what counts as a moral enhancement?

Quite aside from the ethical challenges, mind design will be technically difficult. Given the highly interconnected and interdependent nature of the circuits and functions of human cognition, it seems rather likely that the dimensions of mind aren't all orthogonal: tamper with one and you might find consequences for others. The brain is likely to be a fine-tuned organ in which an 'improvement' or enhancement in one area requires some compensating attenuation in another. If we're going to alter it, we had first better get a good idea of what the design space looks like.

Keeping in mind what matters

Some will surely ask of the notion of designing minds: is it wise?

It's the right question in more ways than one. All advances in human capability bring potential benefits but also ethical dilemmas, and never more so than when the advances may alter our own nature: for example, in technologies of assisted reproduction, cloning, genetic engineering, physical and cognitive enhancement, and human–machine interfaces.

If history teaches us anything about the way our technologies evolve, it is that we are not terribly good at looking in the right direction for what to expect. While developments in cloning and reproductive technologies left us fretting about resurrecting Hitler and totalitarian social control, we didn't think very hard about how those options would play out within an aggressively consumerist free market, where now the concern is that advertising and peer pressure might impel prospective parents on a futile but costly quest for perfection. While we marvelled at the ability of AI to mine and mimic human responses, we failed to anticipate how those algorithms would reproduce our prejudices too. We imagined that social media would open up new channels of dialogue, without much thought to the deluge of misinformation and hate messaging that might flow down them.

I see the same danger in store as we map out Mindspace. As long as computation remains the dominant metaphor for mind and cognition, discourse in neuroscience, animal behaviour, AI, and robotics is likely to go on invoking considerations of inputs and outputs, logic operations, memory capacity, wiring patterns, information integration, optimization schemes. And yet so much of what matters to the human mind lies in other directions. We need empathy, compassion, and most of all, the attribute that I have said nothing about because it is the most neglected of all our features of mind: wisdom.

If our prognoses for the outcomes of advanced AI and alien encounters are often gloomy and dystopian, it seems that may be because we struggle to imagine what attributes like these could mean for non-human minds – and we all too often assume they will therefore be absent. But where are they to be found, and for what reasons? What if other, unfamiliar regions of Mindspace actually embraced more of these affective properties, not fewer – if we weren't after all the exemplar of what is life-affirming and beneficent in minds?

And what if those capacities, rather than our rationality or ingenuity or pattern-recognizing ability, turned out to be the most vital aspects of mind for the future of our species and our planet? That's why we should care about the cartography of minds: because it might show us what we can become.

Acknowledgements

Several years ago, an American science magazine asked me if I might consider writing an article about efforts to open the 'black box' of artificial intelligence: to figure out how today's AI systems are really 'thinking' as they convert input to output. I already knew that, because of the way such AI is designed, the answer is far from obvious; we know it works, and often rather well, but we're not sure how. And this is, after all, not so different from the situation we face in trying to understand our own minds. Being inside the black box doesn't necessarily help; in some ways it makes the question harder to understand and to address.

After reading the paper that triggered the request, I started to dig around and to make inquiries with experts working on AI. But it was frustratingly hard to find advisers who seemed able to comment on the wider issue of how AI 'thinks'. I now know that this was not because no one in AI ponders that question but because I was looking in the wrong places. One lead, however, directed me towards Murray Shanahan, professor of cognitive robotics at Imperial College London, who had advised on Alex Garland's 2014 film *Ex Machina* and has consulted for the AI company DeepMind. Murray was just the person I needed – except that he was too busy at that point to offer comments. But he generously sent me an electronic copy of his seminal 2010 book *Embodiment and the Inner Life: Cognition and Consciousness in the Space of Possible Minds*, which was packed with rich ideas. The subtitle, Murray explained in the book, came from a paper published in 1984 by computer scientist Aaron Sloman of the University of Birmingham titled 'The Structure of the Space of Possible Minds'. I found a copy of Aaron's paper on the web, and managed to get in touch with him, now an emeritus professor. I got more than I bargained for: in response to my questions, Aaron sent me several long emails filled with thoughts, ideas, tangents and much else. I was intrigued – and I still had absolutely no idea what to do with any of this.

Somehow, the trail led me also to a project called Kinds of Intelligence at the Leverhulme Centre for the Future of Intelligence at Cambridge University. (Maybe it was Murray who sent me there: he is a 'spoke leader' of the project.) I contacted one of the senior research fellows in the team, philosopher of cognitive science (what a topic!) Marta Halina, and the next time I was up in Cambridge I arranged to meet her for a chat. So we chatted, and she suggested some other people I might talk to. Still I had no clear picture of where any of this was going, or whether it could ever be an article. But I couldn't let it go.

In 2019 I was invited to spend the summer at Harvard as a visitor in the Medical School's department of systems biology. It was the kind of opportunity you would spend your life regretting if you'd not seized it. I quickly discovered that round the corner from the house where my family and I were staying in Brookline was one of the most wonderful local public libraries I've ever come across, which included a section that sold off excess stock for a dollar a piece. This was a mixed blessing, as I quickly acquired far more titles than my baggage allowance for the return flight would permit. One of the first of these was *The Biological Mind* by neuroscientist Alan Jasanoff. (I never told you this, Alan, but I fear it was personally autographed; let's assume it had been a gift too far for a similarly over-freighted visitor.) I realized that Alan was just up the road at the Massachusetts Institute of Technology – and so I arranged to stop by for a chat.

And suddenly it all became clear. I had never written that article because it wasn't an article at all. There was far, far too much to explore here, and it needed to be a book.

There could have been no better place to start writing it than Boston, a nexus of interdisciplinary research on cognition, neuroscience, psychology, AI, robotics and all things *minded.* It is odd how sometimes things fall into place. No sooner had I realized what my next book had to be about than I visited Neil Gershenfeld at MIT's Media Lab, who, on discovering I had recently written about efforts to put theories of consciousness under experimental scrutiny (another spin-off from my attempts to mine minds), enrolled me to speak on the subject at a forthcoming workshop on animal minds – specifically, on efforts to use information technologies to help us understand the communications and minds of other species, and perhaps to help them understand one another. Neil and colleagues had convened a glittering array of experts for this closed meeting, and I could have wished for no better opportunity to launch me into the world of animal cognition and other minds.

Of course, the subject matter for this topic won't in truth fit even into a book: it demands an encyclopaedia. In *The Book of Minds* I have tried to boil down what seemed to me to matter most, but I don't claim to have done more than start a conversation – or rather, provide pointers to a conversation that is already happening, and has been to some degree ever since Aaron's paper. I believe that conversation is now one of the most exciting, most challenging and ambitious, in all of science. There is an abundance of solid research behind it, but also necessarily still a strong element of speculation, and it is all awash with matters of philosophy and ethics – which I consider to be a good thing.

The process has brought me into contact with scientists,

engineers, thinkers and doers in a wide range of disciplines, all of whom have contributed towards shaping this book whether they knew it or not. I hope I have not omitted anyone from this list of those to whom I am immensely grateful: Björn Brembs, Matthew Cobb, David Cox, Anna Dornhaus, Uta Frith, Neil Gershenfeld, Jeremy Gunawardena, Marta Halina, Thomas Hills, Alan Jasanoff, Arik Kershenbaum, Christof Koch, Marcelo Magnasco, Jennifer Mather, Randolf Menzel, Kevin Mitchell, Melanie Mitchell, Diana Reiss, Anil Seth, Murray Shanahan, Aaron Sloman, Susanne Still, Giulio Tononi, Tomer Ullman and Sara Imari Walker. Melanie, Anil and Anna were particularly generous with their time and advice in reading through the draft manuscript.

My time in Boston was catalytic for this project, and was made possible by the generosity of Galit Lahav and Becky Ward at Harvard Medical School. There I had the pleasure of being able to discuss the ideas at a formative stage with my US editor Karen Merikangas Darling, whose support over several years now I have deeply appreciated. It has been a pleasure too to be working once again with my two editors at Picador, Philip Gwyn Jones and Ravi Mirchandani. All this was made possible, as ever, by the help and encouragement of my agent Clare Alexander, and of my family.

And to Michael at *Quanta*: sorry, I never wrote you that article. It grew up.

Philip Ball
London, December 2021

End notes

CHAPTER 1: MINDS AND WHERE TO FIND THEM

2 'She was nursing': Sacks, *Everything in Its Place: First Loves and Last Tales*, pp. 241–242. Knopf, New York, 2019.
7 'Mind is an immeasurable': Schnell & Vallortigara (2019).
8 'Whatever else': Dennett (1996), p. 4.
9 'A mysterious form': Minsky (1987), p. 55.
10 'takes a set': Ogas & Gaddam, p. 10.
15 'the task of a mind': Dennett (1996), p. 75.
15 'it mines the present': ibid., p. 75.
15 'If our planet': ibid., p. 81.
16 'can have many': Damasio, p. 00.
17 'No organism seems': ibid., p. 00.
20 'segregated field of 'mental': Ryle, p. 00.
23 'we have no clear': Cobb (2020), p. 2.
33 'it can be difficult to tell': Jasanoff, p. 36.
33 'nothing but *reckoning*': Hobbes, *Leviathan* (1651), p. 111. Penguin, London, 1985.
33 'Equating the organic mind': Jasanoff, p. 37.
35 'is the wrong dogma': M. Heisenberg, 'The beauty of the network in the brain and the origin of the mind in the control of behavior'. *Journal of Neurogenetics* 28, 389–399 (2014).
37 'Like evolution itself': Edelman & Tononi (2000), p. 205.

CHAPTER 2: THE SPACE OF POSSIBLE MINDS

39 'Clearly there is not': Sloman (1984).
40 'Many facts are ignored': Sloman, personal communication.
40 'My impression is that': ibid.

40 'We must abandon the idea': Sloman (1984).
41 'This is a classification': ibid.
41 'These explorations can be': ibid.
41 'Instead of fruitless': ibid.
41 'Current AI throws': ibid.
43 'I'm enough of an artist': A. Einstein, interview with G. S. Viereck, 'What life means to Einstein'. *Saturday Evening Post*, p. 117, 26 October 1929.
45 'though clearly a form of perception': Nagel (1974).
45 'It will not help': ibid.
46 'I want to know what': ibid.
46 'For if the facts of experience': ibid.
46 'Martians might learn more': ibid.
49 'what is urgently needed': N. K. Humphrey, in P. P. G. Bateson & R. A. Hinde (eds), *Growing Points in Ethology*, pp. 303–317. Cambridge University Press, Cambridge, 1976.
49 'The essential feature': ibid.
53 'the important question': Wegner & Gray (2016), p. 169.
54 'Perhaps countries are conscious': ibid., p. 224.
54 'Is the United States': ibid., p. 224.
55 'When talking about': D. Hillis, in Brockman (ed.) (2020), p. 173.
55 'to have goals aligned': ibid., p. 173.
57 'A brick scores zero': Shanahan (2016).
58 'To situate human consciousness': ibid.

CHAPTER 3: ALL THE THINGS YOU ARE

63 'Tell me where is fancy': *The Merchant of Venice*, Act 3, ii.
65 'Behaviour itself did not': Pinker (1997), p. 42.
67 'An individual's best': Brosnan et al. (2010).
72 'Every human mind': Dennett (1996), p. 203.
72 'There is no step more': ibid., p. 195.
74 'until a better picture': Wilson (2017), p. 55.
74 'Because the creative arts': ibid., p. 85.
75 'replete with [more] opportunities': Dennett (2018).
76 'Lovers and madmen': *A Midsummer Night's Dream*, Act 5, i.
76 'must have been the single': Dor (2015), p. 209.
76 'Uttering a word is like': ibid., p. vi.
77 'open a venue for human': ibid., p. 2.

79 'We do not know': Damasio (1995), p. 97.
79 'Perception is not simply': A. Gormley, C. Richardson & J. Winterson, *Land*, p. 23. Landmark Trust, Shottesbrook, 2016.
81 'our intuitive sense': Cohen et al. (2020).
81 'controlled hallucination': Seth (2021), pp. 79–80. Seth attributes the phrase to Chris Frith, but tells me he has since been made aware of older origins. It is not clear who first coined it.
82 'recovers the things': Schiffrin et al. (2020).
82 'gild and stain objects': D. Hume (1751), *An Enquiry Into the Sources of Morals*, Appendix 2.
86 'I saw it in my mind's eye': I. McEwan, personal communication.
87 'Cognition's trick is to open': Shanahan (2010), p. 3.
90 'intuitive physics engine': Lake et al. (2016).
94 'You can build a mind': Minsky (1987), p. 17.
97 'I don't think it's right': A. Jasanoff, personal communication.
97 'The emotions are mechanisms': Pinker (1997), p. 373.
98 'Emotions put meat on': Bor (2012), p. 71.
103 'Much of the time': Sacks (1995), p. 248.
103 'We have to learn': Silberman (2015), p. 512.
104 'He can recognize the facts': ibid., p. 99.
106 'I have myself long been': U. Frith, personal communication.
107 'deeper scrutiny of these talents': ibid.
107 'Many of the challenges': Silberman (2015), p. 16.
108 'rich diversity of personal': Stapledon (1930), p. 161.
109 'Taking them out': Dennett (1996), p. 183.
110 'We shouldn't accept a neural': Q&A with Andy Clark, *Nature* 571, S18 (2019).

CHAPTER 4: WAKING UP TO THE WORLD

115 'For someone who claimed': Damasio (1994), p. 248.
115 'It has become customary': Shanahan (201), p. 8.
115 'The one thing, the only thing': Koch, personal communication.
116 'Consciousness is a word': G. A. Miller, *Psychology: The Science of Mental Life*, p. 40. Harper & Row, London, 1962.
116 'As recently as a few years ago': Crick (1994), p. vii.
118 'I knew I was a substance': R. Descartes, *Discourse on the Method* (1637), 6:32–33.

118 'If there is any sense': G. Strawson, *Mental Reality*, p. 53. MIT Press, Cambridge MA, 1994.
123 'The elusive subjective': Koch (2019), p. 4.
123 'a peculiar idea': Dehaene (2014), p. 262.
124 'distinguishing different aspects': Seth, *Aeon* (2016).
126 'Once we have isolated': Pinker (1997), p. 147.
127 'wordless narrative': Damasio (2012), p. 204.
127 'relies extensively': ibid., p. 172.
128 'folded together': ibid., p. 186.
129 'a state of mind': ibid., p. 157.
129 'The brain deserves': Dehaene (2014), p. 236.
132 'Consciousness is global': ibid., p. 163.
133 'riotous parliament': Edelman & Tononi (2000), p. 245.
135 'take possession': Dehaene (2014), p. 21.
135 'To be conscious': Tononi, personal communication.
136 'a system's ability': Koch, personal communication.
138 'Digital computers': ibid.
138 'No matter whether': Koch (2012), p. 131.
138 'I believe that consciousness': Koch, personal communication.
139 'awareness is vividly present': Koch (2019), p. 114.
139 'the problem with panpsychism': J. R. Searle, 'Can information theory explain consciousness?'. *New York Review of Books* 10 January 2013, p. 55. https://www.nybooks.com/articles/2013/01/10/can-information-theory-explain-consciousness/
139 'cannot be spread': ibid.
140 'developed from third': Tagliazucchi.
140 'the hallmark of the conscious': Shanahan (2010), p. 91.
143 'the upper brainstem': Solms, p. 5.
144 'there are many forms of consciousness': Crick (1994), p. 21.
145 'feelings let us mind': Damasio (1995), pp. 159–60.
151 'Those who claim': Baars (2019), p. 13.
151 'Consciousness is filled': Koch (2019), p. 121.
152 'in a conscious mind': Damasio (2012), p. 267.
152 'the organism [becomes]': ibid., p. 176.
157 'Given the astonishing pace': Koch (2019), p. 128.
160 'consciousness has more to do': Seth (2021), p. 9.
160 'The chance that it experiences': Koch (2019), p. 128.
162 'If organised patterns': Bayne et al. (2020).
164 'do not have duties': Brockman (ed.) (2020), p. 252.

CHAPTER 5: SOLOMON'S SECRET

166 'research on animal behavior': LeDoux (2019).
166 'more powerful in leading': Descartes, *Discourse on the Method* (1637), Part V.
167 'The belief that only humans': Koch (2019), p. 31.
168 'the time has come for people': C. Slobodchikoff, personal communication.
170 'biologists . . . have looked for': E. L. Thorndike, 'Animal intelligence: an experimental study of the associate processes in animals', *Psychological Review Monographs Supplement* 2(4), 1–8 (1898), reprinted in *American Psychologist* 53, 1125–1127 (1998).
171 'While the body': Hume (1748), p. 13.
178 'Chimpanzees, like humans': Call & Tomasello (2008).
179 'final nail in the coffin': Caruso (2016).
181 'Humans seem to act': Rosati (2017).
182 'Humans may be unique': ibid.
182 'When a female chimpanzee': de Waal (2016).
184 'The incident can be': de Waal (1997).
184 'The incident at the Brookfield': ibid.
190 'If we imagine': Tomasello & Rakoczy (2003).
193 'Unfortunately it is extremely difficult': Ackerman (2016), p. 32.
195 'acted in nearly the same': Darwin, *The Formation of Vegetable Mould Through the Action of Worms*, p. 97. John Murray, London, 1881.
195 'we should remember': ibid., p. 98.
207 'create new associations': Herculano-Houzel (2020).
210 'hints that entirely novel': Loukola et al. (2017).
215 'I deny sensation to no animal': Boden (2006), p. 70.
218 'We used to think': Abbott (2020).
219 'the gap between ape': Dehaene (2014), p. 247.
224 'Those who consider': K. von Frisch, *Animal Architecture*, p. 244. Hutchinson, London, 1975.
225 'Many animals hide': Dennett (1996), p. 157.
226 'If a lion could talk': L. Wittgenstein, *Philosophical Investigations*, transl. G. E. Anscombe, p. 223. Blackwell, Oxford, 1958.
228 'remove our human glasses': B. Selvitelle, talk at 'Interspecies Internet workshop', MIT Media Lab, 15 July 2019.
228 'The exciting thing': C. Slobodchikoff, personal communication.
229 'cuts too fine': Dennett (1996), p. 56.

229 'They are able to see': M. Magnasco, personal communication.
230 'Dogs are often surrendered': C. Slobodchikoff, personal communication.
230 'made the space to see': D. Reiss, talk at 'Interspecies Internet workshop', MIT Media Lab, 15 July 2019.

CHAPTER 6: ALIENS ON THE DOORSTEP

232 'probably the closest we will come': Godfrey-Smith (2016), p. 9.
232 'If we want to understand': ibid., p. 10.
234 'turn the apparatus around': ibid., p. 52.
234 'They "talk" to you': M. Kuba, personal communication.
235 'a path of lone': Godfrey-Smith (2016), p. 65.
235 'One night, a small octopus': Porcher (2019).
238 'Unity is optional': Godfrey-Smith (2016), p. 87.
240 'When I try to imagine': Godfrey-Smith (2019).
241 'built minds twice over': Godfrey-Smith (2016), p. 9.
245 'impart a certain tone': Ginsburg & Jablonka (2008).
245 'The function of feeling': ibid.
248 'It is hardly an exaggeration': Darwin, *The Power of Movement of Plants*, p. 573. John Murray, London, 1880.
249 'we must believe that in plants': Maher (2017), p. 15.
251 'The ability to learn': Gagliano (2017).
257 'only multicellular animals': Feinberg & Mallatt (2016).
259 'Not just animals are conscious': Ginsburg & Jablonka (2021).
259 'without an internal': Baluška & Reber (2019).
260 'Some level of experience': Koch (2019), pp. 155, 159.
261 'living cells have an intrinsic': Bray (2009), p. 142.
262 'has raised questions about what cognition is': Hernández-Orallo (2017). [4.5]
262 'cognition all the way down': Levin & Dennett (2020).
263 'I've come to think': Keim (2019).
264 'When the bed is thus prepared': L. Taiz & L. Taiz, *Flora Unveiled: The Discovery and Denial of Sex in Plants*, p. 368. Oxford University Press, Oxford, 2017.

CHAPTER 7: MACHINE MINDS

267 'The development of full': S. Hawking, interview for the BBC, 2 December 2014. https://www.bbc.co.uk/news/technology-30290540
267 'biggest existential threat': E. Musk, interview in 2014. See S. Gibbs, 'Elon Musk: artificial intelligence is our biggest existential threat', *Guardian*, 27 October 2014. https://www.theguardian.com/technology/2014/oct/27/elon-musk-artificial-intelligence-ai-biggest-existential-threat
267 'we cannot predict': https://futureoflife.org/ai-open-letter/
268 'has yielded remarkable': ibid.
269 'Highly autonomous AI systems': https://futureoflife.org/ai-principles/
269 'Earth's current environment': Brockman (ed.) (2020), p. 97.
272 'I wish to God': 'A brief history', Computer History Museum, Mountain View, California. https://www.computerhistory.org/babbage/history/
275 'I believe that at the end': Turing (1950).
276 'every aspect of learning': Mitchell (2019), p. 4.
277 'Machines will be capable': ibid., p. 5.
277 'within a generation': Russell (2019), p. 76.
278 'only a clever hack away': C. Koch, personal communication.
277 'science of making machines': Hernández-Orallo (2017), p. 119.
281 'Please write me a sonnet': Turing (1950).
283 'there is probably no elegant': Woolridge (2020), p. 115.
284 'one of the most notorious': ibid., p. 90.
285 'to an appropriate course': Turing (1950).
290 'be conscious of its [own] existence': Mitchell (2019), p. 21.
291 'its teacher will often': Turing (1950).
291 'Any system simple enough': Brockman (ed.) (2020), p. 39.
293 'You could see right there': Mitchell (2019), p. 97.
296 'AI is kind of dumb': D. Cox, personal communication.
299 'It is quite hard to work out': Russell (2019), p. 56.
300 'We all look at a cat': T. Ullman, personal communication.
302 'the absence of common sense': https://www.darpa.mil/program/machine-common-sense
302 'A lot of common sense': D. Cox, personal communication.
302 'The questions people are asking': T. Ullman, personal communication.
302 'We believe that future': Lake et al. (2016).
302 'a kind of idiot savant': Marcus & Davis (2019), p. 64.
303 'We suggest that a current': ibid., p. 25.

306 'involves having a mental model': D. R. Hofstadter, *Gödel, Escher, Bach: An Eternal Golden Braid*, p. 603. Basic Books, New York, 1979.
307 'Arguably all of social cognition': Graziano (2018).
308 'it is surprisingly easy': Winfield (2018).
310 'May not machines carry out something': Turing (1950).
312 'Joe, will you please log out': Block (1995).
311 'a test of human gullibility': J. Markoff, 'Software is smart enough for SAT, but still far from intelligent'. *New York Times* 21 September 2015.
313 'Yet in a circle pallid': https://github.com/jhlau/deepspeare. See Lau et al. (2018).
314 'couldn't believe the expressiveness': personal communication with the author: see https://www.theguardian.com/music/2012/jul/01/iamus-computer-composes-classical-music
319 'We know of no fundamental': C. Koch, http://www.theswartzfoundation.org/abstracts/2001_summary.asp
319 'I am pretty sure': Brooks (2017).
319 'We don't need artificial conscious agents': Brockman (ed.) (2020), p. 51.
320 'one should only create': Schneider (2019), p. 68.
320 'Awareness of the world': Brockman (ed.) (2015), p. 2.
320 'I think it might be a big': Brooks (2017).
322 'To be able to live in the world': I. Rahwan, personal communication.
322 'Let's say you simply ask': ibid.
323 'People are going to have to change': D. Cox, personal communication.
323 'If machines are more vindictive': I. Rahwan, personal communication.
323 'We shape our technologies': the quote sometimes substitutes 'tools' for 'technologies', and does not seem to originate with McLuhan himself. The basic idea appears in J. M. Culkin, 'A schoolman's guide to Marshall McLuhan', *Saturday Review*, 18 March 1967, pp. 51–53 and 71–72. See https://webspace.royalroads.ca/llefevre/wp-content/uploads/sites/258/2017/08/A-Schoolmans-Guide-to-Marshall-McLuhan-1.pdf
324 'As we gain more experience': Russell (2019), p. 248.
324 'can far surpass all the intellectual': Good (1965).
325 'the first ultraintelligent machine': ibid.
325 'the advent of machines that truly think': Brockman (ed.) (2015), p. 46.
325 'Anthropomorphic frames': Bostrom (2014), p. 111.
326 'the AI might find subtle ways': ibid., p. 142.
328 'An artificial intelligence need not': ibid., p. 35.
328 'A characteristic of AI dystopias': Brockman (ed.) (2020), p. 101.
328 'that many of our techno-prophets': Brockman (ed.) (2015), p. 7.

329 'They depend on the premises': Brockman (ed.) (2020), p. 111.
330 'intelligence is the ability': Russell (2019), p. 168.
331 'the one who becomes the leader': ibid., p. 182.
332 'As we move further': Schneider (2019), p. 10–11.
332 'humans and all they've thought': Brockman (2015), p. 9.

CHAPTER 8: OUT OF THIS WORLD

335 'minds that are to our minds': H. G. Wells, *The War of The Worlds*, in *The Science Fiction* Vol. 1, p. 185. J. M. Dent, London, 1995.
337 'The typical Martian organism': Stapledon ([1930] 1972), p. 156.
337 'constituted sometimes': ibid., p. 158.
337 'In so strange a body': ibid., p. 160.
337 'was both far more liable': ibid.
338 'inveterate selfishness': ibid., p. 161.
338 'the older science fiction is': Kershenbaum (2020), p. 323.
339 'a textbook on neutron star': L. David, 'Robert Forward, space futurist, dies at 70', space.com, 23 September 2002. https://web.archive.org/web/20060218003005/http://space.com/news/robert_forward_020923.html
340 'One, they have to have': D. A. Smith, interview with D. C. Denison, *Boston Globe Magazine*, 18 June 1995. See Pinker (1997), p. 61.
341 'Tho' we in vain guess': Huygens (1698), *Cosmotheoros*. Available at https://webspace.science.uu.nl/~gent0113/huygens/huygens_ct_en.htm
341 'not Men perhaps like ours': ibid.
342 'For all this Furniture and Beauty': ibid.
342 'It is idle to try to guess': Jones (1940), p. 250.
342 'such vegetation as now': ibid., p. 211.
343 'a unique objective standard': Cocconi & Morrison (1959).
343 'A sequence of small prime numbers': ibid.
345 'the first sign of intelligent': A. Loeb, *Extraterrestrial: The First Sign of Intelligent Life Beyond Earth*. John Murray, London, 2021.
346 'have done cost–benefit analyses': SETI Institute, https://www.seti.org/seti-institute/project/details/zookeepers-alien-visitors-or-simple-life
346 'the Galaxy is urbanised': ibid.
346 'searching for other versions': Cabrol (2016).
347 'Without some informed analysis': Edmondson (2012).
348 'To find ET we must expand': Cabrol (2016).
348 'will likely have developed': ibid.

354 'well-defined principles that we know': A. Kershenbaum, personal communication.
355 'While Darwinian evolution': S. Bartlett, personal communication.
355 'I'm not convinced': S. Walker, personal communication.
355 'Nowhere in space': Kershenbaum (2020), p. 4.
358 'We can be confident': ibid., p. 171.
361 'There are some planets': Arthur (2020), pp. 316–317.
363 'Ants have dozens of different': A. Dornhaus, personal communication.
364 'On a planet where': Kershenbaum (2020), p. 137.
366 'Writing is the way': A. Dornhaus, personal communication.
367 'wrote a short story': T. Bisson (1991), 'They're made out of meat', https://www.mit.edu/people/dpolicar/writing/prose/text/thinkingMeat.html
369 'The most straightforward': Kurzweil (2005), p. 324.
370 'will be available by the': ibid., p. 199.
371 'Currently, when our human hardware': ibid., p. 325.
373 'Despite the near-religious belief': Koch (2019), p. xiv.
374 'will live out on the Web': Kurzweil (2005), p. 325.
374 'sees the world through': Koch (2019), p. 109.
374 'With sufficient advanced': ibid., p. 111.
375 'a significant fraction': Dyson (1979), p. 207.
375 'When we look into': ibid., p. 210.
376 'biological living space': ibid., p. 211.
376 'Given sufficient time': ibid.
377 'Not only was every solar system': ibid.
377 'within a few hundred years': ibid.
378 'the problem of colonization': ibid., p. 210.
380 'It is easy to imagine': ibid.
380 'be the cosmic version': Brockman (2015), p. 3.
381 'maybe our whole universe': J. Lim, 'Simulation hypothesis: how could we tell if we are living in a computer simulation?'. Futurism, 22 December 2015. https://futurism.com/19945
387 'we need to face': Carroll (2017).
387 'Bob: But everything I know': ibid.
388 'If you reason yourself into believing': ibid.
389 'God is many things': Wegner & Gray (2016), p. 263.
391 'If you comprehend it': Hart (2013), p. 142.
391 'in any of the great theistic': ibid., p. 234.
394 'the deepest truth of mind': ibid., p. 236.
395 'It may actually be': Wegner & Gray (2016), p. xx.

CHAPTER 9: FREE TO CHOOSE

401 'a belief that there is a component': Cashmore (2010).
404 'free will worth wanting': Dennett (1984).
412 'The behavior of large and complex': Anderson (1972).
417 'are not just an interpreter': Hills (2019).
418 'is a biological trait': Brembs (2020a).
419 'the steam whistle': Huxley (1874).
420 'the subjects hand over': K. Smith, 'Brains make decisions before you even know it'. *Nature News* 11 April 2008. https://www.nature.com/articles/news.2008.751
421 'Conscious deliberation': Damasio (2012), p. 271.
430 'Agents are *causes of things*': K. Mitchell, personal communication.
432 'you cannot explain': Dennett (2018).
435 'lawyers, judges, legislators': Damasio (2012), p. 283.
436 'the legal system assumes': Cashmore (2010).
436 'He was not stupid': Damasio (2008), p. 38.

CHAPTER 10: HOW TO KNOW IT ALL

448 'As a human being': A. Einstein, letter to Queen Elisabeth of the Belgians, 19 September 1932.
451 'We are just getting at the rudiments': Haldane (1927).
451 'My suspicion is that the universe': ibid.
451 'Our only hope of understanding': ibid.
452 'we may well be able': O. Campos, M. Ángeles Arráez, M. Moreno, F. Lara, P. Francés & J. R. Alcázar, 'Bioethics and human enhancement: an interview with Julian Savulescu'. *Ilemata* 2(3), 15–25 (2010).
454 'The ongoing digital revolution': Damasio (2012), p. 182.
456 'We may well be able to engineer': Campos et al., op. cit.

Bibliography

A. Abbott, 'What animals really think'. *Nature* 584, 182 (2020)

J. Ackerman, *The Genius of Birds*. Penguin, New York, 2016

S. S. Adams et al., 'Mapping the landscape of human-level artificial general intelligence', *AI Magazine* 33, 25–42 (2012)

K. J. Aitken, 'Are our ideas about octopus life too anthropomorphic to help?'. *Animal Cognition* 26, 257 (2019)

P. Amodio, M. Boeckle, A. K. Schnell, L. Ostojíc, G. Fiorito & N. S. Clayton, 'Grow smart and die young: why did cephalopods evolve intelligence?'. *Trends in Ecology and Evolution* 34, 45–56 (2019)

P. Anderson, 'More is different'. *Science* 177, 393–396 (1972)

X. D. Arsiwalla, R. Sole, C. Moulin-Frier, I. Herreros, M. Sanchez-Fibla & P. Verschure, 'The morphospace of consciousness.' Preprint http://www.arxiv.org/abs/1705.11190 (2017)

W. Arthur, *The Biological Universe: Life in the Milky Way and Beyond.* Cambridge University Press, Cambridge, 2020

B. J. Baars, *On Consciousness*. Nautilus Press, New York, 2019

P. Ball, 'False memories: why we experience tricks of the mind'. *Science Focus* 16 September 2019. https://www.sciencefocus.com/the-human-body/false-memories-tricks-of-the-mind/

— 'Life with purpose'. *Aeon* 13 November 2020. https://aeon.co/essays/the-biological-research-putting-purpose-back-into-life

—, 'Homo imaginatus'. *Aeon* 29 October 2021. https://aeon.co/essays/imagination-isnt-the-icing-on-the-cake-of-human-cognition

F. Baluška & S. Mancuso, 'Deep evolutionary origins of neurobiology'. *Communicative & Integrative Biology* 2(1), 60–65 (2009)

F. Baluška & A. Reber, 'Sentience and consciousness in single cells: How the first minds emerged in unicellular species'. *BioEssays* 41, e1800229 (2019)

D. Baracchi, M. Lihoreau & M. Giurfa, 'Do insects have emotions? Some insights from bumble bees'. *Frontiers in Behavioral Neuroscience* 11, 4 (2017)

S. Baron-Cohen, *The Pattern Seekers: A New Theory of Human Invention*. Allen Lane, London, 2020

Barrett & Henzi, 'The social nature of primate cognition'. *Proceedings of the Royal Society B* 22, 1865 (2005)

S. Bartlett & M. L. Wong, 'Defining lyfe in the universe: from three privileged functions to four pillars'. *Life* 10(4), 42 (2020)

T. Bayne, J. Hohwy & A. M. Owen, 'Are there levels of consciousness?' *Trends in Cognitive Sciences* 20, 405–413 (2016)

T. Bayne, A. K. Seth & M. Massimini, 'Are there islands of awareness?'. *Trends in Neurosciences* 43, 6–16 (2020)

J. Birch, 'The search for invertebrate consciousness'. *Noûs* 2020, 1–21

N. Block, 'The mind as the software of the brain.' In D. N. Osherson, L. Gleitman, S. M. Kosslyn, S. Smith & S. Sternberg (eds), *An Invitation to Cognitive Science*, 2nd edn, vol. 3, pp. 377–425. MIT Press, Cambridge, MA, 1995.

—, 'Neurophilosophy or philoneuroscience'. *Science* 301, 1328–1329 (2003)

—, 'Two neural correlates of consciousness'. *Trends in Cognitive Sciences* 9, 46–52 (2005)

—, 'Comparing the major theories of consciousness'. In M. S. Gazzaniga et al. (eds), *The Cognitive Neurosciences*, pp. 1111–1122. MIT Press, Cambridge, MA, 2009

M. A. Boden, *Mind as Machine: A History of Cognitive Science*, Vols 1 & 2. Clarendon Press, Oxford, 2006

J. Böhm, S. Scherzer, E. Krol, S. Shabala, E. Neher & R. Hedrich, 'The Venus flytrap *Dionaea muscipula* counts prey-induced action potentials to induce sodium uptake'. *Current Biology* 26, 286–295 (2016)

J. Bongard & M. Levin, 'Living things are not (20th century) machines: updating mechanism metaphors in light of the modern science of machine behavior'. *Frontiers in Ecology and Evolution* 9, 650726 (2021)

D. Bor, *The Ravenous Brain: How the New Science of Consciousness Explains Our Insatiable Search for Meaning*. Basic Books, New York, 2012

D. Bor & A. K. Seth, 'Consciousness and the prefrontal pariental network: insights from attention, working memory, and chunking.' *Frontiers of Psychology* 3, 63 (2012)

N. Bostrom, *Superintelligence*. Oxford University Press, Oxford, 2014

D. Bray, *Wetware: A Computer in Every Living Cell*. Yale University Press, New Haven, 2009

B. Brembs, 'Towards a scientific concept of free will as a biological trait: spontaneous actions and decision-making in vertebrates.' *Proceedings of the Royal Society B* 278, 930–939 (2020a)

—, 'The brain as a dynamically active organ'. *Biochemical and Biophysical Research Communications* 564, 55–69 (2020b)

J. Brockman (ed.), *What To Think About Machines That Think*. Harper, New York, 2015

—, *Possible Minds: 25 Ways of Looking at AI*. Penguin, 2020

R. Brooks, 'What is it like to be a robot?', essay, 18 March 2017. https://rodneybrooks.com/what-is-it-like-to-be-a-robot/

S. F. Brosnan, L. Salwiczek & R. Bshary, 'The interplay of cognition and cooperation'. *Philosophical Transactions of the Royal Society B* 365, 2699–2710 (2010)

H. Browning, 'What is good for an octopus?' *Animal Sentience* 26, 243 (2019)

G. E. Budd, 'Early animal evolution and the origins of nervous systems'. *Philosophical Transactions of the Royal Society B* 370, 20150037 (2015)

N. Cabrol, 'What are we looking for: An overview of the search for extraterrestrials'. In J. Al-Khalili (ed.), *Aliens*, pp. 178–187. Profile, London, 2016

—, 'Alien mindscapes – a perspective on the search for extraterrestrial intelligence'. *Astrobiology* 16, 661–676 (2016)

J. Call & M. Tomasello, 'Does the chimpanzee have a theory of mind? 30 years later'. *Trends in Cognitive Sciences* 12, 187–192 (2008)

P. Calvo & A. Trewavas, 'Physiology and the (neuro)biology of plant behavior: a farewell to arms'. *Trends in Plant Science* 25, 214–216 (2020)

S. Carls-Diamante, 'The octopus mind: implications for cognitive science'. *Animal Sentience* 26, 269 (2019)

S. M. Carroll, 'Why Boltzmann brains are bad'. Preprint, https://arxiv.org/abs/1702.00850 (2017)

C. Caruso, 'Chimps may be capable of comprehending the minds of others', *Scientific American* 6 October 2016. https://www.scientificamerican.com/article/chimps-may-be-capable-of-comprehending-the-minds-of-others/

A. Cashmore, 'The Lucretian swerve: the biological basis of human behavior and the criminal justice system'. *Proceedings of the National Academy of Sciences USA* 107, 4499–4504 (2010)

K. C. Catania, 'Tentacled snakes turn C-starts to their advantage and predict future prey behavior'. *Proceedings of the National Academy of Sciences USA* 106, 11183–11187 (2009)

L. Chittka, 'Bee cognition'. *Current Biology* 27, R1037–R1059 (2017)

A. Clark, *Natural-Born Cyborgs: Minds, Technologies, and the Future of Intelligence*. Oxford University Press, Oxford, 2003

—, 'The extended mind', talk at www.hdc.ed.ac.uk/seminars/extended-mind

—, 'Whatever next? Predictive brains, situated agents, and the future of cognitive science'. *Behavioral and Brain Sciences* 36, 181–204 (2013)

N. S. Clayton, T. J. Bussey & A. Dickinson, 'Can animals recall the past and plan for the future?' *Nature Reviews Neuroscience* 4, 685–691 (2003)

M. Cobb, *The Idea of the Brain*. Profile, London, 2000

—, 'The improbability of alien civilizations'. In J. Al-Khalili (ed.), *Aliens*, pp. 156–168. Profile, London, 2016

G. Cocconi & P. Morrison, 'Search for interstellar communications'. *Nature* 184, 844–846 (1959)

M. A. Cohen, T. L. Botch & C. E. Robertson, 'The limits of color awareness during active, real-world vision'. *Proceedings of the National Academy of Sciences USA* 117, 13821–13827 (2020)

C. Cook, N. D. Goodman & L. E. Schulz, 'Where science starts: spontaneous experiments in preschoolers' exploratory play.' *Cognition* 120, 341–349 (2011)

N. C. Cook, G. B. Carvalho & A. Damasio, 'From membrane excitability to metazoan psychology'. *Trends in Neurosciences* 37, 698–705 (2014)

M. C. Corballis, 'Who's in charge? Free will and the science of the brain'. *Laterality* 17, 384–386 (2012)

I. Couzin, 'Collective minds'. *Nature* 445, 715 (2007)

F. Crick, *The Astonishing Hypothesis: The Scientific Search for the Soul.* Simon & Schuster, London, 1994

E. S. Cross & R. Ramsey, 'Mind meets machine: towards a cognitive science of human-machine interactions'. *Trends in Cognitive Sciences* 25, 200–212 (2021)

A. Damasio, *Descartes' Error: Emotion, Reason and the Human Brain.* Picador, London, 1995

—, 'How the brain creates the mind'. *Scientific American* 281, 112–117 (1999)

—, *Self Comes To Mind: Constructing the Conscious Brain.* Vintage, London, 2012

K. Darling, *The New Breed: How To Think About Robots.* Penguin, London, 2021

F. de Waal, 'Are we in anthropodenial?', *Discover* 18(7), 50–53, 19 January 1997. https://www.discovermagazine.com/planet-earth/are-we-in-anthropodenial

—, *Are We Smart Enough To Know How Smart Animals Are?* W. W. Norton, New York, 2016

S. Dehaene, *Consciousness and the Brain.* Penguin, London, 2014

S. Dehaene, H. Lau & S. Kouider, 'What is consciousness, and could machines have it?' *Science* 358, 486–492 (2017)

A. Demertzi et al., 'Human consciousness is supported by dynamic complex patterns of brain signal coordination'. *Science Advances* 5(2), eaat7603 (2019)

D. C. Dennett, *Elbow Room: The Varieties of Free Will Worth Wanting.* MIT Press, Cambridge, MA, 1984

—, *Kinds of Minds: Towards an Understanding of Consciousness*. Weidenfeld & Nicolson, London, 1996

—, *Freedom Evolves*. Penguin, London, 2004

—, *From Bach to Bacteria and Back*. Allen Lane, London, 2017

—, 'Facing up to the hard questions of consciousness'. *Philosophical Transactions of the Royal Society B* 373, 20170342 (2018)

—, 'Herding cats and free will inflation'. Romanell Lecture, *Proceedings and Addresses of the American Philosophical Association* 94, 149–163 (2020)

J. P. Dexter, S. Prabakaran & J. Gunawardena, 'A complex hierarchy of avoidance behaviors in a single-cell eukaryote'. *Current Biology* 29, 1–7 (2019)

S. Dick, 'Bringing culture to cosmos: the postbiological universe'. In S. J. Dick & M. L. Lupisella (eds), *Cosmos and Culture: Cultural Evolution in a Cosmic Context*. NASA, 2009. https://www.nasa.gov/pdf/607104main_CosmosCulture-ebook.pdf

P. Domenici, D. Booth, J. M. Blagburn & J. P. Bacon, 'Cockroaches keep predators guessing by using preferred escape trajectories'. *Current Biology* 18, 1792–1796 (2008)

D. Dor, *The Instruction of Imagination: Language as a Social Communication Technology*. Oxford University Press, New York, 2015

A. Dornhaus, 'Aliens are likely to be smart but not "intelligent": what evolution of cognition on Earth tells us about extraterrestrial intelligence'. [SETI book, ch.6]

C. Douglas, M. Bateson, C. Walsh & A. Bédué, 'Environmental enrichment induces optimistic cognitive biases in pigs'. *Applied Animal Behaviour Science* 139, 65–73 (2012)

R. Dukas, 'Evolutionary biology of animal cognition'. *Annual Reviews in Ecology, Evolution, and Systematics* 35, 37–374 (2004)

F. Dyson, *Disturbing the Universe*. Harper & Row, New York, 1979

G. M. Edelman & G. Tononi, *A Universe of Consciousness*. Basic Books, New York, 2000

Editorial, 'How sociality shapes the brain, behaviour and cognition'. *Animal Behaviour* 103, 187–190 (2015)

W. Edmondson, 'The intelligence in ETI – What can we know?'. *Acta Astronautica* 78, 37–42 (2012)

T. A. Evans & M. J. Beran, 'Chimpanzees use self-distraction to cope with impulsivity'. *Biology Letters* 3, 599–602 (2007)

L. H. Favela, 'Octopus Umwelt or Umwelten?' *Animal Sentience* 26, 238 (2019)

T. E. Feinberg & J. Mallatt, 'The *nature* of primary consciousness. A new synthesis'. *Consciousness and Cognition* 43, 113–127 (2016)

—, *The Ancient Origins of Consciousness: How the Brain Created Experience*. MIT Press, Cambridge, MA, 2016

—, *Consciousness Demystified*. MIT Press, Cambridge, MA, 2018

J. S. Feinstein, R. Adolphs, A. Damasio & D. Tranel, 'The human amygdala and the induction and experience of fear'. *Current Biology* 21, 34–38 (2011)

M.-A. Finkemeier, J. Langbein & B. Puppe, 'Personality research in mammalian farm animals: concepts, measures, and relationship to welfare'. *Frontiers in Veterinary Science* 5, 131 (2018)

M. Gagliano, 'The minds of plants: thinking the unthinkable'. *Communications in Integrated Biology* 10, e1288333 (2017)

M. Gagliano, M. Renton, M. Depczynski & S. Mancuso, 'Experience teaches plants to learn faster and forget slower in environments where it matters'. *Oceologia* 175, 63–72 (2014)

M. Gagliano, V. V. Vyazovskiy, A. A. Borbély, M. Grimonprez & M. Depczynski, 'Learning by association in plants'. *Scientific Reports* 6, 38427 (2016)

M. S. Gazzaniga, *Who's in Charge? Free Will and the Science of the Brain*. Robinson, London, 2016

M. S. Gazzaniga & M. S. Stevens, 'Neuroscience and the law.' *Scientific American Mind*, April (2005)

S. J. Gershman, E. J. Horvitz & J. B. Tenenbaum, 'Computational rationality: a converging paradigm for intelligence in brains, minds, and machines'. *Science* 349, 273–278 (2015)

S. Ginsburg & E. Jablonka, 'The transition to experiencing: I. Limited learning and limited experiencing'. *Biological Theory* 2, 218–230 (2008)

—, 'Sentience in plants: a green red herring?'. *Journal of Consciousness Studies* 28, 17–33 (2021)

P. Godfrey-Smith, *Other Minds: The Octopus and the Evolution of Intelligent Life*. William Collins, London, 2016

—, 'Octopus experience'. *Animal Sentience* 26, 270 (2019)

—, *Metazoa: Animal Minds and the Birth of Consciousness*. William Collins, London, 2020

B. Goertzel & C. Pennachin (eds), *Artificial General Intelligence*. Springer, 2007

I. J. Good, 'Speculations concerning the first ultra-intelligent machine' (1965). Monograph based on talks given in a conference at the University of California, Los Angeles, October 1962, and the Winter General Meetings of the IEEE, January 1963. http://acikistihbarat.com/dosyalar/artificial-intelligence-first-paper-on-intelligence-explosion-by-good-1964-acikistihbarat.pdf

A. Gopnik, T. L. Griffiths & C. G. Lucas, 'When younger learners can be better (or at least more open-minded) than older ones.' *Current Directions in Psychological Science* 24, 87–92 (2015)

H. M. Gray, K. Gray & D. Wegner, 'Dimensions of mind perception'. *Science* 315, 619 (2007)

M. S. A. Graziano, 'The attention schema theory: a foundation for engineering artificial consciousness'. *Frontiers in Neurorobotics* 4, 60 (2017)

D. M. Greenberg, V. Warrier, C. Allison & S. Baron-Cohen, 'Testing the Empathizing–Systematizing theory of sex differences and the Extreme Male Brain theory of autism in half a million people'. *Proceedings of the National Academy of Sciences USA* 115, 12152–12157 (2018)

D. R. Griffin, *Animal Minds: Beyond Cognition to Consciousness*, 2nd edn. University of Chicago Press, Chicago, 2001

R. Gruber et al., 'New Caledonian crows use mental representations to solve metatool problems'. *Current Biology* 29, 686–692 (2019)

A. Guerra, 'The ingenuity of cephalopods'. *Animal Sentience* 26, 241 (2019)

P. Haggard, 'Human volition: towards a neuroscience of will'. *Nature Reviews Neuroscience* 9, 934–946 (2008)

J. B. S. Haldane, *Possible Worlds and Other Essays*. Chatto & Windus, London, 1927. https://jbshaldane.org/books/1927-Possible-Worlds/index.html

S. Hamann, 'Affective neuroscience: amygdala's role in experiencing fear'. *Current Biology* 21, R75-R77 (2011)

A. Hamilton & J. McBrayer, 'Do plants feel pain?' *Disputatio* 12(56), 71–98 (2020)

D. B. Hart, *The Experience of God: Being, Consciousness, Bliss*. Yale University Press, New Haven, 2013

D. Heaven, 'Deep trouble for deep learning'. *Nature* 574, 163–166 (2019)

R. Hedrich, V. Salvador-Recatalà & I. Dreyer, 'Electrical wiring and long-distance plant communication'. *Trends in Plant Science* 21, 376–387 (2016)

M. Heisenberg, 'Is free will an illusion?'. *Nature* 459, 164–165 (2009)

S. Herculano-Houzel, 'Birds do have a brain cortex – and think'. *Science* 369, 1567–1568 (2020)

J. Hernández-Orallo, *The Measure of All Minds*. Cambridge University Prcss, Cambridge, 2017

T. T. Hills, 'Neurocognitive free will.' *Proceedings of the Royal Society B* 286, 20190510 (2019)

D. J. Horschler, E. L. MacLean & L. R. Santos, 'Do non-human primates really represent others' beliefs?' *Trends in Cognitive Sciences* 24, 594–605 (2020)

F. Hoyle, *The Black Cloud*. Penguin, Harmondsworth, 1977

A. C. Huk & E. Hart, 'Parsing signal and noise in the brain'. *Science* 364, 236–237 (2019)

D. Hume (1748), *An Enquiry Concerning Human Understanding*. Oxford University Press, Oxford, 2007

T. H. Huxley, 'On the hypothesis that animals are automata, and its history'. *Fortnightly Review* 95, 555–580 (1874)

J. Jacquet, B. Franks & P. Godfrey-Smith, 'The octopus mind and the argument against farming it'. *Animal Sentience* 26, 271 (2019)

H. F. Japyassú & K. N. Laland, 'Extended spider cognition'. *Animal Cognition* 20, 375–395 (2017)

A. Jasanoff, *The Biological Mind: How Brain, Body and Environment Collaborate to Make Us Who We Are*. Basic Books, New York, 2018

A.-H. Javadi et al., 'Hippocampal and prefrontal processing of network topology to simulate the future'. *Nature Communications* 814652 (2017)

H. S. Jones, *Life on Other Worlds*. English Universities Press, London, 1940

R. Kane (ed.), *The Oxford Handbook of Free Will*. Oxford University Press, Oxford, 2011

S. Kauffman, *Investigations*. Oxford University Press, Oxford, 2000

S. Kauffman & P. Clayton, 'On emergence, agency, and organization'. *Biology and Philosophy* 21, 501–521 (2006)

B. Keim, 'Never underestimate the intelligence of trees'. *Nautilus* 77, 31 October 2019. https://nautil.us/issue/77/underworldsnbsp/never-underestimate-the-intelligence-of-trees

K. Kelly, 'A taxonomy of minds', Blog: *The Technium*, 15 February 2007. https://kk.org/thetechnium/a-taxonomy-of-m/

A. Kershenbaum, *The Zoologist's Guide to the Galaxy: What Animals on Earth Reveal About Aliens – and Ourselves*. Penguin, London, 2020

C. Koch, *Consciousness: Confessions of a Romantic Reductionist*. MIT Press, Cambridge, MA, 2012

—, *The Feeling of Life Itself*. MIT Press, Cambridge, MA, 2019

J. Kondev, 'Bacterial decision making'. *Physics Today* 67(2), 31 (2014)

J. W. Krakauer, A. A. Ghazanfar, A. Gomez-Marin, M. A. MacIver & D. Poeppel, 'Neuroscience needs behavior: correcting a reductionist bias'. *Neuron* 93, 480–490 (2017)

M. Krumm & M. P. Müller, 'Computational irreducibility and compatibilism: towards a formalization'. Preprint, http://www.arxiv.org/abs/2101.12033 (2021).

M. Krützen, J. Mann, M. R. Heithaus, R. C. Connor, L. Bejder & W. B. Sherwin, 'Cultural transmission of tool use in bottlenose dolphins'. *Proceedings of the National Academy of Sciences USA* 102, 8939–8943 (2005)

R. Kurzweil, *The Singularity is Near: When Humans Transcend Biology*. Penguin, New York, 2005

D. Kwon, 'The mysterious, multifaceted cerebellum'. *Knowable* 30 September 2020. https://knowablemagazine.org/article/mind/2020/what-does-the-cerebellum-do

B. M. Lake, T. D. Ullman, J. B. Tenenbaum & S. J. Gershman, 'Building machines that learn and think like people.' Preprint, arxiv.org/abs/1604.00289 (2016)

V. A. F. Lamme, 'Challenges for theories of consciousness: seeing or knowing, the missing ingredient and how to deal with panpsychism'. *Philosophical Transactions of the Royal Society B* 373, 20170344 (2018)

J. H. Lau, T. Cohn, T. Baldwin, J. Brooke & A. Hammond, 'Deep-speare: a joint model of poetic language, meter and rhyme'. Preprint, http://www.arxiv.org/abs/1807.03491 (2018)

A. Lavazza & F. G. Pizzetti, 'Human cerebral organoids as a new legal and ethical challenge'. *Journal of Law and Biosciences* 7, lsaa005 (2020)

J. LeDoux, 'The tricky problem with other minds'. *Nautilus* 75, 29 August 2019. https://nautil.us/issue/75/story/the-tricky-problem-with-other-minds

J. LeDoux, M. Michel & H. Lau, 'A little history goes a long way toward understanding why we study consciousness the way we do today'. *Proceedings of the National Academy of Sciences USA* 117, 6976–6984 (2019)

P. C. Lee, 'Are octopuses special? Mind, sociality and life history'. *Animal Sentience* 26, 255 (2019)

L. Lefebvre, S. M. Reader & D. Sol, 'Brains, innovations and evolution in birds and primates'. *Brain, Behavior and Evolution* 63, 233–246 (2004)

L. Lefebvre, P. Whittle, E. Lascaris & A. Finkelstein, 'Feeding innovations and forebrain size in birds'. *Animal Behavior* 53, 549–560 (1997)

M. Levin & D. C. Dennett, 'Cognition all the way down'. *Aeon* 13 October 2020. https://aeon.co/essays/how-to-understand-cells-tissues-and-organisms-as-agents-with-agendas

B. Libet et al., 'Time of conscious intention to act in relation to onset of cerebral activity (readiness-potential). The unconscious initiation of a freely voluntary act'. *Brain* 106, 623–642 (1983)

O. J. Loukola, C. Solvi, L. Coscos & L. Chittka, 'Bumblebees show cognitive flexibility by improving on an observed complex behavior'. *Science* 355, 833–836 (2017)

T. M. Luhrmann et al., 'Sensing the presence of gods and spirits across cultures and faiths'. *Proceedings of the National Academy of Sciences USA* 118, e2016649118 (2021)

P. Lyon, 'The cognitive cell'. *Frontiers of Microbiology* 6, 264 (2015)

E. L. MacLean, 'Unraveling the evolution of uniquely human cognition'. *Proceedings of the National Academy of Sciences USA* 113, 6348–6354 (2016)

L. McNally, S. P. Brown & A. L. Jackson, 'Cooperation and the evolution of intelligence'. *Proceedings of the Royal Society B* 279, 3027–3034 (2012)

C. Maher, *Plant Minds: A Philosophical Defense*. Routledge, Abingdon, 2017

J. Mallatt, 'The octopus: a beautiful (but disorganized) "mind"'. *Animal Sentience* 26, 250 (2019)

H. Mance, 'Frans de Waal: "We are very much like primates"'. *Financial Times*, 8 March 2019

R. Manzotti & A. Chella, 'Good old-fashioned artificial consciousness and the intermediate level fallacy'. *Frontiers in Robotics and AI* 5, 39 (2018)

G. Marcus, 'What if HM had a Blackberry?' *Psychology Today*, 2008. https://www.psychologytoday.com/us/blog/kluge/200812/what-if-hm-had-blackberry

G. Marcus & E. Davis, *Rebooting AI*. Pantheon, New York, 2019

—, 'Insights for AI from the human mind'. *Communications of the ACM* 64, 38–41 (2021)

E. Margolis & S. Laurence (eds), *The Conceptual Mind: New Directions in the Study of Concepts*. MIT Press, Cambridge, MA, 2015

L. Margulis, 'The conscious cell'. *Annals of the New York Academy of Sciences* 929, 55 (2001)

K. Markel, 'Lack of evidence for associative learning in pea plants'. *eLife* 9, e57614 (2009)

A. Martin, 'Belief representation in great apes'. *Trends in Cognitive Sciences* 23, 985–986 (2019)

J. Mather, 'What is in an octopus's mind?' *Animal Sentience* 26, 209 (2019)

A. Maye, C.-H. Hsieh, G. Sugihara, G. & B. Brembs, 'Order in spontaneous behavior'. *PLoS ONE* 2, e443 (2007)

P. A. M. Mediano, A. K. Seth & A. B. Barrett, 'Measuring integrated information: comparison of candidate measures in theory and simulation'. *Entropy* 21, 17 (2019)

R. Menzel, 'Navigation and communication in insects'. In J. H. Byrne (ed.), *Learning and Memory: A Comprehensive Reference*, 2nd edn, pp. 477–498. Academic Press, Oxford, 2008

—, 'The honeybee as a model for understanding the basis of cognition'. *Nature Reviews Neuroscience* 13, 758 (2012)

—, 'Search strategies for intentionality in the honeybee brain'. In J. H. Byrne (ed.), *The Oxford Handbook of Invertebrate Neurobiology*. Oxford University Press, Oxford, 2015

—, 'The waggle dance as an intended flight: a cognitive perspective'. *Insects* 10, 424 (2019)

G. F. Miller & P. M. Todd, 'Mate choice turns cognitive'. *Trends in Cognitive Sciences* 2, 190–198 (1998)

M. Minsky, *The Society of Mind*. Picador, London, 1987

—, *The Emotion Machine: Commonsense Thinking, Artificial Intelligence, and the Future of the Human Mind*. Simon & Schuster, New York, 2007

K. J. Mitchell, *Innate: How the Wiring of Our Brains Shapes Who We Are*. Princeton University Press, Princeton, 2018

—, *Agents*. Princeton University Press, Princeton, 2022

M. Mitchell, *Artificial Intelligence: A Guide for Thinking Humans*. Pelican, London, 2019

—, 'Why AI is harder than we think'. Preprint, http://www.arxiv.org/abs/2104.12871 (2021)

S. Mithen, *The Prehistory of the Mind*. Thames & Hudson, London, 1996

S. M. Molchanova et al., 'Maturation of neuronal activity in caudalized human brain organoids'. Preprint, https://doi.org/10.1101/779355 (2019)

M. More & N. Vita-More (eds), *The Transhumanist Reader*. Wiley, Chichester, 2013

S. T. Moulton & S. M. Kosslyn, 'Imagining predictions: mental imagery as mental emulation'. *Philosophical Transactions of the Royal Society B* 364, 1273–1280 (2009)

N. J. Mulcahy & J. Call, 'Apes save tools for future use'. *Science* 312, 1038–1040 (2006)

N. Murphy, G. Ellis & T. O'Conner, *Downward Causation and the Neurobiology of Free Will*. Springer, 2009

T. Nagel, 'What is it like to be a bat?' *Philosophical Review* 83, 435–450 (1974)

E. Nahmias, 'Why we have free will.' *Scientific American* 312, 77–79 (2015)

A. W. Needham, *Learning About Objects in Infancy*. Routledge, London, 2016

A. Nieder, L. Wagener & P. Rinnert, 'A neural correlate of sensory consciousness in a corvid bird'. *Science* 369, 1626–1629 (2020)

R. Noble & D. Noble, 'Harnessing stochasticity: how do organisms make choices?' *Chaos* 28, 106309 (2018)

R. O'Doyle, 'Free will: it's a normal biological property, not a gift or a mystery.' *Nature* 459, 1052 (2009)

O. Ogas & S. Gaddam, *Journeys of the Mind: How Thinking Emerged from Chaos*. W. W. Norton, New York, 2022

M. Osvath & H. Osvath, 'Chimpanzee (Pan troglodytes) and orangutan (Pongo abelii) forethought: Self-control and pre-experience in the face of future tool use'. *Animal Cognition* 11, 661–674 (2008)

D. C. Penn & D. J. Povinelli, 'On the lack of evidence that non-human animals possess anything remotely resembling a "theory of mind"'. *Philosophical Transactions of the Royal Society B* 362, 731–744 (2007)

I. Persson & J. Savulescu, 'The perils of cognitive enhancement and the urgent imperative to enhance the moral character of humanity.' *Journal of Applied Philosophy* 25, 162–177 (2008)

T. Pink, *Free Will: A Very Short Introduction*. Oxford University Press, Oxford, 2004

S. Pinker, *How the Mind Works*. Penguin, London, 1997

M. A. Poirier, D. Y. Kozlovsky, J. Morand-Ferron & V. Careau, 'How general is cognitive ability in non-human animals? A meta-analytical and multi-level reanalysis approach'. *Proceedings of the Royal Society B* 287, 20201853 (2020)

M. Pollan, 'The intelligent plant'. *New Yorker*, 23 December 2013

I. F. Porcher, 'The perfecting of the octopus'. *Animal Sentience* 26, 264 (2019)

D. S. Portman, 'The minds of two worms'. *Nature* 571, 40–42 (2019)

N. C. Rabinowitz, F. Perbet, H. F. Song, C. Zhang, S. M. A. Eslami & M. Botvikick, 'Machine theory of mind'. Preprint, http:www.arxiv.org/abs/1802.07740 (2018)

I. Rahwan et al., 'Machine behaviour'. *Nature* 568, 477–486 (2019)

S. M. Reader, Y. Hager & K. N. Laland, 'The evolution of primate general and cultural intelligence'. *Philosophical Transactions of the Royal Society B* 366, 1017–1027 (2011)

S. M. Reader & K. N. Laland, 'Social intelligence, innovation, and enhanced brain size in primates'. *Proceedings of the National Academy of Sciences USA* 99, 4436–4441 (2002)

A. S. Reber, *The First Minds: Caterpillars, 'Karyotes, and Consciousness.* Oxford University Press, Oxford, 2018

A. S. Reber & F. Baluška, 'Cognition in some surprising places'. *Biochemical and Biophysical Research Communications* 564, 150–157 (2021)

A. D. Redish, *The Mind Within the Brain.* Oxford University Press, Oxford, 2013

D. G. Robinson, A. Draguhn & L. Taiz, 'Plant "intelligence" changes nothing'. *EMBO Reports* 21, e50395 (2020)

P. Rochat, T. Striano & R. Morgan, 'Who is doing what to whom? Young infants' developing sense of social causality in animated displays'. *Perception* 33, 355–369 (2004)

A. G. Rosati, 'Chimpanzee cognition and the roots of the human mind'. In M. Muller, R. Wrangham & D. Pilbeam (eds), *Chimpanzees and Human Evolution*, pp. 703–745. Belknap Press, Cambridge, MA, 2017

A. G. Rosati & B. Hare, 'Chimpanzees and bonobos exhibit emotional responses to decision outcomes'. *PLoS ONE* 8, e63058 (2013)

J. S. Rule, J. B. Tenenbaum & S. T. Piantadosi, 'The child as hacker'. *Trends in Cognitive Sciences* 24, 900–915 (2020)

S. Russell, *Human Compatible: Artificial Intelligence and the Problem of Control*. Allen Lane, London, 2019

G. Ryle, *The Concept of Mind*. Penguin, London, 1990

O. Sacks, *An Anthropologist on Mars*. Picador, London, 1995

H. Sakaguchi, Y. Ozaki, T. Ashida, N. Oishi, S. Kihara & J. Takahashi, 'Self-organized synchronous calcium transients in a cultured human neural network derived from cerebral organoids'. *Stem Cell Reports* 13, 458–473 (2019)

A. Sandberg & N. Bostrom, 'Whole Brain Emulation: a Roadmap', *Technical Report #2008.3* (2008). www.fhi.ox.ac.uk/reports/2008-3.pdf

N. Savage, 'Marriage of mind and machine'. *Nature* 571, 515–517 (2019)

D. L. Schachter, D. R. Addis, D. Hassabis, V. C. Martin, R. N. Spreng & K. K. Szpunar, 'The future of memory: remembering, imagining, and the brain'. *Neuron* 76, 677–694 (2012)

R. M. Schiffrin, D. S. Bassett, N. Kriegeskorte & J. B. Tenenbaum, 'The brain produces mind by modeling'. *Proceedings of the National Academy of Sciences USA* 117, 29299–29301 (2020)

S. Schneider, 'Alien minds'. In S. J. Dick (ed.), *The Impact of Discovering Life Beyond Earth*, pp. 189–206. Cambridge University Press, Cambridge, 2015

—, *Artificial You: AI and the Future of Your Mind*. Princeton University Press, Princeton, 2019

A. Schnell & G. Vallortigara, '"Mind" is an ill-defined concept: considerations for future cephalopod research.' *Animal Sentience* 26, 267 (2019)

A. Schurger, P. Hu, J. Pak & A. L. Roskies, 'What is the readiness potential?' *Trends in Cognitive Sciences* 25, 558–570 (2021)

B. L. Schwartz, 'A community of minds'. *Animal Sentience* 26, 240 (2019)

W. A. Searcy & S. Nowicki, 'Bird song and the problem of honest communication'. *American Scientist* 96, 114–121 (2008)

A. Seed, D. Hanus & J. Call, 'Causal knowledge in corvids, primates, and children'. In T. McCormack, C. Hoerl & S. Butterfill (eds), *Tool Use and Causal Cognition*. Oxford Scholarship Online, 2011. DOI:10.1093/acprof:oso/9780199571154.001.0001

T. J. Sejnowski, 'The unreasonable effectiveness of deep learning in artificial intelligence'. *Proceedings of the National Academy of Sciences USA* 117, 30033–30038 (2019)

A. Seth, 'Aliens on earth: what octopus minds can tell us about alien consciousness'. In J. Al-Khalili (ed.), *Aliens*, pp. 47–57. Profile, London, 2016

A. K. Seth, 'The real problem'. *Aeon* 2 November 2016. https://aeon.co/essays/the-hard-problem-of-consciousness-is-a-distraction-from-the-real-one

—, *Being You: A New Science of Consciousness*. Faber & Faber, London, 2021

M. Shanahan, *Embodiment and the Inner Life: Cognition and Consciousness in the Space of Possible Minds*. Oxford University Press, Oxford, 2010

—, *The Technological Singularity*. MIT Press, Cambridge, MA, 2015

—, 'Conscious exotica', *Aeon* 19 October 2016. https://aeon.co/essays/beyond-humans-what-other-kinds-of-minds-might-be-out-there

M. Shanahan, M. Crosby, B. Beyret & L. Cheke, 'Artificial intelligence and the common sense of animals'. *Trends in Cognitive Sciences* 24, 862–872 (2020)

M. Sheldrake, *Entangled Life: How Fungi Make Our Worlds, Change Our Minds, and Shape Our Futures*. Bodley Head, London, 2020

H. Shevlin & M. Halina, 'Apply rich psychological terms in AI with care'. *Nature Machine Intelligence* 1, 165–167 (2019)

J. M. Shine, 'The thalamus integrates the macrosystems of the brain to facilitate complex, adaptive brain network dynamics'. *Progress in Neurobiology* 199, 101951 (2021)

C. M. Signorelli, 'Can computers become conscious and overcome humans?' *Frontiers in Robotics and AI* 5, 121 (2018)

S. Silberman, *Neurotribes*. Allen & Unwin, London, 2015

A. Sloman, 'The structure of the space of possible minds'. In S. Torrance (ed.), *The Mind and the Machine: Philosophical Aspects of Artificial Intelligence*, pp. 35–42. Ellis Horwood, Chichester, 1984

M. Solms, 'The conscious Id'. *Neuropsychoanalysis* 15, 5–19 (2014)

—, *The Hidden Spring: A Journey to the Source of Consciousness.* Profile, London, 2021

C. Solvi, S. G. Al-Khadhairy & L. Chittka, 'Bumble bees display cross-model object recognition between visual and tactile sense'. *Science* 367, 910–912 (2020)

C. S. Soon, M. Brass, H.-J. Heinze & J.-D. Haynes, 'Unconscious determinants of free decisions in the human brain.' *Nature Neuroscience* 11, 543 (2008)

A. Smirnova, Z. Zorina, T. Obozova & E. Wasserman, 'Crows spontaneously exhibit analogical reasoning'. *Current Biology* 25, 256–260 (2015)

A. Sparkes et al., 'Towards robot scientists for autonomous scientific discovery'. *Automated Experimentation* 2, 1 (2010)

E. Spelke, 'Initial knowledge'. *Cognition* 50, 431–445 (1994)

E. S. Spelke, 'Core knowledge, language, and number'. *Language Learning and Development* 13, 147–170 (2017)

E. S. Spelke & K. D. Kinzler, 'Core knowledge'. *Developmental Science* 10, 89–96 (2007)

M. Stacho et al., 'A cortex-like canonical circuit in the avian forebrain'. *Science* 369, eabc5534 (2020)

J. Stamps & T. G. G. Groothuis, 'The development of animal personality: relevance, concepts and perspectives'. *Biological Reviews* 85, 301–325 (2010)

O. Stapledon, *Last and First Men* and *Last Men in London*. Penguin, Harmondsworth, 1972

S. Still, D. A. Sivak. A. J. Bell & G. Crooks, 'Thermodynamics of prediction'. *Physical Review Letters* 109, 120604 (2012)

J. Storrs Hall, *Beyond AI: Creating the Conscience of the Machine.* Prometheus, 2007

T. Suddendorf, D. R. Addis & M. C. Corballis, 'Mental time travel and the shaping of the human mind'. *Philosophical Transactions of the Royal Society B* 364, 1317–1324 (2009)

L. Taiz et al., 'Plants neither possess nor require consciousness'. *Trends in Plant Science* 24, 677–687 (2019)

A. H. Taylor et al., 'Of babies and birds: complex tool behaviours are not sufficient for the evolution of the ability to create a novel causal intervention'. *Proceedings of the Royal Society B* 281, 20140837 (2014)

J. B. Tenenbaum, C. Kemp, T. L. Griffiths & N. D. Goodman, 'How to grow a mind: statistics, structure, and abstraction'. *Science* 331, 1279–1285 (2011)

T. Thiery et al., 'Decoding the neural dynamics of free choice in humans'. *PLoS Biology* 18, e3000864 (2020)

M. Tomasello, *A Natural History of Human Thinking*. Harvard University Press, Cambridge, MA, 2014

—, 'Precis of *A Natural History of Human Thinking*'. *Journal of Social Ontology* 2, 59–64 (2016)

M. Tomasello & E. Herrmann, 'Ape and human cognition: what's the difference?' *Current Directions in Psychological Science* 19, 3–8 (2010)

M. Tomasello & H. Rakoczy, 'What makes human cognition unique? From individual to shared to collective intentionality'. *Mind & Language* 18, 121–147 (2003)

A. Trewavas, 'Intelligence, cognition, and language of green plants'. *Frontiers in Psychology* 7, 588 (2016)

—, 'The foundations of plant intelligence'. *Interface Focus* 7, 20160098 (2017)

A. Trewavas & F. Baluška, 'The ubiquity of consciousness'. *EMBO Reports* 12, 1221–1225 (2011)

A. Trewavas, F. Baluška, S. Mancuso & P. Calvo, 'Consciousness facilitates plant behavior'. *Trends in Plant Science* 25, 216–217 (2020)

A. Tsiaras, I. P. Waldmann, G. Tinetti, J. Tennyson & S. Yurchenko, 'Water vapour in the atmosphere of the habitable-zone eight-Earth-mass planet K2-18 b'. *Nature Astronomy* 3, 1086–1091 (2019)

A. M. Turing, 'Computing machinery and intelligence'. *Mind* 59, 433–460 (1950)

D. A. Vakoch, 'Communicating with the other'. In S. J. Dick (ed.), *The Impact of Discovering Life Beyond Earth*, pp. 143–154. Cambridge University Press, Cambridge, 2015

E. J. C. van Leeuwen et al., 'Chimpanzees behave prosocially in a group-specific manner'. *Science Advances* 7, eabc7982 (2021)

D. Vanderelst & A. Winfield, 'An architecture for ethical robots inspired by the simulation theory of cognition'. *Cognitive Systems Research* 48, 56–66 (2018)

G. Vince, *Transcendence: How Humans Evolved through Fire, Language, Beauty, and Time.* Penguin, London, 2019

J. Vonk, 'Octopi-ing a unique niche in comparative psychology'. *Animal Sentience* 26, 242 (2019)

A. R. Wallace, *Man's Place in the Universe*. Chapman & Hall, London, 1904

D. M. Wegner & K. Gray, *The Mind Club*. Penguin, New York, 2016

A. Whiten & T. Suddendorf, 'Great ape cognition and the evolutionary roots of human imagination'. *Proceedings of the British Academy* 147, 31–59 (2007)

E. O. Wilson, *The Origins of Creativity.* Penguin, London, 2017

A. F. T. Winfield, 'Experiments in artificial Theory of Mind: from safety to storytelling.' *Frontiers of Robotics & AI.* https://www.frontiersin.org/articles/10.3389/frobt.2018.00075/full (2018)

M. Woolridge, *The Road to Conscious Machines*. Penguin, London, 2020

R. V. Yampolskiy, 'The universe of minds'. Preprint, http://www/arxiv.org/abs/1410.0369 (2014)

K. Yokawa, T. Kagenishi, A. Pavlovič, S. Gall, M. Weiland, S. Mancuso & F. Baluška, 'Anaesthetics stop diverse plant organ movements, affect endocytic vesicle recycling and ROS homeostasis, and block action potentials in Venus flytraps'. *Annals of Botany* 122, 747–756 (2018)

Index

Ackerman, Jennifer 192
Addis, Donna Rose 76
Aesop's fables 198
aesthetics, in animals 223–225
affordance 87, 88, 197
agency 409, 423–432, 444, 445
AlexNet 292, 293
aliens 232, 333–381
  biochemistry of 352, 353
  cognition of 357–362
  history of speculations about 334–343
  intelligence of 362–366
  and moral responsibility 440
  technologies 375–381
  see also SETI
AlphaGo 299
altruism 68, 69, 182–185, 359
Anderson, Philip 412
animal cognition 165–240
apes, cognition 166, 172–191
Aristotle 62, 166
Arthur, Wallace 361
artificial intelligence 41, 48, 57, 146, 147, 228, 267–332, 450, 451
  risks of 267–269, 324–332
  and extraterrestrial intelligence 366–374
Asimov, Isaac, *I, Robot* 267, 327, 440
Asperger, Hans 104
Augustine, Saint 391
autism 103–107
autopoesis 258, 259

Baars, Bernard 131
Babbage, Charles 272, 274
baboons 196
Baron-Cohen, Simon 104–106, 189
Bartlett, Stuart 353, 355
Bayne, Tim 162
bees 209–215, 222, 223, 357, 358
  navigation 213, 214
  waggle dance 210–214
Beran, Michael 187
Bongard, Josh 37
Borges, Jorge Luis 112
Bierce, Ambrose 8
biopsychism 11, 257–266
birds 191–209, 223–225
  song 213, 227, 227
Bisson, Terry 367
Block, Ned 13, 14, 16
Boltzmann, Ludwig 386
Boltzmann brains 385–388
bonobos 173, 175, 179, 186, 188, 189
Boogert, Neeltje 193
Bor, Daniel 98, 123, 124, 142, 144, 373
Bostrom, Nick 325–332, 382, 383
bowerbirds 192, 223, 224
Boyajian, Tabetha 379

brain
  anatomy 22–30
  computer analogy 32–35
  and genetics 29, 30
  imaging 31, 124, 420
  simulation of 368, 369
brain organoids 154–164
Brembs, Björn 418
Brooks, Rodney 319, 320
Brownian motion 426

Cabrol, Nathalie 346, 348
Call, Josep 178, 186
Careau, Vincent 446
Carls-Diamante, Sidney 238
Carroll, Sean 387, 388
Carvalho, Gil 256
Cashmore, Anthony 401, 402, 436
cephalopods 143, 197, 231–241
cerebellum 142
cetaceans 227, 229, 230
Chalmers, David 111, 122
Changeaux, Jean-Pierre 132
chimpanzees 173–179, 181–184, 188, 196, 198
Chinese Room argument 120, 121, 306, 310
Chittka, Lars 210, 222, 223
chunking 130, 131
Churchland, Anne 218
Churchland, Patricia 118
Churchland, Paul 118
ciliates 425
Clark, Andy 110
Clayton, Nicola 199, 204
Clayton, Philip 418
cnidarians 244, 246
Cobb, Matthew 23
Cocconi, Giuseppe 343
Cohen, Michael 81
collective intelligence 55, 262–264
compatibilism 399
computation, history of 271–275
consciousness 12, 115–164, 444
  in AI 308–310, 318–321
  in birds 206–209
  and evolution 150–153
  in octopuses 238
convergent evolution 354, 355
ConvNets 294
Cook, Norman 256
cooperation 68
corvids 170, 192, 198
Cox, David 296, 301, 302, 323
creativity, in AI 312–317
Crick, Francis 141, 144
criminal justice 437, 438
Crooks, Gavin 429, 430
ctenophores 244
cultural evolution 71, 72
Cyc (AI) 282–286
Cyrano de Bergerac 334

Damasio, Antonio 16, 17, 79, 115, 118, 152, 219, 278, 421, 454
  on consciousness 143–145
  on free will 435–437
  on neuronal sentience 256
  on the self and 'autobiographical consciousness' 123–130
Darwin, Charles 165, 195, 224, 248, 249
Darwin, Francis 195, 249
Darwinian evolution in aliens 353–357, 359
Davis, Ernest 302, 303
De Waal, Frans 182, 184, 185
deep learning (AI) 292–302, 307
DeepBlue 299, 300

Dehaene, Stanislas 122, 125, 129, 132, 135, 142, 455
Dennett, Daniel 75, 225, 432
  on biopsychism 261
  on consciousness 123, 126, 131, 150
  on definitions of mind 8, 14, 15
  on extended minds 109
  on free will 404, 427, 432
  on language 72, 229
  on machine consciousness 319
Descartes, René 3, 4, 18, 21, 115, 116, 118, 166, 167, 215, 382, 400
Descartes' demon 3, 382, 388
determinism 402, 405–416
Dews, Peter 234
Dickinson, Emily 447
dolphins 220, 227, 229, 230
Dor, Daniel 76, 77
Dornhaus, Anna 360, 363, 366
Douglas, Catherine 220, 221
Drake, Frank 339, 344, 348, 349
  Drake equation 348, 349
Dyson, Freeman 375–381
Dyson, George 291
Dyson spheres 376–381

Earth Species Project 228
earthworm 195
Edelman, Gerald 37, 133, 136, 447
Edmondson, William 348
Einstein, Albert 43, 448
electrical signalling in cells 255, 256
elephants 196
emotions 95–101, 144, 145
  in animals 215–223
ENIAC (Electronic Numerical Integrator and Computer) 273
Epicurus 399
Evans, Theodore 187
evolution of mind 241–247
exoplanets 350, 351
extended mind hypothesis 108–111
extraterrestrial intelligence, *see* aliens *and* SETI

false memories 85, 86
fear 99
Feinberg, Todd 257
Fermi paradox 345, 346
flocking 193, 263
food caching 204–206
Forward, Robert 339, 344
Frankenstein (Mary Shelley) 268
free will 397–441
  and morality 432–441
Frisch, Karl von 212, 224
Frith, Chris 420
Frith, Uta 107
fungi 254,255

Gaddam, Sai 9
Gagliano, Monica 249–251
Gaia hypothesis 5, 265, 266
Galileo 334
Galton, Francis 262
Gardner, Howard 445, 446
Gazzaniga, Michael 403
Gibson, James 87
Ginsburg, Simone 243, 245, 257, 258
global workspace theory 131–135, 140, 141, 147, 152
God, mind of 6, 53, 54, 388–395
Godfrey-Smith, Peter 232, 234, 235, 237, 238, 240, 241
Godwin, Francis 334
Good, I. J. 324, 325
gorillas 173, 184, 185, 196
Gould, Stephen Jay 48
Grandin, Temple 103
Gray, Heather 49–55

Gray, Kurt 49–55, 389, 395
Gray, Russell 197
Graziano, Michael 307

Haldane, J. B. S. 363, 390, 451
Halpern, Abraham 437
Haraway, Donna 111
Hart, David Bentley 391–394
Hawking, Stephen 111, 147, 267, 269, 325, 393, 394
Hayes, Cecilia 71
Hebb, Donald 287
Hebbian learning 287, 288
Hedrich, Rainer 252, 253
Heisenberg, Martin 35
Herculano-Houzel, Suzana 207
Hernández-Orallo, José 262
Herschel, John 272
Herzing, Denise 227
Hillis, Daniel 54, 55
Hills, Thomas 417
Hinton, Geoff 292
Hobbes, Thomas 33
Hoffman, Donald 82
Hofstadter, Douglas 306
honeybees, *see* bees
Hoyle, Fred, *The Black Cloud* 36, 338
human evolution 67, 189
Hume, David 82, 97, 171
Humphrey, Nicholas 49
Husserl, Edmund 10
Huxley, Thomas Henry 419
Huygens, Christiaan 341, 342

Iamus 314
ImageNet 293
imagination 76
imitation game, *see* Turing Test
information theory 303, 304
integrated information theory 136–141, 260
intellectology 58
interoception 145
intuitive physics 88–91, 173–176, 197–202
intuitive psychology 92, 93
IQ 48

Jablonka, Eva 243, 245, 257, 258
Jasanoff, Alan 33, 34, 97,
Jaypassú, Hilton 110
Jones, Harold Spencer 342, 343

Kant, Immanuel 335, 384
Kardashev, Nikolai 377
Kauffman, Stuart 417
Kelly, Kevin 47
Kepler, Johannes 334
Kershenbaum, Arik 338, 354, 355, 358, 364
Koch, Christof 260, 374, 375
  on brain organoids 157–160, 162
  on consciousness 116, 119, 132, 135, 136, 138–141, 147, 148, 151, 154, 167
  on machine consciousness 278, 319, 373
  on the space of minds 56, 57
Krumm, Marius 411
Krupenye, Christopher 179
Kuba, Michael 234
Kurzweil, Ray 369–372, 381

Laland, Kevin 110
Lamarck, Jean-Baptiste 356, 357
Lamme, Victor 137
Lancaster, Madeline 161
language
  in aliens 362–366
  of animals 166, 212, 214, 215, 225–230

human 76, 77, 91, 100, 127
translation by AI 295, 296, 305, 306
Laplace, Pierre-Simon 406
Laplace's demon 406, 407
LeCun, Yann 292
LeDoux, Joseph 100, 101, 166
Lefebvre, Louis 193, 194
Leibniz, Gottfried 272
Lenat, Doug 283, 284
Levin, Michael 37, 262
Li, Fei-Fei 293
Libet, Benjamin 418, 419
Linnaeus, Carl 264
Loeb, Avi 344
Lorenz, Konrad 230
Lovelace, Ada 272, 274
Lovelock, James 5 , 265
Lucian of Samosata 334
Lucretius 400

McCarthy, John 103, 276, 277
McCulloch, Warren 287
McEwan, Ian 85, 86
McGinn, Colin 120
McGurk effect 82
machine behaviour 321–324
machine learning 285–300
McNally, Luke 70
Macphail, Euan 72
Magnasco, Marcello 229, 230
Mallatt, Jon 257
Marcus, Gary 302, 303
Margulis, Lynn 259, 265
Markram, Henry 106
Massimini, Marcello 148
Mather, Jennifer 227, 235, 236
Matrix, The 267, 382, 383
Maturana, Humberto 258
Menzel, Randolf 212, 214
Mettrie, Julien Offray de la 18
Mialet, Hélène 111
Miller, George 116
mind, definition of 7–13
mind uploading 368–374
Minsky, Marvin 94, 277, 291, 383
mirror test 125, 219
Mitchell, Kevin 27, 430
Mithen, Stephen 88
monkeys 172, 180, 190, 191, 220,
moral responsibility 432–441
Morrison, Philip 343
Mulcahy, Nicholas 186
Müller, Markus 411
music perception 84
Musk, Elon 267, 269, 381, 455
mysterianism 120

Nabokov, Vladimir 219
Nagel, Thomas 44–47, 120, 165,
Neher, Erwin 252
nematode worm (*C. elegans*) 256, 257
neural networks (AI) 288–302
neurodiversity 62, 102–107
neurons 23–27
evolution of 242, 243
neurotransmitter 24, 25, 253, 455
neutron stars 339, 344
New Caledonian crows 192, 196, 199–203
Newell, Allan 280
Nieder, Andreas 207, 208
Nikolos, Stanislav 422

octopuses 143, 196, 231–241
Ogas, Ogi 9
optical illusions 84, 85, 133
orangutans 2, 173, 176, 196

pain 219
in plants 254
panpsychism 139

Papert, Seymour 291
Parfit, Derek 128
Pascal, Blaise 272
Pavlov, Ivan 250
Pelagius 400
perceptron 288–292
phase space 443
phenomenology 10
pigs 221
Pinker, Steven 65, 97, 126, 305, 361
on artificial intelligence 328–330
on the 'Chinese Room' 121, 122
Pitts, Walter 287
plants, cognition of 248–258, 263
sentience in 251
Porcher, Ila France 235
primate cognition, *see* apes *and* monkeys
Prisoner's Dilemma 68–70
proprioception 237

qualia 122, 123
quantum mechanics, and determinism 398, 407

Rahwan, Iyad 322, 323
Rakoczy, Hannes 190
randomness, behavioural 426, 427
reductionism 412
Rees, Martin 325, 332
Reiss, Diana 230
Rosati, Alexandra 181, 182
Rosenblatt, Frank 287, 288, 290
Rumbaugh, Duane 166
R. U. R. (Karel Čapek) 268
Russell, Stuart 299, 324, 330, 331
Ryle, Gilbert 7, 19–21, 34, 63, 110,

Sacks, Oliver 1–3, 103
Sagan, Dorion 259
Savage-Rumbaugh, Sue 166
Savulescu, Julian 456
Schiffrin, Richard 82
Schneider, Susan 320, 332, 367, 368, 452
Schnell, Alexandra 7
Schopenhauer, Arthur 404
Schurger, Aaron 420
science fiction 335–341, 380
scrub jays 204–206
Searle, John 116, 120, 121, 139, 310
Seed, Amanda 174, 175
self, awareness of 125–129
Selfridge, John 132
Selvitelle, Britt 228
Seth, Anil 124, 144, 305, 383, 435
SETI (Search for Extraterrestrial Intelligence) 343–348, 362–366, 375–381
sexual selection 71
in aliens 360
SHAKEY (AI) 281
Shanahan, Murray 34, 57, 58, 87, 115, 140, 320
Shannon, Claude 303, 304
Silberman, Steve 103
Simard, Suzanne 263
Simon, Herbert 277, 280
simulation hypothesis 381–385
Skinner, B. F. 167
Slobodchikoff, Con 168, 228–230
Sloman, Aaron 39–42,
Smith, David Alexander 340
Smith, John David 220
social intelligence 66, 67, 70, 188
solipsism 3, 384
Solms, Mark 143, 144
Spearman, Charles 48
Spelke, Elizabeth 91–94
Spock (*Star Trek*) 99, 340

squid 143, 239
Stapledon, Olaf 108, 336–338
stem cells 155
Still, Susanne 429, 430
Strawson, Galen 118
super-intelligence
  and AI 269, 325–330
  and aliens 366–368, 381
symbolic AI 280–285, 294
synapse 24

Tagliazucchi, Enzo 140
Tallinn, Jaan 269
Taylor, Alex 199–202
Tegmark, Max 384
Tenenbaum, Josh 302
Theory of Mind 93, 165, 301
  in AI 301, 307, 308
  in apes 177–182, 188
  in birds 205
Tomasello, Michael 178, 190
Tononi, Giulio 36, 37, 133, 135, 136, 141, 148, 410, 447
tool use 70, 71, 88, 196–202
Tooley, Michael 129
transhumanism 368–375, 381–385, 452, 453
Turing, Alan 274, 275, 291, 310
Turing Test 274, 275, 308–317

Uexküll, Jakob von 169, 259
Ullman, Tomer 300
Umwelt 168–171, 259

Valéry, Paul 14
Vallortigara, Giorgio 7
Varela, Francisco 258
Venus fly-trap 252
vision 28, 78–81, 94
  in cephalopods 240
volition 415–423
Volk, Jennifer 209
Von Neumann, John 32

Walker, Sara 356
Wallace, Alfred Russell 66, 342
Wegner, Daniel 49–55, 389, 395
Weinberg, Steven 115
Weizenbaum, Joseph 311, 312
Wells, H. G. 232, 335, 336
Wilson, E. O. 73, 74
Winfield, Alan 308
Winterson, Jeanette 79
Wittgenstein, Ludwig 76, 226
Wong, Michael 353

Yamanaka, Shinya 155
Yampolskiy, Roman 59

Image credits

Fig. 3.1, p. 85: From S. Barreda, 'Vowel normalization as perceptual constancy', *Language* 96(2), 224–254 (2020). Courtesy of Santiago Barreda.

Fig. 4.3, p. 156: Chris Lovejoy, University College London.

Fig. 5.3, p. 200: From Taylor *et al.*, 2014. Courtesy of Nicola Clayton and Alex Taylor.

Fig 7.4, p. 297: a, From M. Alcorn, Z. Gong, C. Wang, L. Mai, W.-s. Ku & A. Nguyen, 'Strike (with) a pose: neural networks are easily fooled by strange poses'. Conference on Computer Vision and Pattern Recognition, 2019; b, From I. Goodfellow, N. Papernot, S. Huang, R. Duan, P. Abbeel & J. Clark, 'Attacking machine learning with adversarial examples', 24 February 2017. https://openai.com/blog/adversarial-example-research/; c, From A. Nguyen, J. Yosinski & J. Clune, 'Deep neural networks are easily fooled: high confidence predictions for unrecognizable images.' Conference on Computer Vision and Pattern Recognition, 2015. Courtesy of Jeff Clune and Anh Nguyen.